Volcanoes of Europe

Volcanoes of Europe

Alwyn Scarth

formerly University of Dundee

Jean-Claude Tanguy

Université de Paris et Institut de Physique du Globe de Paris

OXFORD

University Press

2001

Oxford University Press

Oxford   New York   Athens   Auckland   Bangkok   Bogotá   Buenos Aires   Calcutta
Cape Town   Chennai   Dar es Salaam   Delhi   Florence   Hong Kong   Istanbul   Karachi
Kuala Lumpur   Madrid   Melbourne   Mexico City   Mumbai   Nairobi   Paris   São Paolo
Shanghai   Singapore   Taipei   Tokyo   Toronto   Warsaw

and associated companies in
Berlin   Ibadan

Copyright © Alwyn Scarth & Jean-Claude Tanguy 2001

Published in the United States of America by
Oxford University Press, Inc.
198 Madison Avenue, New York, N.Y. 10016
http://www.oup-usa.org
Oxford is a registered trademark of Oxford University Press

First published in 2001 by
Terra Publishing, PO Box 315, Harpenden, AL5 2ZD, England
E-mail: Publishing@rjpc.demon.co.uk
Website: http://www.rjpc.demon.co.uk

Library of Congress Cataloging-in-Publication Data
Scarth, Alwyn.
   Volcanoes of Europe / Alwyn Scarth & Jean-Claude Tanguy.
     p. cm.
   Includes bibliographical references and index.
   ISBN 0-19-521754-3
   1. Volcanoes—Europe. I. Tanguy, Jean-Claude. II. Title.

QE526 .S24 2000
551.21'094—dc21   00-053048

Typeset in Melior and Helvetica
Printed and bound by Biddles Limited, Guildford and King's Lynn, England

Contents

Preface   vii
Acknowledgements   ix

PART 1   INTRODUCTION

Introduction   2

PART 2   THE MEDITERRANEAN

Italy   **6**
Campania
Vesuvius   10
The Phlegraean Fields   22
Ischia   30
Aeolian Islands
Stromboli   35
Lìpari   39
Alicudi   42
Filicudi   43
Salina   43
Seamounts   44
Vulcano   44
Sicily
Etna   51
Pantelleria   70
Graham Bank and Foerstner Bank   72
Bibliography   73

Greece   **81**
Santoríni   82
Mílos   92
Kós   92
Méthana   93
Nísyros   94
Bibliography   96

PART 3   THE ATLANTIC

Spain: Canary Islands   **100**
Tenerife   102
Lanzarote   111
Fuerteventura   121
Gran Canaria   122
El Hierro   123
La Palma   124
La Gomera   127
Bibliography   128

Portugal: the Azores   **131**
São Miguel   133
Santa Maria   142
Terceira   143
Graciosa   148
São Jorge   150
Pico   152
Faial   155
Flores and Corvo   161
Bibliography   162

Iceland   **165**
The northern active volcanic zone   171
Shield volcanoes   176
Plinian eruptions in Iceland   177
Subglacial volcanoes   181
Hydrothermal eruptions   183
Island volcanoes   185
Bibliography   189

Jan Mayen   **193**
Bibliography   196

CONTENTS

PART 4   NORTHERN EUROPE

France   **198**
The Chain of Puys   200
Bibliography   210

Germany   **212**
East Eifel   213
West Eifel   214
Bibliography   216

Glossary   217

Vocabulary   222

Eruptions in Europe in
historical times   224

Index of places and features   231

Index of topics and themes   237

Preface

Volcanoes pay no attention to human foibles such as historical periods, political boundaries and scientific definitions. Thus, the title *Volcanoes of Europe* disguises several kinds of arbitrary choices. We have included, for instance, the Canary Islands and the mid-Atlantic islands of Jan Mayen, Iceland, and the Azores within the European umbrella, although two of the Azores and half of Iceland belong to the North American plate, and the Canary Islands belong to the African plate. On the other hand, we do not describe the volcanoes of Turkey and the Caucasus, which many would, no doubt, call European.

It is altogether more difficult to define those volcanoes which are active, dormant, or extinct. Volcanoes do not always display the secrets of their past, nor do they always reveal their future intentions. Several times, even in the course of the twentieth century, expert volcanologists have been puzzled – not to say surprised – when certain volcanoes have suddenly burst into life after a long period of calm.

Clearly, an erupting volcano is active. But such a definition would restrict the European field to Stromboli and Etna. Equally, a volcano such as the Cantal, in central France, has not erupted for many millions of years and is in all probability extinct. But, of course, volcanoes do not all behave in the same way. The eruptions that form cinder cones rarely last for more than a few years, and then fall silent forever, but the violent outbursts of many stratovolcanoes have been limited to short spells between long periods of repose that have occurred, on and off, for thousands of years. Moreover, active volcanoes rarely grow up in isolation. Many now rise from much older volcanic bases. Thus, to set recent activity in its proper context, it is often necessary to extend the volcanic story well back into the geological past. Even hyperactive Stromboli cannot be fully understood if only its recent eruptions are considered. Thus, the ill defined grey areas between the definitions of active, dormant and extinct demand arbitrary choices.

We therefore consider that a volcano can be designated as active if it has had a magmatic eruption during the past 10 000 years. This is the same definition as that adopted by Simkin & Siebert (1994) in the second edition of their *Volcanoes of the world*. This obviously arbitrary date has the advantage of being long enough to encompass most eruptions during postglacial time, but is also short enough to eliminate areas where the only suggestions of eruptions come from barely substantiated legends. But even this definition has its drawbacks. For example, many vents that erupted cinder cones during this period are unlikely to erupt again. May an indulgent reader, then, accept them as the exceptions that prove the rule.

The notion of historical time is also extremely flexible, and historical records count for little within the defined span of 10 000 years. Even within the limited European context, the period during which eruptions could actually be recorded has varied greatly from place to place. Probably no volcano on Earth has a longer recorded history than Etna, where eye-witness accounts have recounted its eruptions, with admittedly varying degrees of fantasy, for 2500 years. However, the Italian volcanoes were in an exceptionally favoured position in the classical world. On the other hand, records in Iceland extend back only to the early centuries after the settlement in AD 874, and no human being even settled in the Azores until 1439.

But there are historical records and historical records. Most reports of eruptions share the defects of all ancient accounts. The series is incomplete, errors are repeated, descriptions are exaggerated, facts are twisted to fit preconceived notions, references are too brief, too vague and often just untrustworthy. Moreover, many volcanoes erupted in what were, for long periods, remote areas; the very mention of an eruption depends on the knowledge of the author, or those from whom the author is copying, on the survival of texts, the spread of news and so on. Repeated mentions do not mean that a volcano was in constant agitation, but neither does the absence of information indicate that a volcano was dormant. Indeed, most historical records achieve only a modicum of reliability at the beginning of the nineteenth century.

Beyond the historical context, accurate dates of eruptions are only just becoming available in many areas. The traditional methods of geological dating by fossils and stratigraphy are very hard to apply to volcanic edifices. The timespan is too short; the volcanic products preserve few animal or vegetal remains; and the erosion of valleys and their subsequent occupation by further lavas make large and active volcanoes a stratigrapher's nightmare. In many cases, too, the most recent eruptions have masked the products of their predecessors to such an extent that the story of the volcano can scarcely be elucidated at all. At least, monogenetic cinder cones and lava flows invite the gratitude of volcanologists by their broad simplicity.

In recent decades, new techniques of absolute dating have done much to overcome these handicaps. Radiocarbon dates have been calibrated with greater precision, and volcanic rocks can be dated by thermoluminescence, potassium–argon, and palaeomagnetic and archaeomagnetic studies. A whole range of these techniques is now being applied, especially to those more dangerous volcanoes whose tempestuous past must be discovered before their future furies can be predicted with accuracy. Nevertheless, the absolute dates of many European eruptions have yet to be established.

A geomorphologist is often tempted to assess the age of volcanic features by their appearance and by the relative amounts of weathering and erosion that they seem to have undergone. However, the speed of degradation depends upon many conflicting factors, including the varying power of the atmospheric elements, and the different strengths of the rocks that they attack. In general, as with human beings, young volcanoes have fresh features and svelte, sharply defined outlines, whereas older volcanoes are blunted and scarred by innumerable gullies. Sometimes such analysis is valid, as many instances in the text might demonstrate, but judging the age of a volcano in this fashion is fraught with difficulty and it involves an unavoidable element of subjectivity. Such morphological exercises that have been tried here should be regarded as a last resort in the attempt to provide a coherent history of a volcanic area.

The availability of information is a further element that imposes its own limitations on any treatment of European volcanoes. In spite of the boom in volcanological research during the past few decades, some volcanoes are still imperfectly known. Thus, several Italian volcanoes are in intensive care, whereas those in the Azores, for instance, have undeservedly progressed little beyond the waiting list. Consequently, the balance and the treatment of active European volcanoes is, in part at least, influenced by the amount of the scientific literature that is available. Our work therefore indicates not only the broad state of knowledge at the end of the twentieth century but also highlights where some fruitful research could be accomplished. But the chief aim of this study is to stimulate a wide range of readers – to encourage them to take an active, informed interest in some of the most sublime and fascinating features in the natural world – and, especially, to go and see them. Aesthetic rewards will also enhance their scientific pilgrimages, because the European volcanoes embellish landscapes beyond compare.

Authorship and acknowledgements

Jean-Claude Tanguy was primarily responsible for the sections on Etna, Vesuvius and Pantelleria. Alwyn Scarth bears the responsibility for the remaining chapters; he gratefully acknowledges the invaluable assistance of Juan-Carlos Carracedo, Victor Hugo Forjaz, Harry Hine, Maxime Le Goff, and especially Anthony Newton.

The photographs without explicit credits were taken by the authors.

Acknowledgements

The authors and publisher would like to thank the following for permission to make use of copyright material in this book on the pages indicated in grey type. All maps are based on the originals cited, but have been much modified for the purposes of the present text.

Académie des Sciences à Paris
63: Tanguy, J. C., M. Le Goff, V. Chillemi, A. Paiotti, C. Principe, S. La Delfa, G. Patanè 1999. Variation séculaire de la direction du champ géomagnétique enregistrée par les laves de l'Etna et du Vésuve pendant les deux derniers millénaires. *Comptes Rendus* 329: 557–64, figs 1–4.

Acta Vulcanologica
7: Ferrari, L. & P. Manetti 1993. Geodynamic framework of Tyrrhenian volcanism: a review. 3: fig. 1, p. 2.

Bulletin of Volcanology
173: Sigvaldason, G. E., K. Annertz, M. Nilsson 1992. Effect of glacier loading/unloading on volcanism: postglacial volcanic production rate of the Dyngjufjöll area, central Iceland. 54: fig. 2, p. 387.

Consejo Superior de Investigaciones Cientificas (Estación Volcanológica de Canarias)
125: Carracedo, J. C., S. J. Day, H. Guillou, P. J. Gravestock 1999. Geological map of Cumbre Vieja volcano (La Palma, Canary Islands).

Earth and Planetary Science Letters
33: Beccaluva, L., G. Gabbianelli, F. Lucchini, P. L. Rossi, C. Savelli 1985. Petrology and K/Ar ages of volcanics dredged from the Eolian seamounts: implications for geodynamic evolution of the southern Tyrrhenian basin. 74: fig. 1a, p. 189.

Geological Society of London
144: Self, S. 1976. The recent volcanology of Terceira, Azores. *Quarterly Journal* 132: fig. 13, p. 661.

Harvard University Press
20: Pliny [Gaius Plinius Secundus] 1969. *Letters and Panegyricus*, vol. VI: *letters 16 and 20* [translated by B. Radice]. [Loeb Classical Library]

Journal of Volcanology and Geothermal Research and Elsevier Science Publishers
67: Barberi F., M. L. Carapezza, M. Valenza, L. Villari 1993. The control of lava flow during the 1991–1992 eruption of Mt Etna. 56: fig. 1, p. 3.
124: Carracedo, J. C. 1994. The Canary Islands: an example of structural control on the growth of large oceanic island volcanoes. 60: fig. 1, p. 226.
113: Carracedo, J. C., E. Rodríguez Badiola, V. Soler 1992. The 1730–1736 eruption of Lanzarote, Canary Islands: a long, high-magnitude basaltic fissure eruption. 53: fig. 1a, p. 240.
136: Cole, P. D., G. Quieroz, N. Wallenstein, J. L. Gaspar, A. M. Duncan, J. L. Guest 1995. An historic subplinian/phreatomagmatic eruption: the AD 1630 eruption of Furnas volcano, São Miguel, Azores. 69: fig. 1, p. 119.
31: Orsi, G., M. Piochi, L. Campajola, A. D'Onofrio, L. Gianella, F. Terrasi 1996. ^{14}C geochronological constraints for the volcanic history of the island of Ischia (Italy) over the last 5000 years. 71: fig. 1, p. 250.
17: Rosi, M., C. Principe, R. Vecci 1993. The 1631 Vesuvius eruption: a reconstruction based on historical and stratigraphical data. 58: fig. 4, p. 159.

Jökull
169: Saemundsson, K. 1979. Outline of the geology of Iceland. 29: fig. 1, p. 8.

The Holocene
178: Larsen, G., A. J. Dugmore, A. J. Newton 1999. Geochemistry of historical-age silicic tephras in Iceland. 9: fig. 3, p. 465.

International Association of Volcanology
90: Georgalás, G. C. 1962. Catalogue of the active volcanoes of the world including solfatara fields, part XIII: Greece: fig. 5, p. 22.

Nature
174: Larsen, G., K. Grönvold, S. Thórarinsson 1979. Volcanic eruption through a geothermal borehole at Námafjall, Iceland. **278**: fig. 1, p. 707.
153: Woodhall, D. 1974. Geology and volcanic history of Pico Island Volcano, Azores. **248**: fig. 1, p. 664.

Observatório Vulcanológico Geotérmico dos Açores
143: Forjaz, V. H. 1997a. *Alguns vulcões da Ilha de São Miguel*: fig. 113, p. 137.

Parc Naturel Régional des Volcans d'Auvergne
211: de Goër de Hervé, A., P. Boivin, G. Camus, A. Gourgaud, G. Kieffer, J. Mergoil, P. M. Vincent 1991. *Volcanologie de la Chaîne des Puys* (3rd edn): fig. 3, p. 37; pp. 112–13.
202: de Goër de Hervé, A., P. M. Vincent, P. Boivin, D. Briot, A. Gourgaud, G. Kieffer, G. Camus 1995. *Volcanisme et volcans de l'Auvergne*, no. 8, 1–43: fig. 22, p. 21.

Société Géologique de France
53, 70: Kieffer, G. & J. C. Tanguy 1993. L'Etna: évolution structurale, magmatique et dynamique d'un volcan polygénique. *Pleins feux sur les volcans* [Mémoire 163], R. Maury (ed.): fig. 1, p. 254; fig. 6, p. 262; fig. H, p. 268.

Springer Verlag
22: Barberi, F. & M. L. Carapezza 1996. The Campi Flegrei case history. In *Monitoring and mitigation of volcano hazards*, R. Scarpa & R. I. Tilling (eds): fig. 1, p. 773.

Thor Thordarson
170: Thordarson, T. 1995. *Volatile release and atmospheric effects of basaltic fissure eruptions*. PhD thesis, University of Hawaii, Manoa.

Yale University Press
From: Scarth, A. 1999. *Vulcan's fury: man against the volcano.*
20: fig. on p. 32,
Plate 3b: fig. on p. 46.

United States Geological Survey
138, 139: Moore, R. B. 1992. *Geological map of São Miguel, Azores.*

PART 1 INTRODUCTION

1 Introduction

The distribution of the volcanoes of Europe is perhaps more difficult to understand than on any other continent because of the complications caused by the collision between the Eurasian and the African **plates**.* Most of the volcanoes occur on the margins of the European continent: in the Mid-Atlantic Ridge, the Canary Islands, southern Italy, and the Aegean Sea. The remaining areas of volcanic activity are broadly associated with old rifts in France and Germany. Thus, some **eruptions** are related to the growth of the Eurasian plate along the Mid-Atlantic Ridge; others are related to collision and subduction linked to the clash between Europe and Africa; some eruptions seem to be associated with **hotspots**; and others to the presence of deep fractures transecting the Earth's **crust**. But, in spite of all the remarkable advances made in volcanology in recent decades, it is still sometimes difficult to explain the exact position of some quite important European volcanoes.

The volcanoes on the Mid-Atlantic Ridge are clearly linked to the growing edge of the Eurasian plate. Eruptions are largely submarine and continuous along the whole length of the Ridge, and they were unseen and, indeed, largely unsuspected, until research in the past few decades revealed their enormous importance in the dynamics of the Earth. These eruptions occur chiefly from multitudes of **fissures** that produce the basalts that make the world's oceanic crust. In Jan Mayen, Iceland and the Azores, the crest of the Mid-Atlantic Ridge and part of its flanks have been built up above the waves, so that this

vital volcanic activity can be inspected and analyzed at close range. However, the emissions on these islands are not wholly basaltic, and they have also included eruptions of more evolved **magmas** after reservoirs have developed. But, although the Canary Islands, which lie on the African plate, seem to have been generated by a complex hotspot, no fully satisfactory explanation for the details of their formation has yet emerged. Practically all the volcanic islands in the North Atlantic Ocean represent considerable accumulations of **lavas**. Their bases often lie more than 2000 m deep on the sea floor, and several volcanic peaks rise more than 2000 m above the waves. Thus, Beerenberg in Jan Mayen, Öraefajökull in Iceland, Pico in the Azores, and Teide in the Canary Islands, form some of the most prominent mountains in the North Atlantic Ocean – and are, at least, on a par with any volcanic piles in the rest of Europe.

The volcanoes in Italy and Greece are closely linked to the prolonged collision of the African and Eurasian plates, during which several microplates detached themselves from their parent masses and pursued varying and independent courses. At the same time, the edges of the microplates and the Eurasian plate were smashed, fractured and crumpled as the African plate advanced broadly northwards. Thus, continental sediments carried on the plates were thrust up and contorted to form the Atlas Mountains, the Alps, the Apennines, and the chains of the Balkans, Greece, and Turkey; and magma made its way to the land surface up major faults that transect the Earth's crust. Etna and Vesuvius might have formed in this way. Parts of the forward edges of the African plate were also

---

* Words in bold type are explained in the Glossary (p. 219).

subducted beneath the Eurasian plate and the adjacent microplates. This subduction caused the eruptions that formed the Aeolian Islands and the volcanic islands in the Hellenic Arc in the Aegean Sea. Subduction is thus probably responsible for the most violent European eruption during the past 4000 years, on the Greek island of Santoríni, and for the world's most diligent volcanic performer in modern times, Stromboli. But these Mediterranean volcanoes are as varied as the tectonic conditions that have given them birth. Some, such as Etna and Stromboli have long histories of moderate and mainly basaltic eruptions; others, such as the Fossa cone at Vulcano, have erupted tuffs in more vigorous outbursts; yet others, such as Santoríni and Vesuvius, have erupted huge volumes of **fragments** (pyroclasts) during episodes of great violence that buried whole cities.

The third main group of European volcanoes formed broadly in relation to the discontinuous rifts that traverse the continent from Oslo, in the north, to the Rhine Rift Valley and on to the Limagnes of central France. Eruptions have given rise to many cones and **maars**, in both the Eifel Massif in Germany and the Chain of Puys in central France.

The Mediterranean volcanoes

The volcanoes and the major fold structures of the Mediterranean area have been caused fundamentally by the collision between the Eurasian and African plates (e.g. Dercourt et al. 1985). However, this generalization hides a great complexity of events that perhaps has no equal anywhere else on Earth. Although the Mediterranean area has been intensively studied, much work still needs to be done before the intricacies of the scenario of its development can be truly unravelled. It is thus difficult to explain the distribution and the causes of the Mediterranean volcanoes without making generalizations that may prove to be misleading, inadequate or even inaccurate as research progresses.

The collision takes place between two plates carrying continents that are themselves directly involved in the impact, which has shattered their edges, formed microplates that have moved in different directions, and crumpled and faulted the rocks for millions of years. Thus, collision has not only brought about subduction and deep faults that transect the whole crust, but also

areas of crustal extension. As a result, individual Mediterranean volcanoes can have several different causes or, indeed, combinations of causes. In broad terms, subduction seems to be responsible for the Greek volcanoes on the Hellenic arc in the Aegean Sea, and for the volcanoes in the Aeolian Islands, whereas Etna and Vesuvius perhaps owe their growth to eruptions at the intersection of deep major faults. But the relationships are far from simple and the specialists rarely agree about the exact details of the course of events.

With this varied tectonic background, it is not surprising that the Mediterranean volcanoes have displayed virtually the complete range of eruptive styles. Thus, Vesuvius has been extremely violent, often erupting **Plinian columns** and **nuées ardentes**, but was largely effusive from 1631 to 1944; Etna has been chiefly effusive, but had some violent outbursts about 2000 years ago; Stromboli has been mildly explosive for many centuries; and Santoríni has produced only moderate eruptions since its great explosion in the Bronze Age. And all the while, thousands of less spectacular eruptions have formed **cinder cones**, lava flows, **domes**, **fumaroles** and mudpots, which have been feared ever since antiquity. Italy is the zone of greatest tectonic complexity. And it has also been the forum of the greatest, most varied and most closely studied volcanic activity in Europe.

The Mid-Atlantic Ridge

The Mid-Atlantic Ridge is the most clearly defined, and probably the best known, of all the **mid-ocean ridges**. It forms a sinuous curve bisecting the Atlantic Ocean from the Arctic to Antarctica, in a continuous chain of volcanic accumulations rising 2 km or more from the ocean floor. A longitudinal rift runs along its crest, which marks the site of continual volcanic eruptions, where oceanic crust is generated as the North American and Eurasian plates diverge. The main source of these eruptions is the multitude of fissures and **dykes** that run parallel to the trend of the crest. They have provided the basaltic **pillow lavas** and the black smokers that are characteristic of this environment. Generally speaking, the youngest volcanic rocks occur at the higher, central parts of the ridge, whereas increasingly older rocks are found in roughly parallel strips farther and farther from the crest.

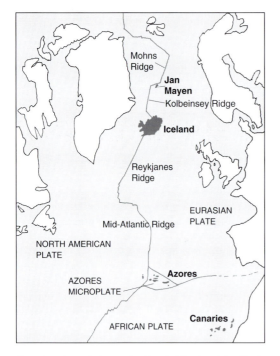

The plate boundaries and volcanic islands of the North Atlantic Ocean.

The ridge is mostly submerged and it is only in exceptional circumstances that volcanic eruptions are so frequent as to have built it above sea level. The most common explanation for these exceptional conditions is that a **mantle** hotspot lies beneath the mid-ocean ridge and this seems to be the most likely reason for the emergence of Jan Mayen, Iceland, and the Azores as some of the culminating points on the Mid-Atlantic Ridge. Jan Mayen lies on the Eurasian flank of the ridge; Iceland is transected by the ridge so that its western part belongs to the North American plate and the eastern part to the Eurasian plate; and the ridge divides the Azores.

Finally, the Canary Islands are apparently not related to the Mid-Atlantic Ridge, but may have been initiated in part by the collision between the African and Eurasian plates. But they bear no evidence of subduction and seem to have developed chiefly in response to one or more hotspots beneath the African plate.

The study of the islands also shows significant variations from this simplified pattern. Major offsets develop in the trend of the Mid-Atlantic Ridge that are associated with notable transform faults and fracture zones more or less at right angles to the trend of its crest. One of their broad effects is to create further fissures up which

magma can then rise. They also tend to facilitate activity on the flanks of the ridge, where eruptions can build up and widen the ridge itself. Thus, the Azores rise from a broad submerged platform on the flanks of the ridge. The rocks on these flanks usually increase in age with their distance from its crest, but the materials emitted during the flank eruptions are much younger than the rocks upon which they lie and they do not usually increase in age with their distance from the crest of the ridge.

The morphology of the Mid-Atlantic Ridge is further complicated near the Azores by the development of a triple junction. Rifting occurs not only between the Eurasian and North American plates but also between the Eurasian and African plates. A zone of secondary spreading seems to have developed and may have formed a microplate supporting the central and eastern Azores.

The fourth rather abnormal feature of the ridge is the eruptions from central clusters of **vents**, which are often related to flank volcanism. In Iceland, for instance, they often form large basaltic **shields** and **pahoehoe** surfaces, but sometimes more explosive eruptions take place if the magma has undergone some evolution in a reservoir. In these conditions, the prevalent **basalts** are replaced by intermediate lavas such as andesites or even **rhyolites**. At the same time, lava flows become less numerous and fragments become increasingly important as a **stratovolcano** is constructed. In this way, Hekla and Oraefajökull, for instance, have grown up in Iceland, and Beerenberg in Jan Mayen. In the Azores, most of the stratovolcanoes have also undergone a markedly explosive phase that led to the formation of large **calderas** on their summits. Iceland, too, has more than a dozen calderas, some of which, like Grímsvötn, are hidden beneath ice caps.

Iceland is by far the largest emerged zone of any mid-ocean ridge in the world, and it is the best studied of the three European components of the Mid-Atlantic Ridge. The growth of the much smaller and less complex island of Jan Mayen has also been broadly elucidated. In the Azores, where the mixture of eruptions is greater, several important detailed studies in recent years have begun to clarify the picture of their development.

PART 2 THE MEDITERRANEAN

2 Italy

Vulcano, Stromboli, Etna and Vesuvius are the most famous active volcanoes on Earth. It is the Italian volcanoes, more than any others, that have given the Western world its views, fears, fascination and preconceived notions about volcanic activity for over 2000 years. Entangled with myths, deities, devils and saints, their varied and often spectacular eruptions have always inspired terror and fascination. The continual eruptions of Stromboli fully justify its nickname, "Lighthouse of the Mediterranean"; Etna probably served as a model for the Polyphemus story in the Odyssey, and its emissions have often been televised in the past few decades. The awesome powers of Vesuvian outbursts were immortalized by Pliny the Younger in AD 79 and, ever since, have forced many a Neapolitan sinner to the confessional. Vulcano itself, the forge of Vulcan, has stamped its name on practically every European language. The renown of the Italian volcanoes has sprung from centuries of intimate contact and observation by a large population that has clustered, in spite of all the dangers, on the rich soils of their flanks. The Seven Hills of Rome, too, were carved from old volcanic products, whereas Pompeii and Herculaneum were later buried by others.

The volcanoes of Italy take a whole gamut of forms, from calderas and stratovolcanoes to cinder cones, domes and mudpots. Their eruptions have ranged from vast **Plinian** outbursts with **nuées ardentes**, to lava flows and **solfataras**. Their products similarly cover a wide volcanic spectrum, from a predominance of basalts in Sicily, to the concentrations of potassium-rich rocks, **trachybasalts**, **tephrites**, **trachytes**, **latites** and **phonolites**, as well as **rhyolites**, in peninsular Italy. The active volcanoes have themselves strikingly different personalities and behaviour. Etna, Vesuvius, Stromboli and Vulcano are distinctive members of the volcanic family and, as a result, their activity has often been used to establish archetypes for a basic classification of styles of volcanic eruption. Long familiarity and study has also given Italy many type-localities, and Italian is a major source of technical terms. **Strombolian**, **Vulcanian**, **lapilli**, scoria, fumarole, solfatara, latite, atrio, not to mention volcano, provide just a few examples out of many. It is not surprising, therefore, that Italy has become the scene of some of the most vigorous contemporary volcanic research from several European countries. The progress of science has generated a vast bibliography. But, although the histories of many active and recently extinct Italian volcanoes have been elucidated, many intriguing problems about them have yet to be resolved, not least of which are those concerned with the very causes of Italian volcanic activity.

In the present state of knowledge, Italy seems to be an unusual volcanic environment: an area of continent-to-continent collision, manifest in large and frequent earthquakes; with widespread, but not always clear, subduction; with crustal extension following upon compression and the development of deep fractures; with magma often of high potassic content, and also perhaps derived from the crust in some areas and from the mantle in others, with the opportunity for differentiation as well as contamination en route. The uncommon combination of these features is a major warning against taking the Italian volcanoes as archetypes and extrapolating their

The volcanic zones of southern Italy.

characteristics to the different outside world. In any case, they have enough individuality to merit all, and more, of the close study that they have always received.

The volcanoes of Italy stretch in a broad but discontinuous band along the western flanks of the Apennines from Tuscany to Campania alongside the Bay of Naples and on, through the **seamounts** of the Tyrrhenian Sea, to the Aeolian Islands and eastern Sicily. In addition, isolated outposts occur in Sardinia, the Euganian Hills in the Po Valley, in Monte Vulture in Basilicata and in the island of Pantelleria in the Straits of Sicily. In a very general way, volcanic activity started in the north and spread irregularly southwards. Thus, the northern areas seem to be wholly extinct, whereas many of the southern areas are clearly still active.

The intense orogenic crumpling of the Apennines was followed by tensional movements that were concentrated on the western margins of the Apennines and were probably connected to the opening of the Tyrrhenian Sea and the continuing collision of the African and Eurasian plates, and of the microplates between them. Deep fractures could then sometimes provide an easy path for rising magma that perhaps played a major role in the eruptions of Campania, Monte Vulture and eastern Sicily. It was probably only

southeast of Calabria that the collision was accommodated by the development of a **subduction zone** that gave rise to the volcanoes of the Aeolian Islands. The intensity and types of fracturing, the varying natures of the magmas and their different rates of ascent, then produced distinct regional patterns of volcanic activity. Thus, the volcanism that is only recently extinct or is still active at the present time falls into five main zones: Tuscany, Latium, Campania, the Aeolian Islands and Sicily.

Most of the volcanic areas of Tuscany are small, scattered and eroded, and were formed between 5 million and 2.3 million years ago. However, Monte Amiata, much the youngest and largest volcanic area in Tuscany, dates from less than 400 000 years ago. Although Tuscany has no active volcanoes at present, the magmas still warm the waters circulating in the upper layers of the crust. This heat is harnessed at Larderello and Monte Amiata to such an extent that Tuscany is one of the most important sources of geothermal energy in Europe.

The volcanoes of Latium, or Lazio, stretch in a broad band, 100 km long, on the western margins of the Apennines and form the largest outcrop of volcanic rocks in Italy. Compared with Tuscany, the volcanoes are larger, more varied and more recent. They form distinct stratovolcanoes, calderas and vast sheets of **pumice**, as well as maars, domes and cinder cones. The eruptions are often associated with the formation of lakes such as Bolsena, Vico, Bracciano, Albano and Nemi, which embellish the rolling outlines of the Latium countryside. But the chief volcanic legacy in Latium is the series of hills dominating its skyline: the Monti Vulsini, Montefascione, Vico, Monti Cimini, Monti Sabatini to the north of Rome, and the Colli Albani (Alban Hills) to the south. For example, the Monti Vulsini were active from 400 000 to 60 000 years ago, the Monti Sabatini from 500 000 to 100 000 years ago, and the Colli Albani from 580 000 to 19 000 years ago. The eruptions produced rocks of a high potassic content and most are rich in **leucite**. **Leucitites**, phonolites and tephrites, are perhaps the most common, with smaller quantities of latites and trachytes.

The volcanoes of Campania are concentrated around the northern shores of the Bay of Naples, although Monte Vulture (in Basilicata) and Roccamonfina (on the border of southern Latium) also belong to this group. Like those in Latium, the Campanian volcanoes are associated with

The Colli Albani (Alban Hills)

The Via Appia links Rome to the Colli Albani, 25 km away to the southeast. The Colli Albani form a vast cone, some 2500 km² in area, which is covered with evergreen oaks and chestnuts, and scattered with many villages surrounded by citrus groves. They enclose a large central hollow, 10 km across, which is open to the west. This depression, the Cinta Esterna caldera, itself encloses a smaller hollow, the Cinta Interna, which is about 2.5 km in diameter and is also open to the west. The whole region is covered with small cinder cones, innumerable lava flows, craters and hot springs. Three craters pockmark the southwest of the Cinta Esterna; the Ariccia crater is dry, but its two neighbours are partly occupied by the famous beauty spots of Lago Albano and Lago Nemi.

Lago Albano is 6 km² in area and has a maximum depth of 150 m. It is in fact an elliptical hollow formed by two intersecting craters. The smaller northeastern crater was partly destroyed when the larger, deeper southeastern crater exploded on its flanks. Lago Nemi lies 3 km to the southeast of Lago Albano. It is 35 m deep and occupies the southernmost of two intersecting hydrovolcanic craters, which together cover 1.6 km². Both craters are fringed by grey peperini tuffs and were exploded through the leucite lava flows covering the outer flanks of the old stratovolcano. Nemi village was built on one of these lava flows. The Ariccia crater, which is clearly visible from the Via Appia Nuova, lies 2 km west of Lago Nemi. It used to be occupied by a lake that has now drained away, leaving a nearly circular hollow 100 m deep and 3 km² in area. It is now dominated by the Castel Savelli cinder cone, which rises to 325 m and was no doubt formed during the last phase of activity that created the Ariccia crater.

The Colli Albani were formed in response to tensional movements on the western margin of the Apennines. Their eruptions produced some 290 km³ of volcanic rocks. Activity began about 580 000 years ago and lasted until about 19 000 years ago, although they could also, perhaps, have been the source of the stones, noted by Livy, that reputedly fell around Rome in 640 BC and 212 BC (Stothers & Rampino 1983). The earliest eruptions formed a broad basement that is now largely masked by subsequent emissions. The tephritic leucite lava flows at Acquacetosa, for example, belong to this initial phase. They were followed by five successive violent eruptions of the "lower tuffs", which are composed of fragments of leucite, tephritic leucite or phonolite lavas in variously coloured layers.

Then began the construction of a large stratovolcano, which probably rose to a height of about 2000 m and was composed chiefly of leucitites. It produced widespread lava flows, including the 10 km-long Capo di Bove flow. These effusions were accompanied by eruptions of welded cinders, and red and yellow lapilli, that have blan-

keted wide areas around the Colli Albani. This phase climaxed with the collapse of the Cinta Esterna caldera.

More than 200 000 years of calm then elapsed before the final, and perhaps most violent, episode in the history of the Colli Albani took place. Explosions blasted away the western sectors of the calderas. These explosions, or subsequent ones, formed the crater of the Valle Ariccia and the intersecting craters that are now partly occupied by Lago Nemi and Lago Albano. These eruptions also expelled the upper grey tuffs (known as peperini) that cover more than 60 km² on the western half of the Colli Albani. They were apparently hydrovolcanic in character, because they formed maars and diatremes and expelled lithic fragments up to 1 m in diameter among the tuffs. The last of these eruptions took place from Lago Albano about 19 000 years ago.

It is possible that the Colli Albani are only dormant. Earthquakes have been recorded in the area ever since Roman times, and studies have shown that contemporary earthquakes are concentrated on the western flanks of the Colli Albani, where the most recent eruptions occurred. Seismic tomography investigations have indicated that a hot, or perhaps molten, mass lies more than 6 km below the craters. It seems also that the continuing arrival of fresh supplies of magma could be responsible for an increase in the level of activity since 1980, and the uplift of the area, which has reached a maximum of 40 cm in that period. However, the earthquakes are centred above the top of the magma reservoir and might merely indicate the continued adjustment of the overlying rocks to the changes in the reservoir below (Chiarabba et al. 1997). It is thus uncertain whether this contemporary unrest truly represents a warning of renewed activity in the Colli Albani.

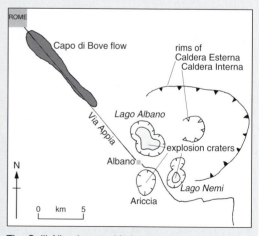

The Colli Albani, central Italy

crustal stretching and the development of deep fractures delimiting elongated fault troughs and fault blocks. Some Campanian volcanoes have erupted during **historical times**, notably in Vesuvius, and the Phlegraean Fields and Ischia. The areas on the fringe are older. The Ponza Islands were formed between 1 and 2 million years ago. The imposing mass of Roccamonfina began to erupt its leucitites, tephrites and phonolites about 1.2 million years ago and its last sign of activity occurred 25 000 years ago. In the east rises the equally large mass of Monte Vulture, which erupted leucitites, tephrites, phonolites and trachytes between about 800 000 and 400 000 years ago.

The Phlegraean Fields form the core of Campania. Here, many vents are concentrated from which chiefly trachytes and phonolites were emitted, often in notably explosive eruptions that produced extensive ashflows and blankets of pumice such as the Campanian **Ignimbrite**. Similar eruptions have also marked the offshore islands of Procida, Nisida and Ischia. **Hydro-volcanic eruptions**, solfataras and fumaroles brought the activity into historical times. In the 1980s, vertical movements in the Pozzuoli area were substantial enough to give rise to fears of a renewed eruption, which has yet to materialize.

The glory of Campania is, of course Vesuvius – perhaps the only volcano in the world that needs no introduction. At the risk of provoking the most dangerous volcano in Europe, suffice it to say that it has been dormant since 1944, and that it usually erupts varieties of leucite–tephrite and phonolite.

The Aeolian Islands are probably a volcanic **island arc**, caused by the subduction of the African plate or the Ionian microplate. They are mostly less than a million years old and have erupted calc-alkaline rocks, and latterly more potassic products, culminating in the **shoshonitic lavas** now emitted by Stromboli and Vulcano. The islands and their volcanoes are quite varied in age, appearance and style of activity. Several of them seem to be extinct, Lìpari may be still active, Vulcano erupts infrequently and Stromboli has been one of the most continual volcanic performers of modern times.

Etna and the Monti Iblei constitute the contrasting volcanic areas on the island of Sicily, and offshore eruptions formed Pantelleria and Linosa to the south and Ustica to the northwest. Ustica in the Tyrrhenian Sea may be related to the Aeolian subduction zone. Pantelleria and Linosa erupted in submarine fault troughs. In 1831, they gained an ephemeral companion when Graham Island rose above the waves in a **Surtseyan eruption**. The Monti Iblei also erupted just below sea level. Their **tholeiitic** and **alkali basalts** erupted on a limestone platform on the outer margin of the African plate, where it had been broken by major faults.

Etna, by its size and origin, is in a class of its own in relation to the other Italian volcanoes. It seems to have formed where at least three major fractures intersect and usually allow magma to rise rapidly and relatively unhindered to the surface. It is a spectacular cone, more than 3000 m high and covering an area of 1200 km^2. Its eruptions began with tholeiitic basalts and continued with alkaline lavas, such as trachybasalts (hawaiites), **trachyandesites** (mugearites) and **benmoreite**. Although Etna has experienced two caldera-forming eruptions in the past few thousand years, its predominant activity has usually been persistently effusive and mildly explosive. Its frequent eruptions have produced many lava flows and **satellite cones**.

CAMPANIA: VESUVIUS

Vesuvius is irresistible. It is the focus of one of the world's most beautiful landscapes, sweeping up from the Bay of Naples in a sleek cone of exquisite proportions in front of the protecting arm of Monte Somma. The view from Naples has been painted and photographed so often that Vesuvius has been instantly recognizable for centuries. It is by far the most violent volcano that has erupted in Europe in historical times, and its outbursts have been observed for nearly 2000 years, because they repeatedly destroyed the crops, property and lives of those attracted to the fertile soils around it. Vesuvius also has great methodological importance. In his two letters to the historian Tacitus, Pliny the Younger described the great cataclysm of AD 79 with such accuracy and vivid clarity that they constitute the first scientific account of any eruption in the world, and this outburst rightly became the prototype of Plinian eruptions (Sheridan et al. 1981a, Walker 1981).

As if the aesthetic quality of the site, the exceptional length of its historical records, and the violence of its major eruptions were not enough, Vesuvius has also played an equally outstanding archaeological role. It is the most famous volcano in the world because it buried Pompeii and Herculaneum. It is still by far the most dangerous volcano in Europe in terms of risks to human life. It has been at rest ever since it erupted to greet the advancing Allied armies in March 1944, but after such a long interval of repose, the next eruption will be the most violent since 1631. The outburst is likely to be sudden

and directed to the large population living on its southern flanks. No practical steps could be taken to stop the rain of pumice or the nuées ardentes. Nearly a million people would have perhaps less than a day to save their lives (Santacroce 1996, Civetta 1998).

Today, the main cone, or Gran Cono, of Vesuvius, rising to 1281 m, is enclosed on the north and east by the semi-circular rampart of Monte Somma, which is 1132 m high. Between the two lies the Valle dell'Inferno, which extends westwards to valleys of the Atrio del Cavallo and the Fosso della Vetrana, which many lava flows have invaded in historical times. Monte Somma is the remains of the old Somma stratovolcano, which covers an area of about 480 km^2. The crest of Somma probably collapsed piecemeal during a series of violent eruptions that began about 18000 years ago. They left behind an asymmetrical caldera, whose northern and northeastern rims now form Monte Somma. Vesuvius grew up within this caldera. The much lower southern and southwestern sectors of the Somma caldera have been all but completely covered by eruptions of Vesuvius, although at the start of the twentieth century they still formed a shoulder on the flanks sloping towards the Bay of Naples. Thus, the southern, highly populated, sectors of the volcano have been wide open to the dangerous effects of virtually all the eruptions of Vesuvius for several thousand years. Both Vesuvius and Somma have had a strong central focus of activity and, although lateral fissures have erupted lava flows since 1631, few satellite cones appear to have formed. On the lower southern slopes, the prehistoric Camaldoli della Torre

Vesuvius from Naples. with the ridge of Monte Somma on the left and extensive settlements on the right, which would be in danger of destruction during any new eruption.

now forms a cone, 80 m high, that is now almost completely swamped by lava. The two cones of Il Viulo and Fossa Monaca erupted nearby in about AD 1000, and another followed in 1760. The lateral vents that developed in the southwest in 1794 and 1861 were no more than wide fissures and they are already disappearing beneath the vegetation, but higher vents extruded the cupolas of the Colle Margherita in 1891 and the Colle Umberto in 1895.

The base of Somma–Vesuvius lies below sea level, so the volcano is larger than it seems. Nevertheless, Vesuvius itself is a relatively small stratovolcano, in spite of its strong focus and its long history of copious output. If the testimony of the many landscape painters is to be believed, the almost persistent activity from 1631 to 1944 produced remarkably few changes beyond the crater. The volcano reached its maximum recorded altitude of 1335 m on 22 May 1905, after six months of persistent explosive and effusive activity that had itself formed part of the continual agitation lasting since December 1875 (Perret 1924, Carta et al. 1981). The eruption in April 1906 decapitated this summit, and a subsequent landslide from the crater rim reduced it to 1186 m, before it reached its present height of 1281 m after the eruption in 1944.

Historical records of eruptions

Vesuvius has been accessible and easily visible in the midst of a well populated area for more than 3000 years and its behaviour has been scrutinized by literate observers for longer than any other volcano in Europe except Etna (Albore Livadie et al. 1986, Gasparini & Musella 1991, Scandone et al. 1993).

Vesuvius was calm for a long period before AD 79, although there may have been an eruption in the eighth century BC and another, perhaps, in 217 BC (Stothers & Rampino 1983, Principe, personal communication 1998). In sharp contrast, hour-by-hour events of the eruption of AD 79 have now been reconstructed (Sigurdsson et al. 1985). Thereafter, information is vague, although some kind of activity, probably persistent, occurred between AD 80 and 120, 170 and 235 (especially in 203), and in the late fourth century. The details of the Pollena eruption that occurred in AD 472 have been deduced from documentary and geological evidence, as well as from a broad range of radiocarbon dates. Activity

The ridge of Monte Somma and the cone of Vesuvius from the air

was also noted, or claimed, in 536, 685, 787, 968, 991, 999, 1006–1007, 1037, 1139, and possibly about 1350 and 1500 (Alfano & Friedlander 1929, Carta et al. 1981). But it is not clear what erupted, or whether these dates represent separate eruptions that were part of longer periods of activity. Nevertheless, current archaeomagnetic studies indicate that, between about 750 and 1150, many copious lava flows erupted, which now form most of the coast between Portici and Torre Annunziata (Principe et al. 1998, Tanguy et al. 1999).

Vesuvius fell into repose well before the great eruption in 1631 initiated the revival of the pulsating activity that then lasted until 1944. After 1631, too, the more accurate scientifically based analyses can often be compared with paintings and literature to provide a much more extensive picture. There were notable eruptions in 1707, 1737, 1754, and especially 1760, when vents opened low down on the southern slopes of the cone. Lava flows spread in several directions in 1767 and an immense lava fountain marked the eruptive climax in 1779. Further notable eruptions occurred in 1794, 1822, 1839, 1850, 1855, 1858, 1861 and 1872. The period between 1875 and 1906 was the most complex episode of persistent activity in modern times, although it continued on a slightly lesser scale until the eruption in 1944 brought this phase to an end.

The growth of Vesuvius

Vesuvius rises at the eastern end of the Campanian volcanic province in one of the most fractured and unstable zones of the Italian Peninsula. The volcanic activity arises where it is caught between the westward encroachment and

The climax of the eruption in April 1906 seen from close range by Frank Perret

Towards 10 p.m. [22.00, 7 April] masses of cone material again began to be mixed with the incandescent jets, with consequent renewal of powerful, brilliant electrical discharges in wondrous contrast to the golden spearheads of the ejected lava piercing the dark detritus clouds. Then all would again be cleared of ash, and the pillar of liquid fire – maintained continuously at a height of several kilometers by multiple projections from all parts of the magma column within the conduit – illumined the Gulf of Naples from Capri to Miseno. [At 22.20 and 22.40] two brilliant fountains of fire upon the flank of the mountain seemed at first to indicate the opening of new craters, but they proved to be a violent renewal of activity at the lava vents of Bosco di Cognoli and the Valle dell'Inferno. From these there issued veritable torrents of lava, which in a few hours reached the first houses of Boscotrecase and later crossed the town and the circum-Vesuvian railway. . . .

At 12.37 a.m. [00.37] on 8 April, a strong shock was felt, followed soon after by visibly greater explosive activity at the crater. At the altitude of the Observatory the ground was in constant motion. There was, in a word, a continuous earthquake, and for some hours (which constituted the period of dynamic culmination) it was impossible to stand quite still. Within the building it was difficult to cross a room without steadying oneself with a hand against the wall. To make sure of this not being merely an oscillation of the structure, the writer went outside and braced himself against a stone wall; but the effect was the same – like the shell of a humming boiler, the mountain was pulsing and vibrating continuously, but with a period and an amplitude proportioned to its size. . . .

The obvious peril to the Observatory drove us out of doors, where at only 2.5 km from the radiant column of incandescent pasty fragments – at this time already 300 m in diameter and 3000 m in height – the air was so cold as to constrain us to build a fire and sit close about it for warmth. This condition was caused by the aspiration of cold air from the sea by the tremendous updraft of the pillar of fire. . . .

The most alarming feature at this time was the continuous increase – each earthshock felt above the regular pulsation was stronger than its predecessor; each wavecrest on the sea of sound was louder than before;

the jets of the great fiery geyser shot ever higher into the dark overhanging pall of blackness that extended over our heads and fell westwards in a thick veil, through which, from Naples, could be seen only fitful gleams. But between the Observatory and the crater all was clear, and it becomes increasingly difficult to describe the events of the great culmination in words befitting a scientific book. . . .

With increasing amounts of detritus, the electrical manifestations now reached an appalling intensity. . . . At 3.30 a.m. [03.30], April 8, there began the truly dynamic culmination of the great eruption, with a literal *unfolding* outwardly of the upper portions of the cone in all directions, like the petals of a flower. Those who speak of the mountain having fallen *in* were not eyewitnesses of what occurred. No mass of matter, however great, could descend against the mighty uprush of gas that was now liberated from the depths. . . . This colossal column, with ever-increasing acceleration, was actually coring out and constantly widening the bore of the volcanic chimney. . . .

The sound, also throughout the entire culmination, was an uninterrupted compound note not unlike the roar of Niagara, but with a recurrent crescendo–diminuendo effect giving to this phenomenon, along with all other manifestations of the great eruption, the wave form. Above the curve of dynamic intensity – itself a slow wave synchronizing with the still slower one of luni–solar influence – there were superposed these rolling cadences of sound, the seismic pulsing of the ground, and the exceedingly rapid undulations of the electric flashes in the cloud. The great eruption was a sublime manifestation of rhythm.

Strongest of all impressions received in the course of these remarkable events, greatest of all surprises, and most gratifying of all features to record, was, for the writer, that of an infinite dignity in every manifestation of this stupendous releasing of energy. No words can describe the majesty of its unfolding, the utter absence of anything resembling effort, and the all-sufficient power to perform the allotted task and to do it majestically. Each rapid impulse was the crest of something deep and powerful and uniform which bore it, and the unhurried modulation of its rhythmic beats sets this eruption in the rank of things which are mighty, grave, and great. (Perret 1924)

subduction of the Apulian and Adriatic microplates, on the one hand, and the eastward movements opening up the Tyrrhenian Sea basin on the other. Recent activity of Somma–Vesuvius has also been helped by reactivation of major regional faults along a fault trough running from northeast to southwest (Luongo et al. 1991, Scandone et al. 1991).

Although volcanic products from its southern basement have been dated to 0.3–0.5 million

years ago, most of the Somma stratovolcano, which had a total volume of 100–150 km³, grew up after the expulsion of the Campanian Ignimbrite, about 37 000 years ago (Andronico et al. 1998). It seems that repeated explosive and effusive eruptions of potassic lavas built Somma into a cone rising perhaps to 1800 m above sea level. Then came the great eruptions and asymmetrical caldera collapses that decapitated Somma and left the caldera rim much higher on

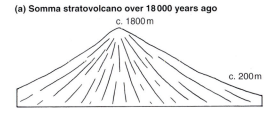

(a) Somma stratovolcano over 18 000 years ago

c. 1800 m

c. 200 m

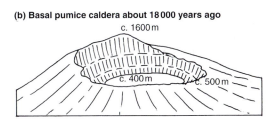

(b) Basal pumice caldera about 18 000 years ago

c. 1600 m

c. 400 m

c. 500 m

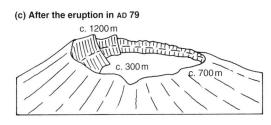

(c) After the eruption in AD 79

c. 1200 m

c. 300 m

c. 700 m

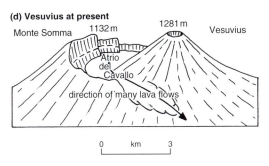

(d) Vesuvius at present

Monte Somma   1132 m   1281 m   Vesuvius

Atrio del Cavallo

direction of many lava flows

0   km   3

The growth of Monte Somma and Vesuvius.

the north than on the south. The date of the formation of the caldera, and consequently the date of birth of Vesuvius within it, is one of the main structural and morphological problems on the volcano. It was often believed that the caldera was formed in AD 79, but there is evidence to suggest that Vesuvius could be older. Most of the lavas found on the northern slopes of Somma are more than 18 000 years old. Therefore, the caldera and Monte Somma must have started to form at that date and prevented lavas from reaching beyond the ridge. Subsequent eruptions have remodelled the caldera and could have built up

a cone that can be considered to be the direct ancestor of modern Vesuvius. Thus, the cataclysm of AD 79 was only the last of the major outbursts that have altered the form of the caldera (Santacroce et al. 1987, Cioni et al. 1999).

Vesuvius emits potassic (sometimes highly potassic) materials generated when basic magma ascends continually from the depths, but halts intermittently and undergoes spells of differentiation in shallow reservoirs (e.g. Métrich 1985, Santacroce et al. 1987). These reservoirs develop in dolomitic layers beneath the volcano, and it was long believed that assimilation of these carbonates played a crucial role in the composition of the magma (Rittmann 1962). Today, such a role tends to be denied or minimized. Before the eruption in AD 79, a shallow reservoir of at least 5 km^3 was probably lying at about 3–6 km deep (Barberi et al. 1981, Cioni et al. 1995), but it was largely emptied by that eruption. At present, there does not seem to be a shallow reservoir, but recent seismic studies have detected the roof of a mass that could be magma about 10 km deep (Zollo et al. 1998). In general terms, when dark, rather primitive magma is able to reach the surface relatively unimpeded, **basanites** and tephrites are emitted in brief, but quite frequent, periods of persistent activity. When the magma halts a while in reservoirs and undergoes some differentiation, then both primitive and intermediate lavas, tephritic phonolites and phonolites are given off in Plinian and sub-Plinian eruptions. A relationship exists between the length of the repose periods, the power of the ensuing eruptions, and the volumes of materials ejected (Santacroce et al. 1987, Civetta et al. 1992). Intervals of repose are therefore crucial to understanding and predicting its behaviour.

Thus, during the past 18 000 years or so, Vesuvius has undergone several major cycles, usually lasting for a few millennia (Santacroce 1983, Andronico et al. 1998). These major eruptive cycles were all started by vast Plinian outbursts that are indicated by dated phonolitic pumice layers. They had all been preceded by long periods of repose that are revealed by the presence of ancient soils. The eruption of the basal pumice (pomici di base) initiated the caldera and the first cycle 18 000 years ago. The eruption of the Mercato Pumice marked the start of the second cycle about 8000 years ago, and the Avellino Pumice, dated to 3800 years ago, began the third cycle. The Pompeii pumice, erupted in AD 79, initiated the latest cycle.

A few sub-Plinian eruptions of medium intensity, and many eruptions of lesser intensity, probably occurred during each cycle, before a long period of repose supervened. These periods usually lasted several centuries while the next cataclysm was brewing in the magma reservoir. The **magmatic differentiation** began when tephritic magma at about 1100°C halted in a shallow reservoir, and tephritic–phonolitic and even phonolitic magmas slowly developed. The upper reaches of the reservoir became increasingly evolved and gaseous as the repose period lengthened. The stratification developed by the crystal fractionation and slow cooling was emphasized by the regular influx of fresh basic magma into the bottom of the reservoir. These changes continued until the pressures generated caused the magma to rise and overcome the strength of the confining crust under the volcano. The longer the magma remained in the reservoir, the greater the explosion eventually unleashed by Vesuvius. The work of centuries could then culminate in almighty explosions usually lasting less than three days.

Semi-persistent activity

Small-scale pulsating activity was typical of the behaviour of Vesuvius between 1631 and 1944, and especially between 1694 and 1872. The pulsations were similar but more prolonged after 1875 (Carta et al. 1981). Phases of calm were followed by persistent mild activity, then by larger eruptions with bigger lava flows, and lastly by a short so-called final eruption that could almost reach sub-Plinian proportions. The flows produced by dark, rather primitive magmas, composed mainly of the basanites and tephrites that rose fairly rapidly from the depths (Santacroce 1983). These eruptions completely emptied the vent, which then became blocked when its walls crumbled and withstood with increasing efficiency the magmatic pressure from below. Thus, a phase of calm, expelling only fumaroles, always followed one of these final eruptions.

Persistent Strombolian activity
The phase of calm was followed by a phase of mild continuous activity, which included Strombolian explosions and sometimes lava fountaining, that formed a small **spatter** and cinder cone, as well as minor lava flows inside the crater. The crater thus gradually filled. This activity

occurred for more than half the total period between 1631 and 1944. It lasted for varying periods, often of several years, and even exceeded 30 years between 1875 and 1906, and between 1913 and 1944 (Imbò 1949). Except for a doubtful case in 1682, no such phase was ever followed by a return to calm conditions: eruptive activity always increased.

Intermediate eruptions
The intermediate eruptions were Strombolian on a grander scale and they were often superimposed upon the periods of persistent activity. During these phases, more lava, with a higher gas content, was expelled at a faster rate and built up a large cinder cone, and also flooded the Atrio del Cavallo and beyond with tephritic–leucititic lava flows. Forty-four such phases have been identified between 1631 and 1944. Until the end of the nineteenth century, they usually lasted only a week or a month or so. Then, as the vent remained continuously open from 1875 to 1906, long periods of Strombolian activity alternated with intermediate eruptions for years at a time: 1881–4, 1885–8, 1891–4, 1895–9 and 1903–1904. These intermediate eruptions were never followed by a return to repose. They could be succeeded either by a return to the mild phase or by a sudden change of gear to a final eruption.

Final eruptions closing periods of persistent activity
The final eruptions were more vigorous, much shorter and less frequent than any other form of activity between 1631 and 1944. The 23 such phases each lasted only from one to two weeks. They often began by hydrovolcanic explosions, followed by an eruption of lava fountains and fine fragments that formed an eruptive column that rose into the stratosphere and showered down **ash** and pumice and often obliterated the Sun more than 100 km away. Vast fluid lava flows emerged from both the central crater and lateral fissures. Although the final eruptions are several orders of grandeur down the eruptive scale from the Plinian eruption of AD 79, they have been sufficiently imposing to impress painters such as Tomasso Ruiz, poets such as Lamartine in 1822, scientists such as Poullet-Scrope in 1822 and Perret in 1906, as well as the Allied troops in 1944. In 1794, for example, lava surged from a fissure that opened just above Torre del Greco, destroyed three quarters of the town within a few hours, and trapped and killed

15 people (Hamilton 1795). In 1872, flows spread over the northwestern slopes of the Gran Cono from a fissure that suddenly opened and fired out volcanic **bombs** that caused 12 deaths (Palmieri 1872). These final eruptions bring the cycles of persistent activity to a close, because they are invariably followed by a period of calm.

The eruption in 1906

The final eruption of 1906 lasted an unusually long time, from 4 April to 23 April, second only to the 23-day final eruption in 1794. It had an eminent eye witness, Frank Perret, who stayed at the Vesuvian Observatory with the director, Matteucci, throughout its climax. The eruption began when moderate explosions intensified at the crater, and lava flows emerged from new vents on the upper southern slopes of the main cone. On 7 April, as tremors increased in intensity, lava fountains soared 3 km skywards, and torrents of lava gushed from a fissure and partly destroyed Boscotrecase on the lower southeastern flanks of Vesuvius. Early on 8 April, the eruption column shot obliquely to the northeast and showered the lower flanks of Monte Somma with so many lapilli and bombs that many roofs caved in at San Giuseppe and especially at Ottaviano, where 100 people were killed when the church roof collapsed on them. Then, a sub-Plinian column rose 13 km into the air as if it had been generated by a giant steam engine. It lasted for 18 hours and eventually decapitated the summit of the cone. From 9 April until the eruption ended on 23 April, large amounts of fine, often pale, ash were expelled. Heavy rain also caused **mudflows** that severely damaged Ottaviano. Beneath the falling ash in Naples, processions of panic-stricken inhabitants filled the streets invoking the intercession of the Saints. This eruption killed at least 218 people. Molten lava trapped and killed three old men at Boscotrecase, and the rest died when roofs fell upon them. Calm duly returned and remained for seven years. Vesuvius had lost 115 m from its summit and the crater was now an abyss 700 m across and more than 600 m deep (Mercalli et al. 1907, Perret 1924).

The eruption in 1944

When Vesuvius resumed activity on 5 July 1913, it began a long period of persistent activity that filled the crater formed in 1906 to such an extent that the summit of the inner cone that nested within it could be seen from Naples rising above the main crater rim. The final eruption of 1944,

which was witnessed by Giuseppe Imbò, lasted from 18 to 29 March, as the Allied armies advanced through Campania. From 13 March, oscillations of the magmatic column destroyed the inner cone, which collapsed into the main vent. At 16.30 on 18 March, lavas emerged from the crater, channelled westwards along the Atrio del Cavallo, and caused the greatest damage during the eruption when they engulfed San Sebastiano and Massa within the next three days. On 20–21 March, eight lava fountains, each lasting 20–40 minutes, gushed 2000 m above the crater, with smaller lava fragments reaching as high as 4 km (Imbò 1949). They were followed on 22 March by violent hydrovolcanic explosions that ejected ash-laden clouds and small nuées ardentes. The wind blew great quantities of ash eastwards, where, for instance, 80 cm accumulated 5 km away at Terzigno. Gradually, the ash explosions waned and they ceased altogether on 29 March. The eruption of 1944 removed most of the material that had filled the crater during the previous 31 years of persistent emissions. The new crater had shifted 200 m to the south and was only one third of the size of its predecessor. It was 300 m deep, 580 m long and 480 m wide,

Vesuvius erupting in March 1944.

and has been the scene of variable mild fumarole activity throughout the ensuing period of repose.

Repose periods

Since 1631, repose periods have usually lasted for months or years after the final eruptions. Half of these calm phases lasted only between one and three years, and none lasted for longer than seven years (Carta et al. 1981, Santacroce 1983). However, the calm since 1944 has now lasted so long that it far exceeds any period in modern times and may signal a complete change in the rhythm of activity of Vesuvius. If so, it may indicate either the close of the period that began in 1631, or even the end of the major Pompeian cycle that started in AD 79. In the first case, the calm will last only decades or a few centuries, as happened, for instance, before the sub-Plinian eruptions in AD 472 and 1631; Vesuvius will, therefore, most probably erupt with sub-Plinian violence, perhaps between 50 and 250 years hence. In the second case, if the repose since 1944 marks the last phase of the Pompeian cycle, then complete calm should reign at Vesuvius for at least several centuries before a Plinian eruption initiates a new major cycle. At all events, because of the relationship between the length of the repose period and the violence of the eruption that concludes it, the next outburst will be the most intensely explosive since 1631.

Sub-Plinian eruptions

Sub-Plinian eruptions are smaller versions of Plinian eruptions, but are just as dangerous to those living nearby. They seem to be set in motion by the upsurge of unusual volumes of fresh magma from the depths that bring a long period of repose to an end. The eruptions in AD 472 and 1631, for example, were preceded by at least several dormant decades. These sub-Plinian eruptions consisted of violent explosions of pumice and nuées ardentes, and they discharged fairly large volumes of 0.3–1 km^3 of material from intermediate or evolved magmas, usually ranging from tephrites to phonolites. In some cases, the sub-Plinian eruptions seemed to clear the vent for a long period of smaller-scale semi-persistent activity.

The Pollena eruption in AD 472
The sub-Plinian Pollena eruption that occurred in AD 472 deposited more than 30 m of nuée ardente fragments at Pollena on the northwestern slopes of Monte Somma, and 8 cm of ash on Constantinople, 1200 km away. On 5–6 November, "Vesuvius . . . vomited up its completely consumed inner parts and turned day into night, covering the whole surface of Europe with fine dust. Every year, on 6 November, the people of Constantinople celebrate the memory of those terrifying ashes" (Stothers & Rampino 1983). Cassiodorus also clearly described the nuées ardentes as "rivers of dust and sterile sand that a raging blast had raised into liquid flows". The Pollena deposits indicate a large eruption with a high eruptive column, with nuées ardentes on the northwest and pumice falls that accumulated thickly, especially to the east-northeast of Vesuvius. As the eruption progressed, evolved phonolitic magmas were expelled first; and they were followed by phono-tephritic magmas from the lower reaches of the reservoir. Eventually, hydrovolcanic activity brought the eruption to a close. Some 0.32 km^3 of fragments were ejected in the course of about a week (Rosi & Santacroce 1983). It was probably the most violent of all the outbursts of Vesuvius after AD 79.

The eruption of 1631
The eruption of 1631 is of great importance in the history of Vesuvius because it not only initiated the semi-persistent activity that lasted until 1944, but also because it was the most violent eruption since AD 472, and it ended the long interlude of Renaissance calm (Rosi et al. 1993). Vesuvius had been shaken by some earthquakes during the preceding five months, but they became more frequent on 10 December when the ground rumbled and wells dried up. The shocks reached alarming proportions during the night of 15–16 December. The eruption began with a violent explosion at 07.00 on 16 December 1631. A sub-Plinian column soared to the stratosphere for 11 hours. The volcano then roared throughout the night. At 10.00 the following morning, nuées ardentes burst from the cone, swept to the sea between Portici and Torre Annunziata, and more or less engulfed Torre del Greco, Pugliano, Portici, as well as Resina, which had been built, although no-one yet knew it, over part of Herculaneum. These nuées ardentes (pyroclastic flows or surges) killed most of the victims of the eruption; and at least 20 000 people fled in terror to Naples, where they were housed in the leper colony to prevent the spread of disease (Principe 1998). Soon, violent explosions of ash also

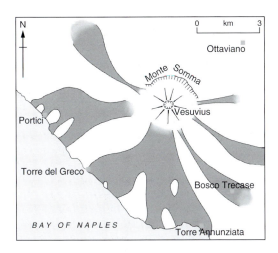

Distribution of nuées ardentes erupted from Vesuvius in December 1631

blanketed the volcano. Even at Naples, 12 km away to the west, the ash accumulated to a depth of 30 cm in total darkness that lasted for over two days. Torrential rain mixed with the ash on Monte Somma and generated mudflows that swamped Pollena, Massa and Ottaviano, and devastated the plain of Nola. After about a week, the power of the eruption waned considerably, but carried on into January 1632.

Contrary to long-established opinion, no lava flows seem to have been expelled during this eruption; the supposed fast-moving lava flows were, in fact, nuées ardentes. The lavas that have been attributed to this eruption really emerged between about 750 and 1139 (Principe et al. 1998, Tanguy et al. 1999). In 1631, the initial eruptions discharged phonolitic leucitites from the upper parts of the evolving magma reservoir, and tephritic leucitites were expelled from its lower reaches as the eruption concluded. At the end of the eruption, at least 4000 people, and perhaps as many as 10 000, had been killed (Gasparini & Musella 1991, Nazzaro 1997). Vesuvius had lost 470 m from its crest, the sub-Plinian gas blast had widened the crater to 1600 m across; and the whole area around the volcano had been devastated. Calm then returned until the first phase of semi-persistent effusive activity in modern times began on 15 April 1638.

Large-scale Plinian eruptions

The Plinian outbursts of Vesuvius were gas-blast eruptions of enormous proportions and their deposits all reveal a similar pattern. They began with explosions of trachytic or phonolitic ash and pumice that rose over 20 km high in an eruptive column and spread out into a typical umbrella-pine tree shape. The fragments were first white and then grey. As the column collapsed from time to time, nuées ardentes – the really lethal elements of the eruption – rushed down the volcano. At length, when most of the contents of the reservoir had been ejected, the summit of the volcano collapsed into a caldera, which every subsequent outburst then widened. In all, up to 5–10 km³ of fragments (Sigurdsson et al. 1985, Cioni et al. 1999), covering about 500 km² in area, were probably expelled in periods of two or three days. Thus, these Plinian eruptions occupied phases totalling scarcely 15 days in 18 000 years, but enough material was discharged at sufficient speed to bury several towns completely when the occasion presented itself in AD 79.

The cycle that began with the eruption of the Avellino Pumice about 3800 years ago included at least six sub-Plinian eruptions (Andronico et al. 1998), the last of which might have been the doubtful eruption in 217 BC. However, scholars such as Diodorus Siculus, Vitruvius and Strabo recognized that Vesuvius was a volcano because of the burnt stones strewn across its summit. Nevertheless, the main perceived hazard in the district was undoubtedly the earthquakes that often shook Campania. Most of them caused no more than a shudder in the ground, but the earthquake on 5 February AD 62 had caused considerable damage in and around Pompeii, which had not been entirely repaired 17 years later (Albore Livadie et al. 1986).

The eruption in AD 79

In AD 79, Pompeii was probably one of the largest towns in Campania, with a population of about 20 000, 10 km southeast of Vesuvius. Herculaneum was a smaller seaside town of about 5000 people, 7 km due west of Vesuvius. Naples, which was probably then, as now, the chief centre, and Stabiae, Oplontis and the naval base of Misenum in the west, were the other major settlements on the Bay of Naples. All were built on recent volcanic rocks, and Pompeii, Herculaneum and Oplontis were constructed on rocks expelled by Vesuvius within the previous 10 000 years. Being on the southern side of the volcano, they would thus be vulnerable if Vesuvius were ever to revive (Plate 1a).

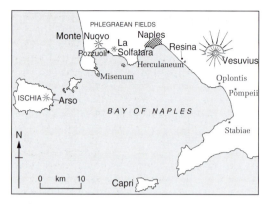

Vesuvius and the Phlegraean Fields, with Monte Nuovo, Arso and La Solfatara.

In the summer of AD 79, the 17-year-old Pliny the Younger was staying with his mother and her brother, Pliny the Elder, at Misenum, 32 km west of Vesuvius. His first letter to Tacitus describes his uncle's journey to his death at Stabiae, and the second his own experiences at Misenum (e.g. Scarth 1994, 1999). There is no mention of Pompeii or Herculaneum in Pliny's narratives.

The tectonic earthquake that damaged much of Campania in AD 62 could have been a precursor of the eruption 17 years later. Two features, recounted by Seneca (VI: 1 and 27), suggest that it could have been related to an increase in magmatic pressure beneath the volcano: the earthquake was centred on the area around Vesuvius, and 600 sheep were poisoned in the Pompeii region, "when, at ground level, they inhaled some baleful air," probably carbon dioxide.

Earth tremors of increasing frequency began several days before and gave premonitory but unheeded signs of the impending eruption. Early on the morning of 24 August, the initial hydrovolcanic explosions cleared the conduit and distributed fine ash, especially to the north and east (Sigurdsson et al. 1985, Scarth 1999).

Just before 13.00 on 24 August, the Plinian climax began and its eruptive column of gas and mainly phonolitic fragments rose from the upper layers of the magma reservoir and soon reached a height of 27 km, as the wind winnowed out the ash and white pumice, raining them down in a southeasterly direction. Pompeii was thus situated in the zone of maximum accumulation. Throughout the next seven hours, a thickness of about 1.40 m of ash, pumice and fist-size lithic fragments fell on Pompeii. Roofs caved in and, by the evening, most of the people had probably panicked, fled, and saved their lives.

From about 20.00 on 24 August, darker tephriphonolitic magma exploded from the middle zones of the magma reservoir. The Plinian column rose to a height of 33 km. The debris expelled became coarser and changed to greenish-grey. By midnight, a thick blanket covered an area southeast of Vesuvius from Terzigno to Oplontis. On top of the white pumice, 1.30 m of grey pumice eventually fell on Pompeii that night. However, the western areas around Misenum remained unaffected on 24 August, and Herculaneum, much closer to Vesuvius, had been spared all but showers of fine ash (Plate 1b).

At 01.00 on 25 August came the major change in the Plinian eruption. Magma that was rather poorer in explosive gases began to be emitted. Thus, the column could not always sustain its upward impetus, and its lower parts collapsed from time to time, crumbling like a pillar of fire, to form nuées ardentes. The first nuée ardente was funnelled westwards by the Atrio del Cavallo at a speed of at least 100 km an hour and reached Herculaneum in less than 4 minutes. Its remaining inhabitants had time only to flee to the shore, where they died choking in the swirling

The fuming eastern wall of the crater of Vesuvius.

The eruption that killed Pliny the Elder

"At that time [24 August AD 79] my uncle was at Misenum in command of the fleet. About one in the afternoon, my mother pointed out a cloud with an odd size and appearance that had just formed. From that distance it was not clear from which mountain the cloud was rising, although it was found afterwards to be Vesuvius. The cloud could best be described as more like an umbrella pine than any other tree, because it rose high up in a kind of trunk and then divided into branches. I imagine that this was because it was thrust up by the initial blast until its power weakened and it was left unsupported and spread out sideways under its own weight. Sometimes it looked light coloured, sometimes it looked mottled and dirty with the earth and ash it had carried up. Like a true scholar, my uncle saw at once that it deserved closer study and ordered a boat to be prepared. He said that I could go with him, but I chose to continue my studies.

Just as he was leaving the house, he was handed a message from Rectina, the wife of Tascus, whose home was at the foot of the mountain, and had no way of escape except by boat. She was terrified by the threatening danger and begged him to rescue her. He changed plan at once and what he had started in a spirit of scientific curiosity he ended as a hero. He ordered the large galleys to be launched and set sail. He steered bravely straight for the danger zone that everyone else was leaving in fear and haste, but still kept on noting his observations.

The ash already falling became hotter and thicker as the ships approached the coast and it was soon superseded by pumice and blackened burnt stones shattered by the fire. Suddenly the sea shallowed where the shore was obstructed and choked by debris from the mountain. He wondered whether to turn back, as the captain advised, but decided instead to go on. "Fortune favours the brave", he said, "take me to Pomponianus". Pomponianus lived at Stabiae across the Bay of Naples, which was not yet in danger, but would be threatened if it spread. Pomponianus had already put his belongings into a boat to escape as soon as the contrary onshore wind changed. This wind, of course, was fully in my

uncle's favour and quickly brought his boat to Stabiae. My uncle calmed and encouraged his terrified friend and was cheerful, or at least pretended to be, which was just as brave.

Meanwhile, tall broad flames blazed from several places on Vesuvius and glared out through the darkness of the night. My uncle soothed the fears of his companions by saying that they were nothing more than fires left by the terrified peasants, or empty abandoned houses that were blazing. He went to bed and apparently fell asleep, for his loud, heavy breathing was heard by those passing his door. But, eventually, the courtyard outside began to fill with so much ash and pumice that, if he had stayed in his room, he would never have been able to get out. He was awakened and joined Pomponianus and his servants who had sat up all night. They wondered whether to stay indoors or go out into the open, because the buildings were now swaying back and forth and shaking with more violent tremors. Outside, there was the danger from the falling pumice, although it was only light and porous. After weighing up the risks, they chose the open country and tied pillows over their heads with cloths for protection.

It was daylight everywhere else by this time, but they were still enveloped in a darkness that was blacker and denser than any night, and they were forced to light their torches and lamps. My uncle went down to the shore to see if there was any chance of escape by sea, but the waves were still running far too high. He lay down to rest on a sheet and called for drinks of cold water. Then, suddenly, flames and a strong smell of sulphur, giving warning of yet more flames to come, forced the others to flee. He himself stood up, with the support of two slaves, and then he suddenly collapsed and died, because, I imagine, he was suffocated when the dense fumes choked him. When light returned on the third day after the last day that he had seen [on 26 August], his body was found intact and uninjured, still fully clothed and looking more like a man asleep than dead." (Pliny VI: 16)

cloud. Hundreds of victims were discovered there in the excavations in 1982. As fewer than a dozen victims had previously been discovered in Herculaneum, it was assumed that the townspeople had been able to flee to safety, and thus that the town must have been among the last casualties of the eruption. Herculaneum, in fact, was the first major victim of Vesuvius in AD 79.

Meanwhile, as the grey pumice continued to fall, a second nuée ardente at about 02.00 completed the burial of Herculaneum. A third nuée ardente, at about 06.30, buried Oplontis and spread as far as the northern walls of Pompeii, and may have prompted many other citizens to

leave. A great increase in tremors about that time, marking perhaps the beginning of caldera collapse, can only have reinforced their fears. Pompeii was already blanketed in 2.4 m of ash, pumice and even larger rock fragments, and they were still falling in the stifling darkness.

At 07.30 on 25 August, a fourth nuée ardente completely overwhelmed Pompeii and all the 2000 people remaining in the city. Some were killed by falling columns or tiles and bricks ripped from the buildings, others were baked, but most were asphyxiated by a mixture of hot ash and mucus inhaled with their last breaths. The plaster casts made during the excavations

Pliny at Misenum: a report of human reactions to a violent eruption 2000 years ago

"Meanwhile, my mother and I had stayed at Misenum. After my uncle left us, I studied, dined and went to bed, but slept only fitfully. We had had earth tremors for several days, which were not especially alarming, because they happen so often in Campania. But that night they were so violent that everything felt as if it were being shaken and turned over. My mother came hurrying to my room and we sat together in the forecourt facing the sea.

By six o'clock, the dawn light was still only dim. The buildings around were already tottering and we would have been in danger in our confined space if our house had fallen down. This made us decide to leave town. We were followed by a panic-stricken crowd that chose to follow someone else's judgement rather than decide anything for themselves. We stopped once we were out of town and then some extraordinary and alarming things happened. The carriages we had ordered began to lurch to and fro, although the ground was flat, and we could not keep them still even when we wedged their wheels with stones. Then we saw the sea sucked back, apparently by an earthquake, and many sea creatures were left stranded on the dry sand. From the other direction over the land, a dreadful black cloud was torn by gushing flames and great tongues of fire like much-magnified lightning.

The cloud sank down soon afterwards and covered the sea, hiding Capri and Capo Misenum from sight. My mother begged me to leave her and escape as best I could, but I took her hand and made her hurry along with me. Ash was already falling by now, but not very thickly. Then I turned around and saw a thick black cloud advancing over the land behind us like a flood. "Let us leave the road while we can still see", I said, "or we will be knocked down and trampled by the crowd." We had hardly sat down to rest when the darkness spread over us. But it was not the darkness of a moonless or cloudy night, but it was just as if the lamps had been put out in a completely closed room.

We could hear women shrieking, children crying and men shouting. Some were calling for their parents, their children, or their wives, and trying to recognize them by their voices. Some people were so frightened of dying that they actually prayed for death. Many begged for the help of the gods, but even more imagined that there were no gods left and that the last eternal night had fallen on the world. There were also those who added to our real perils by inventing fictitious dangers. Some claimed that part of Misenum had collapsed or that another part was on fire. It was untrue, but they could always find somebody to believe them.

A glimmer of light returned, but we took this to be a warning of approaching fire rather than daylight. But the fires stayed some distance away. The darkness came back and ash began to fall again, this time in heavier showers. We had to get up from time to time to shake it off, or we would have been crushed and buried under its weight. I could boast that I never expressed any fear at this time, but I was only kept going by the consolation that the whole world was perishing with me.

After a while, the darkness paled into smoke or cloud, and the real daylight returned, but the Sun shone as wanly as during an eclipse. We were amazed by what we saw, because everything had changed and was buried deep in ash like snow. We went back to Misenum and spent an anxious night switching between hope and fear. Fear was uppermost because the earth tremors were still continuing and the hysterics still kept on making their alarming forecasts." (Pliny V: 20)

Left: the Roman waterfront at Herculaneum, with Resina in the background and Vesuvius on the horizon.

Below: victims in the southern part of Pompeii.

Emergency plans for the area around Vesuvius

In 1991, the Italian Minister of Defence appointed a commission to develop guidelines to assess the volcanic risk in the area around Vesuvius, based on the distribution of deposits from the major eruptions during the past 20 000 years and on numerical simulations. The commission established two hazard zones. In the red zone, which covers 236 km^2, many sectors could be almost entirely destroyed by nuées ardentes, mudflows or heavy ballistic falls. Nuées ardentes severely damaged 40 per cent of this area in AD 472 and 20 per cent in 1631. The yellow zone (1125 km^2) could be affected by heavy falls of ash and lapilli, as well as mudflows. In 1631, 10 per cent of this area was badly damaged. This zone also includes a blue sector (98 km^2) that could be subject to flooding.

In 1993, the Minister of Defence appointed another commission to prepare a plan for the evacuation of the area surrounding Vesuvius, based on the vulnerability of the hazard zones, the establishment of alert levels, and the civil defence action that should then be undertaken. The Vesuvius Observatory is responsible for issuing scientific alerts, based on eight danger levels, which should be transmitted to the local and national administration and thence to the threatened population. It is planned to evacuate the 600 000 people living in the red alert zone, and accommodate them in other Italian regions, even before the eruption actually starts. (L. Civetta 1998)

reveal their terrible fate and now make gruesome archaeological exhibits. Five minutes later, a fifth and even larger nuée ardente swept over Pompeii and carried onwards to the outskirts of Stabiae, 17 km from the volcano. Earthquakes, probably related to further collapse of the caldera, reached a frightening pitch throughout the region. About 08.00 on 25 August, the sixth and largest nuée ardente swept onwards to Stabiae and killed Pliny the Elder. Another branch of the same nuée ardente surged westwards across the waters of the Bay of Naples and halted, 32 km away, within sight of Pliny the Younger and his mother above Misenum.

Vesuvius released other nuées ardentes throughout the morning of 25 August and still more pumice rained down incessantly. Then, as the day went on, the nuées ardentes diminished in vigour, the Plinian column reduced in height, and the discharge of the most gaseous and evolved phonolitic parts of the magma reservoir was completed. The final hydrovolcanic phase ensued as water entered the vent when the magma reservoir collapsed. Many lithic fragments, in a fine ash replete with accretionary lapilli, were scattered, 1 m thick, all around the flanks of Vesuvius. Eventually, glimmers of daylight returned to the more distant areas such as Misenum in the west, the tremors declined in number and size, and Pliny the Younger and his mother returned home. The end of Pliny's first letter is uncharacteristically ambiguous about when the eruption ended. Some authorities have suggested 27 August, or even 28 August, but specialist interpretation of the text indicates that eruption probably drew to a close on the morning of 26 August (Hine, personal communication).

When the daylight returned, all Campania was covered with a thick blanket of fine grey pumice in undulating dunes. About 300 km^2 around Vesuvius was completely devastated and the farms, villages, towns and cities of the plain had vanished. According to Sigurdsson et al. (1985), Vesuvius had discharged 1 km^3 of white pumice, 2.6 km^3 of grey pumice, 0.37 km^3 of nuée ardente deposits, and 0.16 km^3 of hydrovolcanic materials, expressed as **dense rock equivalent**. The magma reservoir thus delivered 4.13 km^3 of dense rock, or about 10 km^3 of fragments, although these figures have recently been reduced by more than half (Cioni et al. 1999).

Preventive techniques may well have improved when the next Plinian eruption of Vesuvius occurs. If such an eruption were to occur tomorrow in defiance of all trends, no known palliative technique could save the many towns on the southern flanks of Vesuvius, and the only effective defence would seem to be immediate and comprehensive evacuation.

CAMPANIA: THE PHLEGRAEAN FIELDS

The Phlegraean Fields form an array of cones, craters, calderas and occasional domes and lakes, covering a low-lying area of about 100 km², due west of Naples. They do not really live up to their name, which is derived from the Greek *phlegein* (to burn), although they have always been notorious for the sinister fumes that the Greeks and Romans believed were the exhalations of the underworld. Odysseus (*Odyssey* XI) and Aeneas (*Aeneid* VI) both visited Hades from here and, at the Mare Morto, Charon was reputed to row the souls of the dead across the Styx to the underworld. Innumerable hot springs made Baia a leading spa, the sulphurous emanations from La Solfatara were known as the Forum Vulcani, and the old crater forming the harbour at Misenum was a major base of the Roman Fleet.

Many eruptions have taken place in the Phlegraean Fields in the past 12 000 years and it is possible that some of them could have been witnessed by the Greek colonists, who first named and settled in the area in the eighth century BC.

However, the formation of Monte Nuovo in AD 1538 is the only eruption recorded in historical times, and radiocarbon dating suggests that it followed 3000 years of quiescence. On the other hand, the recent activity of the Phlegraean Fields has been concentrated on the emissions at La Solfatara and, more dangerously, on the **bradyseismic** movements that are caused by oscillations in the levels of the subterranean magma. They are now focused on Pozzuoli, but they have been common in the Phlegraean Fields at least since the ground first rumbled beneath the feet of Aeneas. Bradyseismic uplift increased markedly again during the two years before Monte Nuovo erupted in 1538. In modern times, uplift has occurred at Pozzuoli in 1969–72 and between June 1982 and December 1984 (Orsi et al. 1996a). Disquiet was followed by panic, departure and evacuation, as the population, the administration and scientists prepared for a renewal of volcanic activity that fortunately did not take place. But the incidence of volcanic hazards is high in this area, where some 1.5 million people live (Rosi & Santacroce 1984). Although any future eruption is most likely to form a cone like Monte

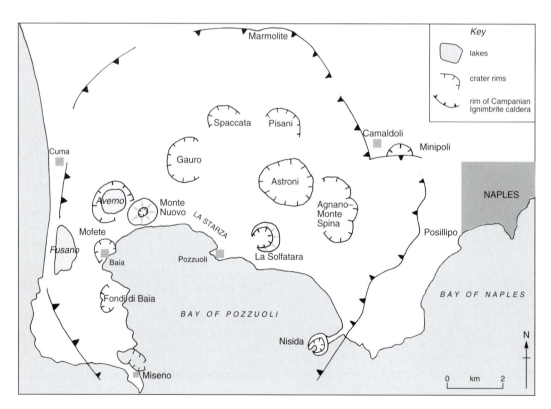

The Phlegraean Fields.

Nuovo, it will most probably take place in the densely populated area centred on Pozzuoli.

The volcanic rocks of the Phlegraean Fields belong to the potassic series, like most of those in Campania. They range from rare trachy-basalts, through latites, trachytes and alkaline trachytes to phonolitic trachytes (Civetta et al. 1998). Most of the magma has now cooled and solidified, but a reservoir of liquid trachyte, esti-mated at a mere $1.4 km^3$ in volume, lies at a depth of 4–5 km below the hub of the Phlegraean cal-dera at Pozzuoli. However, its thermodynamic pulsations are enough to cause the modern bradyseismic movements. If this magma reser-voir is ever refilled from the depths, then a large explosive eruption is likely to take place, but there is no evidence as yet that any such refilling has occurred in the past few thousand years.

The Phlegraean Fields have been the scene of many predominantly trachytic explosive erup-tions. Ashflows, Plinian eruptions and small cal-dera collapses have accompanied the formation of dozens of cones, **tuff rings** and craters. Explo-sive intensity was often increased by hydro-volcanic interplay from underground waters, ephemeral lakes and the sea. Hence fragments, especially pale ash and pumice, predominate among the materials ejected, and little effusive or extrusive activity has taken place, even during the waning phases of the past few millennia. Thus, lava domes are rare and eruptions of lava flows have been almost entirely absent.

The Phlegraean Fields have excited scientific curiosity for more than 200 years (Hamilton 1776/9), but it is only recently that a simpler pat-tern of development has been revealed than once seemed likely. Eruptions began below sea level and eventually formed a stratovolcano more than 50 000 years ago (Rosi et al. 1983, Rosi & Sbrana 1987, Luongo & Scandone 1991, Barberi & Cara-pezza 1996, Orsi et al. 1996a, Rosi et al. 1996, Orsi et al. 1999, de Vita et al. 1999, De Vito et al. 1999).

The Campanian Ignimbrite

The arrival of fresh magma in the reservoir provoked perhaps the greatest eruption in the Mediterranean area during the past million years. The summit of the stratovolcano collapsed as the grey Campanian Ignimbrite was expelled. Its volume has been estimated at $80–150 km^3$ of magma (Fisher et al. 1993), which represents up to $500 km^3$ of ignimbrites.

The materials were possibly expelled from a ring fracture fed by a subterranean system of cone sheets. The enormous momentum of the explosion spread the finer fragments in a rapidly expanding incandescent cloud, which formed the Campanian Ignimbrite. Its trachytic pumice and lithic fragments are embedded in an ashy matrix and it thickens to as much as 60 m in lower-lying areas. This enormous **ashflow** not only blanketed the whole of the Campanian Plain, swept across the Bay of Naples into the Tyrrhenian Sea and rode over hills 1000 m high in the Sorrento peninsula, but covered $30 000 km^2$ and much of southern Italy as well. It has been dated most recently to 36 000–37 000 years ago (Fisher et al. 1993). It is usually sup-posed that it burst out in a single unit in less than about a week. However, cores taken from the Tyrrhenian sea floor have revealed five separate trachytic ash layers that suggest that the Cam-panian Ignimbrite was ejected in five distinct stages (Paterne et al. 1988). The chief episodes occurred 36 000, 33 500 and 26 900 years ago, with smaller emissions about 38 700 and 24 000 years ago. If these correlations are substantiated, then the ignimbrite would have been erupted in brief individual episodes extending over an unusually long period. The irregular scalloped outline of the associated caldera, which could indicate piecemeal foundering, gives some sup-port to the concept of five-stage eruptions.

The collapse of the caldera was the key event in the history of the Phlegraean Fields. Its floor foundered by at least 700 m and formed a hollow 12–15 km in diameter, making it one of the larg-est calderas in Europe. But it has only recently been properly identified, mainly because sub-sequent eruptions have masked both its nature and its limits. Part of its floor and its southern rim sank below the Bay of Naples, but remnants of its scalloped rim can be seen at the Monti di Procida in the south, at Cuma in the west, at Marmolite in the north, and especially at Camaldoli in the east. They demarcate the zone in which all subsequent volcanic activity took place in the Phlegraean Fields. Thereafter, the eruptions decreased in intensity and became increasingly concentrated on the centre of the caldera.

The Neapolitan Yellow Tuff

Soon, eruptions began to fill the sea-flooded caldera, and perhaps about $23 km^3$ of magma erupted altogether (Gilberti et al. 1984), but the chronological sequence of the trachytic tuffs lying between layers of latitic and trachytic flows is hard to establish. However, this period concluded with a major event: the eruption of the Neapolitan Yellow Tuff, whose colour comes from the diagenetic alteration of trachyte erupted in shallow water. The outburst occurred about 12 000 years ago and it has been linked to an important ash layer in the Tyrrhenian Sea that erupted about 12 300 years ago (Paterne et al. 1988). In all, some 20–50 km^3 of these hydrovolcanic trachytic tuffs were expelled (Scarpati et al. 1993, Orsi et al. 1996a, Rosi et al. 1996), either in a single eruption, or from multiple emissions with their main source in the east, or from vents arranged in small arcs around the inner rim of the caldera as well as in its centre. The arcs could well represent a persistence of ring fractures fed by cone sheets. The vents in the arc stretch from Miseno to Monticelli and those belonging to the eastern arc extend along Posillipo Hill. The central vent formed the Gauro volcano, the largest within the caldera. These explosively hydrovolcanic, or even hydro-Plinian, eruptions produced pumice, accretionary lapilli and ash. Sometimes the yellow tuff is stratified, as at Mofete and Miseno; sometimes it is chaotic, as at Camaldoli. This major eruption was accompanied by the formation of the Neapolitan Yellow Tuff caldera, which occupied the centre of the Campanian Ignimbrite caldera and foundered some 60 m below sea level.

Activity during the past 12 000 years

Activity in the Phlegraean Fields during the past 12,000 years was concentrated in three periods (Di Vito et al. 1999).The first period lasted from 12 000 to 9500 years ago, when 34 mainly hydrovolcanic eruptions took place, in or near the sea, around the perimeter of the Neapolitan Yellow Tuff caldera. The resultant cones now form a circle stretching from Capo Miseno volcano, in the west, to Montagna Spaccata in the north and round to Nisida volcano in the east. A thousand years of quiescence then ensued, during which a paleosol developed on the land areas within the caldera. The second period of activity was marked by six relatively mild eruptions that began 8600 years ago with the formation of the Fondi di Baia cone in the west and ended 8200 years ago with the eruption of the San Martino

cone in the north. A long period of calm followed this brief eruptive episode and a major paleosol developed widely over the land surface. Towards the end of this calm period, the central part of the caldera was uplifted by some 40 m and formed the La Starza terrace, which now stretches northwestwards from Pozzuoli. The third and latest period of volcanic activity lasted from 4800 to 3800 years ago. It witnessed 16 explosive and 4 effusive eruptions, all of which took place in the northeastern sector of the Neapolitan Yellow Tuff caldera, except the explosion of the Averno volcano in the west. Most of these eruptions were trachytic. The main events of this episode included the eruptions of Agnano–Monte Spina volcano (which began the series 4800 years ago), the explosion of Averno volcano about 4500 years ago, the extrusion of the dome of Monte Olibano and the eruption of La Solfatara 3800 years ago, which were immediately followed by the explosions of the Astroni tuff ring and the Fossa Lupara, which brought the third period to a close.

Most of these eruptions had some quite important hydrovolcanic components. Nuées ardentes of ash and pumice radiated from some of the vents, and accretionary lapilli and vesiculated pumice were common. Thus, the lava dome that formed Monte Olibano was most untypical. On the other hand, Astroni and Averno volcanoes are typical tuff rings, produced by shallow explosions, which have wide craters surrounded by tuff ramparts almost 200 m high. However, it was only much later that water filled the Averno crater, although Agnano crater contained a warm shallow lake until it was drained in 1870. Some explosions, such as that at Agnano–Monte Spina, were clearly violently hydrovolcanic in character. Many cones, such as Senga or Monte Nuovo, are squat and broad in the typical style of hydrovolcanic forms. It seems that small lakes, or perhaps subterranean waters, interfered with the more effusive eruptions that would otherwise have taken place. The eruptions were mostly sporadic and brief. The volume of volcanic material expelled during the past 12 000 years amounts to no more than 4.5 km^3, which indicates how the magma reservoir has continued to shrink and cool. However, although the formation of Monte Nuovo in 1538 marks the only subsequent magmatic event, the Phlegraean Fields are far from extinct, and both hydrothermal emissions and bradyseismic movements have been common.

Bradyseismic movements at Pozzuoli

Pozzuoli is built over the hub of the Phlegraean
Fields caldera in the zone most sensitive to
recent bradyseismic movements, which are
probably related to a resurgence within the caldera linked to thermodynamic oscillations in the
magma reservoir situated at a depth of between
4 km and 5 km. Thus, although its molten parts
are now only small, its fluctuations are easily
propagated to the land surface (Orsi et al. 1996a).
As the area around Pozzuoli has been quite
densely populated since antiquity, even small
movements have been noticed. In fact, bradyseismic movements can be traced back to when
the terrace called La Starza was uplifted 5400
years ago near Pozzuoli.

Some of the bradyseismic movements of
antiquity were registered on the columns of the
old Roman market, built about 100 BC, which
early archaeologists mistook for the Temple of
Serapis. Soon afterwards, the pavement had to be
built 2 m higher, after sinking ground had taken
the original floor below the waves. Subsequent
fluctuations were recorded on its columns.
When the Roman Empire fell, the building was
abandoned, and the lower 3.6 m of its columns
were covered by earth and rubble. Apparently

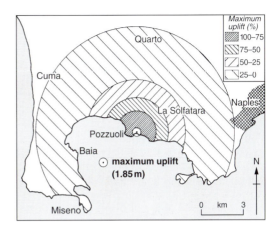

Bradyseismic uplift around Pozzuoli, 1982–4.

the area sank by about 5.8 m during the tenth
century and flooded the ruins. The submerged
parts of the columns were then attacked by the
stone-boring mollusc, *Lithodomus lithofagus*.
Later, bradyseismic movements elevated the
land, and the waters drained away to reveal the
pitted surface of the pillars. The depredations of
Lithodomus revealed the old depth of the sea
water very accurately. The earth and rubble had
protected the lower parts of the columns from the
Lithodomus and, when it was removed, the

The Roman market at Pozzuoli, known as the Temple of Serapis. After the destruction and flooding of the edifice, the
lowest parts of the columns were protected by rubble; the parts exposed to the water were attacked by the stone-boring
mollusc *Lithodomus lithofagus* and appear rough and eaten; the middle and upper parts were exposed to the air and
remained relatively unaffected by atmospheric weathering.

smooth bases of the columns were also revealed, still intact. These upward movements, amounting to about 6 m in all, probably occurred in the decades preceding the eruption of Monte Nuovo. After the eruption the ground sank again, perhaps by as much as 4 m. All these displacements were of course independent of worldwide **eustatic** changes of sea level and contemporaneous Italian tectonic movements.

The first 60 years of the twentieth century witnessed an irregular overall sinking of about 60 cm, punctuated by minor uplifts and periods of stability. This trend was reversed in mid-1969 by the onset of a phase of uplift that continued until mid-1972. It produced a maximum uplift of 1.70 m, which was accompanied by tremors and caused considerable anxiety at Pozzuoli. During the next two years the ground sank by about 20 cm, thus giving a net overall maximum rise of 1.50 m from the levels of 1969 (Barberi et al. 1984a,b, Barberi & Carapezza 1996). However, the feared earthquake or volcanic eruption fortunately did not occur. The crisis did highlight the need for civil-defence plans, if evacuation were needed, as well as systematic monitoring of the whole area. It is astonishing that these facilities were not available in 1969, although a variety of political, social and scientific explanations may account for their absence.

These omissions were largely rectified during the ensuing period of relative calm. Then, further uplift occurred from August 1982 until December 1984. The greatest rise (1.85 m) was registered at the harbour of Pozzuoli. Thus, central Pozzuoli has risen 3.35 m since mid-1969. The area had an established network of seismic, levelling and tidal-gauge stations, gravity and geothermal monitoring – all coordinated in a research centre at the Vesuvius Observatory. The ground deformation occurred throughout the Phlegraean Fields, but was most marked at the hub of the caldera at Pozzuoli. Analysis of the hydrothermal emissions, especially close to the hub at La Solfatara, showed an initial increase of activity, but it was not maintained and it suggested that the magma was not itself rising.

The earthquakes did not begin until November 1982 after uplift had been under way for some months. The initial magnitude 3.5 earthquake occurred at La Solfatara on 15 May 1983 and a magnitude 4.0 earthquake occurred at Pozzuoli on 4 October 1983. As a result, 40 000 people were evacuated from the centre of Pozzuoli. This was a sound safety measure, because many of the buildings in the town were damaged and some later crumbled during the swarms of small tremors between October 1983 and April 1984. In all, some 8000 buildings were damaged and the harbour is now too shallow for large boats, and lower quays have been built alongside their now uselessly high predecessors. But the earthquake foci remained at the same depth of between 3 km and 4 km, which again suggested that the magma was not rising and, therefore, that a volcanic eruption was unlikely unless there was a radical change. The crisis ended when both uplift and the earthquakes stopped in December 1984. In spite of the panic, the media exaggeration, and the evacuation, which with hindsight was perhaps not entirely necessary, the crisis was handled quite successfully. However, the crucial symptoms indicating an imminent eruption in these conditions are still not fully known. Pozzuoli is still in the forefront of volcanotectonic analysis. The vital need to solve the problems is emphasized by the great vulnerability of a population of 400 000 living in the potential danger zone.

La Solfatara

As its name indicates, La Solfatara is famous for its sulphurous emissions, for which it has become the type locality. La Solfatara lies only 1.5 km east of Pozzuoli, very close to the centre of the most recent volcanic and tectonic activity in the Phlegraean Fields, and it may be wrong to assume that these emissions represent the dying phases of activity on this vent. La Solfatara forms a horseshoe cone, wide open on its southwestern side. Its crater, the Piano Sterile, is 750 m long and almost 600 m broad; its almost rectangular outline seems to have been determined by two sets of faults, one trending from northwest to southeast, the other from northeast to southwest.

La Solfatara from the air, with the outskirts of Pozzuoli.

most active area currently

The crater of La Solfatara.

The latest episode of eruptions began in the district less than 3800 years ago with the extrusion of the trachytic dome of Monte Olibano. Immediately afterwards, eruptions from a new vent, situated only 300 m to the north, shattered the northern half of the dome, and built a cone of ash and white pumice alongside it. This was the cone of La Solfatara. Violent explosions then opened up all its southwestern flank, and small nuées ardentes surged forth and covered almost 1 km². In the final act of the eruption, a piston of viscous trachyte plugged the vent and intruded the northern wall of the crater. Thus, the Piano Sterile of La Solfatara is encircled by a wall composed of a plug in the north, an ash and pumice cone in the centre, and a broken dome in the south. The fumaroles have greatly altered all the surrounding rocks. Some of the fumaroles reach temperatures of 160°C and are composed of over 80 per cent of steam and water, but also exhale carbon dioxide, hydrogen sulphide and small amounts of nitrogen, hydrogen and carbon monoxide (Chiodini et al. 1997). The basic cause of all this fumarole activity is the recycling of atmospheric waters that are heated by the high temperatures that prevail just below the surface.

Fumarole activity has probably been more or less continuous at La Solfatara ever since the crater was formed. Various small vents emitting mud, or creating small hot muddy lakes in the Piano Sterile, were known to the Romans and were also recorded as early as the fifteenth century, although activity seems to have declined since the beginning of the eighteenth century. Perhaps the greatest event recorded in the crater in historical times occurred in 1198. It has sometimes been described as a hydrovolcanic eruption (Bonito 1691, in Rosi & Santacroce 1984), but was most probably a muddy geyser that rose about 10 m into the air after a small earthquake. A similar event occurred in 23 July 1930, when mud clots were thrown 30 m skywards.

The crater of La Solfatara forms a wide arena floored with hardened grey mud, pockmarked with bubbling muddy pools and hissing vents exhaling steam and sulphurous fumes. At present, several small active fumaroles cluster on the eastern half of the Piano Sterile and the adjacent crater walls. Many seem to be related to fractures following the regional trends: northwest–southeast and northeast–southwest. Each cluster has its individual characteristics: the Forum Vulcani is small but vigorous, the Bocca Grande is large and hottest of all. The Fangaia area, nearest the centre of the crater, has hot fumaroles and, as its name suggests, pools of hot mud that are usually 2 m deep and cover up to 500 m². Most of the fumaroles are encrusted with beautiful but fragile sulphur crystals, derived from the oxidation of hydrogen sulphide when it reaches the air. This sulphur then reacts with water to produce sulphuric acid, which contributes greatly to the chemical alteration of the rocks, which are then often bleached to form the hardened mud that makes little pale-grey mounds between the new muddy pools that

The eruption of Monte Nuovo according to Scipione Miccio

"In the year 1538, the city of Pozzuoli and all the Terra di Lavoro were much tried by strong earthquakes. On the 27th of the month of September, they ceased neither by day nor night in that city. The plain that stretches from Lago Averno to Monte Barbaro arose by a certain amount, and opened up in many places. Water surged forth from these openings and, at the same time, the sea alongside that plain dried up over a distance of 200 paces (370 m). As a result, the fish were left stranded high and dry, and fell prey to the local inhabitants.

On the 29th of the aforementioned month, about the second hour of the night [20.00], the ground opened near the lake. It exposed a most horrifying abyss, from which smoke, fire, stones and ashy mud then issued furiously. The opening made a noise like a roll of thunder, which was heard as far away as Naples. The fire coming out of the abyss ran close to the walls of the unhappy city [Pozzuoli]. The smoke was both black and white: the black part covered the land in darkness, and the white part resembled the whitest bombax [cotton]. Having risen into the air, this smoke eventually touched the sky. When the stones were thrown out, the all-devouring flames converted them into pumice. Some say that the largest of those that were thrown a great distance were the size of an ox. These stones rose as far as a crossbow shot into the air, and then fell back down, sometimes into the abyss, and sometimes onto its margins. In fact, the expulsion of many of them could not be seen because of the obscurity of the smoke. But they were clearly evident when they emerged from the smoking fog; and they had no little odour of fetid sulphur, just as the cannonballs of a bombardment can be seen when the smoke produced by the powder has dispersed.

The mud had the colour of ash, which was very liquid at first, but, little by little, it eventually became drier. It was expelled in such quantities, along with the aforementioned stones, that a mountain a mile high was created in less than 12 hours. Not only Pozzuoli and the neighbouring area, but also the city of Naples were covered by this ash, where it spoilt much of the beauty of its palaces. Transported by the rage of the winds, it swirled and burned the green vegetation and trees through which it passed, and broke down many trees under its weight.

Apart from this, many birds and various animals that were coated with this mud left themselves prey to man.

This eruption continued incessantly for two nights and two days. In fact, sometimes it was more powerful than others. When it was at its most powerful, a great din, noise and booming resonance could be heard in Naples – like the thundering noise when two armed enemy forces enter into conflict. The eruption stopped on the third day, when the mountain was exposed without its cover, causing no little amazement among all those who saw it. From the summit of the said mountain, a round concavity was seen stretching down [within it] to its feet. It was a quarter of a mile wide, and the stones that had fallen back into it could be seen boiling in its midst – like a great cauldron of water boils on top of the naked flames.

The citizens of Pozzuoli abandoned their homes and fled, with their wives and children – some by sea, and others over land. The Viceroy [Pedro de Toledo] at once took horse in the direction of the city, and halted on the Monte San Gennaro to observe the fearsome spectacle and the misery of the city, which was so completely covered with ash that any traces of the houses could scarcely be seen. These appalling ruins terrified the citizens of Pozzuoli, and they determined to abandon the city. But the viceroy, who was unwilling to consent to the desolation of such an ancient city that was of such use to the world, decreed that all the citizens should be repatriated and exempted from taxes for many years. To demonstrate his good faith, he himself built a palace with a fine strong tower, and erected public fountains and a terrace a mile long, with many gardens and springs. He reconstructed the road to Naples and widened the tunnel so that it could be traversed without lights. He built the San Francesco church at his own expense. He also had the satisfaction of completing his own palace, and of seeing that many Neapolitan gentlemen had built mansions there too. He also restored the hot baths as successfully as possible, and had the city walls rebuilt. And, to stimulate interest in the city, he decided to spend half the year in Pozzuoli, although ill health subsequently enabled him to stay there only in the spring".
Scipione Miccio 1600. *Vita di Don Pedro da Toledo*. Archivio Storico Italiano ix (1846).

appear from time to time. In contrast, the north-western part of the crater floor is calm enough to support a campsite and bar for those who appreciate sulphurous aromas. The inactive areas are coated with greyish-white hardened mud that resonates when walked upon. Sometimes, however, the hardened crust that masks the boiling waters below is dangerously thin and visitors are well advised to follow the designated safe paths.

Monte Nuovo

The formation of Monte Nuovo marked the latest, but almost certainly not the final, eruption in the Phlegraean Fields. It took place in the space of a week, between Sunday 29 September and Sunday 6 October 1538, 5 km west of Pozzuoli, at the hamlet of Tripergole, which had been reputed for its hot springs since Roman times. The eruption formed a small squat cone, 130 m high, on the La Starza terrace (Di Vito et al. 1987, Dvorak & Gasparini 1991, Orsi et al. 1996a).

Monte Nuovo and Lago Averno.

Throughout the Middle Ages, the area had been generally subsiding, and the Roman Lago Averno had become an arm of the Bay of Naples. The precursory phase reversed this trend: the land began to rise and increasingly frequent earthquakes started to shake the district around Pozzuoli, especially from 1537 onwards. Rapid but very localized uplift, accompanied by stronger earthquakes, took place on 27 and 28 September 1538. The Temple of Serapis at Pozzuoli apparently rose by some 5 m, and a tract of the sea bed, some 350 m wide, was exposed along the coast. By this time, the earthquakes had reduced much of Pozzuoli to ruins. The first and most important (hydrovolcanic) eruptive phase began on the evening of 29 September at 19.00 when first water, then gas, and finally magma gushed from a fracture that quickly progressed towards Tripergole. Soon, wet trachytic ash rained down as far as Naples, and black and white clouds rose about 4 km high. The hydrovolcanic eruption produced a mixture of pumice, yellow tuffs and old lava fragments in a coarse ash matrix that piled up near the vent to build up the bulk of the cone of Monte Nuovo by the end of 30 September 1538. It emerged for the first time from its pall of thick black smoke on the next afternoon, 1 October. The fragments also

A contemporary view of the eruption of Monte Nuovo in 1538.

The northern part of the fuming crater of La Solfatara.

again built up the isthmus separating Lago Averno from the Bay of Naples. During the calm interlude on 2 October, observers climbed the new cone and saw fuming lava bubbling in the crater. Then, when water was no longer invading the vent, mild Strombolian explosions covered the cone with about 3 m of typical **cinders**. However, a short violent explosion on the afternoon of 3 October expelled a nuée ardente that rushed about 7 km southwards into the Bay of Naples. It was followed by three days of calm. The final eruption was the shortest, most restricted in extent, and the most disastrous. It lasted little more than a few minutes on 6 October 1538. On that Sunday afternoon, the sudden explosion of a nuée ardente killed 24 people who were climbing up the southern flanks of the cone. The bodies of several victims were never recovered. It was the last fling of Monte Nuovo. The new mountain was built by both Surtseyan and Strombolian types of eruption; the former gave it its characteristic broad squat cone and wide crater 125 m deep; the latter gave it its covering of dark cinders (Parascandola 1946, Scarth 1999).

CAMPANIA: ISCHIA

Rich green with its mantle of pines, olives and vineyards, Ischia stands guard over the entrance to the Bay of Naples and the Campanian volcanic region along its northern shores, and, with its predominantly trachytic eruptions, it resembles the Phlegraean Fields nearby. It lies on a major fault that can be traced through Procida, the Phlegraean Fields and on to the Apennines at Benevento. But the island has also developed a clear individuality, in which the landscape owes much to local volcanotectonic uplift. Ischia covers an area of about 45 km² and stretches 9 km from east to west and 6 km from north to south. It rises to 787 m at Monte Epomeo, which marks the crest of the uplifted block dominating the island and, indeed, is higher than any volcanic summit in the adjacent Phlegraean Fields. It is thought to have been uplifted, faulted, and fractured by an increase of magmatic pressure in the shallow magma reservoir below (Rittmann 1930, Vezzoli et al. 1988, Orsi et al. 1998). This mass of magma has not yet completely cooled. After the uplift began, eruptions took place along these fractures, so that perhaps 50 volcanic vents can be distinguished on the island. Eruptions occurred for many thousands of years and they continued after the Greeks settled on the island in about 770 BC, through the Roman period and into the Middle Ages, when the emission of the Arso lava flow in 1302 marked the latest episode of activity. Eruptions on Ischia were noted in about 470 BC by Pindar and Strabo, in about 350 BC by Strabo and Pliny the Elder, and again in 19 BC by Pliny the Elder (Stothers & Rampino 1983); however, it is difficult to link these references with specific vents on the island.

The island is still tectonically unstable; but, rather paradoxically in an island that owes much of its altitude to uplift, the net trend of movements since Roman times has been downwards. Ischia also suffered a major earthquake on 28 July 1883, which destroyed much of Casamicciola and cost over 2000 lives. This was only the latest and most lethal of several such earthquakes recorded during historical times. The island has many hot springs that support a tourist industry, notably at Casamicciola. These relatively recent volcanic, seismic and hydrothermal features suggest that Ischia may not have witnessed its final volcanic eruption.

The rocks expelled were chiefly trachytes and alkali-trachytes, together with some latites and

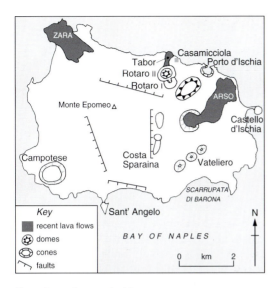

Key

- recent lava flows
- domes
- cones
- faults

ZARA
Casamicciola
Tabor
Porto d'Ischia
Rotaro II
Rotaro I
ARSO
Monte Epomeo △
Castello d'Ischia
Costa Sparaina
Vateliero
SCARRUPATA DI BARONA
Sant' Angelo
BAY OF NAPLES
N
0   km   2

Recent eruptions on Ischia.

phonolites. It seems that an original reservoir of trachybasaltic magma not only underwent gravitational differentiation and fractional crystallization, but was also subject to periodic influxes of new magma, which precipitated a succession of eruptions.

Ischia began life as a submarine volcano. Its base lies at a depth of about 500 m in the Tyrrhenian Sea, except where it faces the Phlegraean Fields, and it now forms a much-eroded basal complex that outcrops on the southeast of the island (Orsi et al. 1998). When the complex had been built above sea level, explosive eruptions expelled predominantly trachytic fragments. As is shown by the Scarrupata di Barona cliff face on the southeast of the island, prolonged activity was punctuated by episodes of repose, when fossil soils had time to develop before eruptions resumed (Poli et al. 1989). Most of this basal complex was then eroded away and the episode was brought to a sudden close more than 150 000 years ago by the collapse of a caldera, of which little direct morphological evidence remains.

The second phase saw the growth of trachytic cones, and trachytic and phonolitic domes, along the faulted rim of this caldera. By coincidence, the rim follows the general outline of the present coasts of Ischia. The Castello of Ischia crowns one of their remnants, and on the south coast the Sant'Angelo dome forms another. This episode may well have lasted from about 150 000 to 74 000 years ago, but activity seems to have been concentrated around 130 000 years ago. At

least 20 000 years of repose – in many places considerably more – then followed.

The third phase, lasting from about 55 000 to 30 000 years ago, began with the influx of new magma into the Ischian reservoir (Civetta et al. 1991). It immediately provoked the vast eruption of trachytic ignimbrite that deposited the Monte Epomeo Green Tuff about 55 000 years ago. It not only filled the centre of the island with a blanket 200 m thick, but it might also have extended into the Phlegraean Fields and covered about 300 km². The tuff could have resulted from one single overwhelming ashflow, or it may have been formed in stages, represented by three distinct ash layers on the adjacent Tyrrhenian sea floor, which have been dated to about 60 000, about 55 000 and about 51 000 years ago, with another possibly related layer about 47 000 years ago (Paterne et al. 1988). Subsequent eruptions from the west and southwest took place and might also be linked to ash layers on the nearby sea floor, which have been dated to about 40 000, about 36 000 and about 35 000 years ago. At all events, these eruptions generally declined in importance and they reveal an increase in potassium and calcium content, and a progressive evolution towards alkali trachytic magmas, probably by fractional crystallization, which formed highly differentiated magma in the upper part of the reservoir. Thus, these most explosive of all the eruptions on Ischia produced the most voluminous fragments, but did not form one large volcano.

The fourth phase in the growth of Ischia lasted from about 28 000 to 18 000 years ago. It was initiated by the incursion of new magma into the Ischian reservoir. The new magma probably caused the beginning of the uplift of the Monte Epomeo fault block, which was to increase greatly in the succeeding phase. Nevertheless, the fourth phase is crucial to the development of the present morphology of Ischia, because many eruptions of this phase occurred around the rising Monte Epomeo block. At this time too, the Campotese volcano erupted in the southwest of the island and produced trachytic lava flows before ending with a great explosion 18 000 years ago, which breached the crater and expelled the Citara–Serrara tuffs.

The fifth, and current, phase of the growth of Ischia began about 10 000 years ago (Buchner 1986, Vezzoli 1988, Orsi et al. 1996b). It was marked by the increasingly rapid uplift of the Monte Epomeo fault block, accompanied from

The eruptions of Rotaro

Perhaps the most varied eruptions in Ischia occurred on the short Rotaro fissure near the northeastern coast of the island, 1.5 km south of Casamicciola (Buchner 1986). The vents erupted at irregular time intervals, at regular distances of 300 m. The main cone, Monte Rotaro I, grew up first in the south as a small stratovolcano in about 600 BC. Then, a violent hydrovolcanic explosion, some 300 m along the fissure to the north, destroyed the northern flank of Monte Rotaro and formed a crater 500 m wide. During the Roman period, a trachytic dome, Rotaro II, rose to a height of 175 m within this crater, almost filling it. Very soon afterwards, another 300 m farther north along the fissure, an explosion blew a hole, 40 m deep, into the northern base of this dome. This explosion may have occurred as the dome was solidifying, because viscous alkali trachytes emerged from the hole and flowed out to reach the shore in a small promontory, the Punta della Scrofa, almost 1 km away. Finally, a nearly solid piston of viscous lava pushed like toothpaste up from a vent some 300 m north of the source of the flow. This piston, forming Monte Tabor, still carries a cap of older tuffs taken upwards as the piston rose. It now rises 76 m and is 150 m in diameter, which is probably scarcely wider than the original vent from which it emerged. The formation of Monte Tabor seems to have brought the eruptions on the northern part of this fissure to a close.

time to time by eruptions around its edges, especially in the lower area to the east of Monte Epomeo. Thus, at least 14 eruptions have taken place during the past 5000 years, and some of these have been dated. About 3800 years ago a violent Surtseyan eruption occurred 2 km off the southeastern coast at the Secca di Ischia and deposited the Piano Liguori unit of surge layers over much of southeastern Ischia. Between 3000 and 4500 years ago, trachytic lavas oozed out in a wall, nearly 1 km long and 200 m high, which forms the Costa Sparaina. Other eruptions also took place in the far northwest and formed the pair of partly collapsed trachytic domes and lava flows comprising the Zara complex. It reaches a height of 200 m and constitutes the northwestern peninsula of the island. At least one of the flows is older than the fourteenth century BC, because a Bronze Age settlement of that date was built upon it (Buchner 1986). About 930 BC, an eruption probably took place from the Cannavale vent, in the east of the island. A Surtseyan eruption built the tuff ring surrounding the maar forming the Porto. It occurred possibly as late as the fourth century BC, because the tuffs lie on top of Greek ware from the fifth century BC. (A channel was cut through the tuff ring to the open sea in 1834 to make the present harbour.) Additional explosive eruptions occurred about AD 60, and some time later the contiguous domes of Montagnone and Maschiatta probably extruded from the same vent. The little cones of Molara, Vateliero and Cava Nocelle erupted at 500 m intervals along a northeast-trending fissure. The cinders from Vateliero cover a fossil soil containing Roman pottery from the second and third centuries AD, and these eruptions could perhaps be related activity dated to AD 430 (Vezzoli 1988, Orsi et al. 1996b). A further explosive eruption,

which laid down the Fiaiano Pumice, occurred from about AD 670 to AD 890 from the vent that was to produce the Arso eruption in 1302.

The eruption of the Arso cone marked the latest episode of activity on Ischia. It began on Friday 18 January 1302 and lasted for about two months. It broke out on the eastern flanks of Monte Epomeo, probably on an extension of the Rotaro fissure. The eruption began with a violent hydrovolcanic explosion of pumice that choked the sea nearby, and ash spread in a sulphurous cloud all over Campania and even covered Avellino, 75 km away, in a blanket like snow. It was soon followed by emissions of olivine trachytes that formed a crescent-shape ridge of spatter, 50 m high and 450 m wide, and by a lava flow that set fire to the vegetation, and spread 2.7 km to the sea in a broad band between 200 m and 1 km wide. Many terrified islanders are said to have fled to Procida and Capri, as well as to the mainland, and some commentators even claimed that many people and their animals had been killed.

Enumeration of the main volcanic features formed during this latest phase inflates its importance. Activity declined throughout this period; the magmas became increasingly basic, and eruptions tended to be effusive, small and restricted to the northeast of the island during historical times. Ischia joins the nearby Phlegraean Fields, therefore, in experiencing waning and more localized eruptions.

AEOLIAN ISLANDS

The Aeolian Islands, north of Sicily, are the most famous volcanic archipelago in the Mediterranean Sea. Seven volcanic islands and some islets form an arc with a north–south trending pendant, making a T-shape group about 100 km across. Vulcano, Lìpari and Salina, forming the pendant, are the largest islands, but Stromboli is now by far the most active. The islands are only the summits of bulky volcanic accumulations, the bases of which lie on the floor of the Tyrrhenian Sea at depths of 1000–2000 m, and whose crests rise a further 500–900 m above the waves. They are usually composed of one or more stratovolcanoes, with several calderas, smaller domes, cones or lava flows.

During a volcanic history stretching back more than a million years, the focus of activity has shifted from island to island (Beccaluva et al. 1985, Santo et al. 1995). The Aeolian Islands have given the volcanic world the Strombolian and Vulcanian types of eruption, which for many years gave a useful shorthand means of describing two contrasting forms of activity (Mercalli 1907, Lacroix 1908). Stromboli is deservedly famous as the "lighthouse of the Mediterranean", and is the most diligent volcano on Earth at present. It was probably a latecomer to the Aeolian scene, but has apparently been in almost continual activity for at least 200 and perhaps for 2500 years. Vulcano, believed in antiquity to be the home of Vulcan (the God of Fire), has several calderas and a beautiful **tuff cone** in the Fossa, as well as hot, bubbling fissures and mudpots. Its efforts seem to have been more episodic than

Stromboli's, and it has been resting since the Fossa last erupted in 1888–90. Lìpari last had a major eruption at Monte Pilato, probably in AD 729, and its fumaroles and seismic activity are lethargic, but it has the most varied scenery in the archipelago that reaches a blinding white climax in the pumice and the turbulent **obsidian** flow structures of the Rocche Rosse. The remaining Aeolian Islands, Alicudi, Filicudi, Salina and Panarea, are generally less populated and more isolated, and erosion has already smoothed their old and extinct volcanic features. Islets such as Basiluzzo, Lisca Blanca, Strombolicchio, and the various Faraglioni between Vulcano and Lìpari are extinct, and their original landforms have been eroded so much that they cannot be recognized. However, Vulcanello, between Vulcano and Lìpari, may still be considered an active vent. Several seamounts, which broadly prolong the arc formed by the upper part of the T shape of the archipelago, could well be extinct, because rocks dredged from their surfaces seem to be among the oldest in the area.

Although there are certain local difficulties with the hypothesis, it seems generally agreed that the Aeolian Islands form a volcanic island arc in the Tyrrhenian Sea, which was generated when the African plate, or an adjacent microplate, was subducted northwestwards beneath the Eurasian plate, or an adjacent microplate (Barberi et al. 1974, Dercourt et al. 1986, Frepoli et al. 1996). Nevertheless, there is often disagreement about where these actual plate boundaries should be drawn. And, although the activity has been distinctly calc-alkaline, as befits a subduction zone, trace-element study, for example, also reveals intraplate influences on the emissions (Ellam et al. 1988). Strike-slip faults transecting the arc are also associated in a general way with potassic volcanism in the Aeolian Islands. However, in terms of time, space or chemistry, there is no simple transition from calc-alkaline to potassic activity in the islands. Salina, for example, is composed of calc-alkaline tholeiitic basalts, basaltic **andesites**, and andesites that date from between 13 000 and 430 000 years ago. The lavas of Vulcano erupted during the past 120 000 years are potassic, with higher potassic contents in the younger rocks, which range from leucite-tephrites to rhyolites.

It seems that the islands have been formed from individual magma reservoirs, with perhaps several separate reservoirs beneath the larger islands such as Lìpari and Vulcano, where they

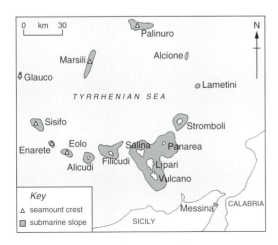

The Aeolian Islands and accompanying seamounts.

have been established long enough for rhyolites to evolve by fractional crystallization. It also seems that the potassic lavas may have been formed as the extreme end-product of otherwise normal subduction-related enrichment. With the notable exception of Stromboli and Vulcanello, silicic explosive magmas have dominated the recent eruptions of the Aeolian Islands. Thus, fine fragments usually far outweigh lava flows; many islands are crowned by steep cones; lava domes are numerous; but lava flows are often small, viscous and silicious, although, like the Rocche Rosse on Lìpari and the Pietre Cotte on Vulcano, they tend to form distinctive features in the landscape.

Generally speaking, the first phase of volcanic activity built up the islands of Alicudi, Filicudi and Panarea, as well as the older parts of Salina and Lìpari. At the close of this phase, first Panarea became extinct, then Alicudi and Filicudi. After a long period of repose, the second phase began when potassic materials erupted, which culminated in the present eruptions at Stromboli and Vulcano. The marine platforms around the older islands, at altitudes ranging from 105 m above to 4 m below the present shores, register both sea-level changes and volcanotectonic displacements. They are best preserved on Filicudi, Panarea, Salina and in western Lìpari. Their absence from both Stromboli and Vulcano indicates the extreme youth of these islands.

The earliest historical eruptions in the Aeolian Islands are suggested when Homer describes the home of Eolus, the god of the winds (*Odyssey* X: 1–5). Stromboli is usually considered to be the home of Eolus, but the description is probably based on several islands. Eolus and his family feast continually and their home resounds to the noise of their festivities, which probably refers to the crater of Stromboli, and its repeated explosions, vibrations, rumblings, roaring, smoke, hisses and the pyrotechnic displays from incandescent ash and lapilli. Other Aeolian eruptions are mentioned or described with varying degrees of detail and accuracy by a succession of classical authors (Stothers & Rampino 1983). In general, the eruptions on Vulcano are noted most often in antiquity, and the possible birth and some subsequent eruptions of Vulcanello might also be inferred from these writings. Curiously enough, the first explicit reference to an eruption of Stromboli dates only from the second century BC (Polybius in Strabo VI: 2.10). Strabo also saw the difference between the eruptions of Stromboli and Vulcano. Stromboli, he said, was given over to the fire, but its flames were brighter than those of Vulcano, although they were less powerful. But references to what were remote islands are scarce and unreliable until the beginning of the nineteenth century, and most detailed scientific studies date from only the later parts of the twentieth century.

AEOLIAN ISLANDS: STROMBOLI

Stromboli is the northernmost and most famous of the Aeolian Islands, and it is the only volcano among them in continual activity. The island is the emerged summit of a large stratovolcano, 75 km from the Sicilian coast, which has grown from a depth of 2000 m on the floor of the Tyrrhenian Sea to reach an altitude of 924 m. Stromboli is elongated along the regional tectonic trends, stretching almost 5 km from northeast to southwest and about 3 km across, with a total emerged area of 12 km². The island forms a pyramid with overall slopes of 13–15°, with similar gradients continuing to the submarine base of the volcano (Gabbianelli et al. 1993). Apart from the active summit vents, its most notable feature is the Sciara del Fuoco ("scar of fire"), the steep black swathe more than 1 km broad that scars the northwestern flanks of the island. The main settlements, San Vincenzo and San Bartolo, lie on an esplanade on the northeast coast. They face the islet of Strombolicchio, an eroded remnant of an early cone that rises to 49 m almost 2 km off shore. The only other settlement, Ginostra, clings more precariously to the steeper down-faulted western coast dominated by other small satellite cones including the Timpone del Fuoco and the Vigna Vecchia. Stromboli has a double crest: Vancori at 924 m is separated by a saddle from the Pizzo sopra la Fossa, at 918 m, but neither marks an active vent. Together, they form a natural balcony overlooking the active craters, lying 750 m above sea level, which erupt near the top of the Sciara del Fuoco. During the past 200 years, when eruptions have been described with

Stromboli from the northeast.

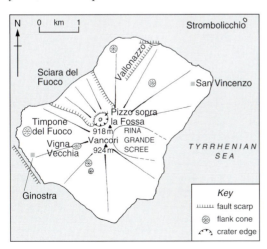

Chief features of Stromboli.

reasonable accuracy, at least three, often five, and sometimes up to eight vents have functioned more or less regularly and often in unison, at intervals ranging from a few minutes to several hours. Although firm evidence is lacking, this kind of moderate, persistent and repeated activity is reputed to have operated for 2000 years or more, which would make Stromboli the most diligent, reliable and consistent volcanic performer in the world during that period, and most likely ever since the Sciara del Fuoco was formed about 5000 years ago (Kokelaar & Romagnoli 1995). Thus, for centuries, the summit of Stromboli has apparently been the place to visit in Europe to be sure of seeing a volcanic eruption.

It is perhaps appropriate to introduce a note of caution into this generalization. None of the ancient texts specifically states that Stromboli was in continual eruption at that time and there are no direct references in the official lists to eruptions before 1558. Indeed, the first detailed observation of the activity dates only from 1788 (Spallanzani 1792). Reliable observations of eruptions have been made only during the past two centuries and many of those do not explicitly mention the alleged continual activity. Even now, the activity of the three groups of vents is independent and, within a single group, the style of eruptions and the intervals between them can vary. Such irregularity has surprised untrained observers and has even caused several fatal accidents in recent years. Periods of magmatic activity have also been succeeded by phases when only fumes escaped from the craters. Thus, for instance, no magmatic eruptions at all seem to have occurred between June 1907 and May 1910 (Barberi et al. 1993a). On the other hand, eruptions during the later eighteenth century seem to have been not only rather more violent but much more frequent than at present. For example, Spallanzani, the Italian savant and traveller, stressed that the eruptions usually occurred at interludes of less than three minutes, although they varied greatly in vigour: on some occasions ash exploded less than 15 m into the air, but on others jets of fragments rose 800 m skywards, and, more rarely, glowing blocks 1 m across were hurled 1500 m off shore.

Sudden outbursts still occasionally cause damage over the island. On 22 May 1919, blocks weighing several tonnes crashed onto houses and killed four people and injured many more. On 11 September 1930, another vigorous explosion, accompanied by a small nuée ardente, caused six deaths (Abbruzzese 1936). An average of one or two brief but unusually vigorous explosions have occurred every year during the past 110 years, and 15 have been recorded since 1993. However, averages are misleading because they happen irregularly: the six explosions that took place in 1998, for example, occurred on 16 January, 23 August, 8 September, 24 November, 26 December, and 28 December. Several craters produced a variety of lava fountains, incandescent ash, pumice and bombs (Bertagnini et al. 1999). On the other hand, lava flows have been known to cascade down the Sciara del Fuoco for several weeks at a time, with or without any concomitant explosions from the craters above. Thus, close monitoring has revealed that Stromboli behaves less monotonously than has been supposed.

Nevertheless, most of the present **Strombolian eruptions** have modest vigour, modest dispersal and great frequency, but limited danger, forming cinder cones and, more rarely, lava flows. This kind of activity is, perhaps, the most common type of eruption on land. It has been associated with Stromboli ever since it was designated as a major eruptive type (Mercalli 1907, Lacroix 1908). But Stromboli is not as typical as the type should be. Elsewhere, such eruptions rarely last for more than a few years – enough to build up a cinder cone perhaps 100 m to 200 m in height and emit copious lava flows. These eruptions do not form stratovolcanoes, but Stromboli is composed of massive accumulations of lava flows, cinders and deposits of nuées ardentes that were emitted from a central cluster of vents that form a pile 3000 m tall. Paradoxically, then, Stromboli is by no means a typical result of Strombolian eruptions (Plate 4b).

The growth of Stromboli

Stromboli lies on a basement of continental crust about 18 km thick (Rosi 1980, Hornig-Kjarsgaard et al. 1993). The island grew up at the intersection of two faults, the most important of which runs from northeast to southwest. It is the predominant tectonic influence on Stromboli, dictating not only the elongation of the island but also marking the boundary between the older and younger volcanic eruptions. An older vent forming Strombolicchio rises at its northeastern end and the present active craters are aligned on it in the centre. Several spatter cones above

---

Spallanzani peers into the crater, 4 October 1788

Spallanzani watched the eruptions for over an hour from the shelter of a cave overlooking the crater. "The walls of the crater form a confused mass of lava, cinders, and sands . . . that narrow sharply downwards like a truncated, upturned cone . . . The crater is filled up to a certain level with liquid, incandescent matter, like molten bronze. This is the lava itself. It is agitated by two very distinct motions. One is circular, tumultuous and internal, and the other works up and down. The liquefied material rises inside the crater at varying speeds. When the lava reaches 25–30 feet [8 m] from the upper lip, it bursts like a clap of thunder. At once, part of this matter is torn into a thousand pieces, and is thrown into the air at an incredible speed, while the crater overflows with fumes, sparks and sand. A few moments before the explosion, the surface of the lava balloons up and forms large bubbles, some of which are several feet across. These bubbles burst with a bang [give off fumes] and throw out the materials . . . Most of the fragments fall back near, or into, the crater, but exceptional eruptions rain debris onto the slope leading down to the sea. After the eruptions, the lava falls back almost noiselessly in the crater and, whenever it rises again, it sounds like liquid boiling furiously in a pot." (Spallanzani 1792)

Eruptions on Stromboli

The famous moderate explosive eruptions are the glory of Stromboli and, especially at night, are among the most spectacular natural performances on Earth. The first sign of another eruption is a low rumbling and a slight trembling of the ground. Then a red glow lights up first one and then another of the vents. The rumbling becomes a deep echoing roar of escaping gases. The red brightens inside the vents to pale vermilion, even yellow at times, as the molten rock is hurled to the surface. One small vent hisses out a blue flame like a huge Bunsen burner. The roar becomes high-pitched, raucous and hoarse. Incandescent fragments rise 100 m or more into the air in a constellation of sparks, hot ash and cinders that twist and twirl in the steamy air. Then they fall back gracefully and lie as glowing embers on the cones around the vents. Their vermilion deepens to red and the larger fading clots slide slowly down the pyre. The roaring calms to a subdued moan as the last cinders crash back down into the vent. Then silence returns. The whole sequence takes perhaps five minutes. Often, there is less than half an hour to wait for the next display. These mild and almost rhythmic eruptive spasms depend on a constant supply of ascending lava that is driven upwards by the gases that break out from the magma as the pressures upon it are released. The resulting explosions, which usually take place near the top of the vent, break up the lava into molten spatter, cinders, lapilli and ash, which are then thrown into the sky (Scarth 1999).

Occasionally eruptions are slightly different and very occasionally are more dangerous. A lucky observer might see a lava flow erupt. All the flows for at least the past 200 years have been confined to the Sciara del Fuoco. Some flows make their way downwards from the active craters, but many also rise from the higher reaches of the Sciara del Fuoco itself. Some solidified lava boulders break into angular rubble that becomes part of the scree slipping slowly down the 40° slope. But often, the molten stream cascades down the Sciara del Fuoco, and,

A nocturnal eruption from the southwestern vent of Stromboli.

as its surface solidifies, blocks break off and roll down into the Tyrrhenian Sea, where they hiss like tempered steel in billowing clouds of steam.

Ginostra mark the southwestward continuation of the fault, and two submarine seamounts, representing satellite vents, reveal its extension some 4 km off shore.

Stromboli is a young stratovolcano: it has few deep gullies and no marine terraces. The only deep valley, the Vallonazzo, clearly follows a fault down to the north coast. The main features of degradation on Stromboli are the great screes, especially those flanking the southeastern summit area. The largest, the Rina Grande, marks the speediest path by which the visitor can descend from the crest. These morphological indications are corroborated by absolute dating (Condomines & Allègre 1980, Gillot & Keller 1993). A potassium–argon age of 204 000 years has been

Strombolicchio, the marine-eroded remains of a much larger volcanic pile.

Stromboli from the northeast and the villages of San Vincenzo (left) and San Bartolo (right) The Rina Grande scree lies on the upper left slopes of the volcano.

obtained for Strombolicchio, the ancestor of modern Stromboli. The past 100000 years of the history of Stromboli can be divided into seven phases (Rosi 1980, Hornig-Kjarsgaard et al. 1993).

In the first phase, which lasted until about 65000 years ago, high-potassium, calc-alkaline, hornblende-bearing andesites and basaltic andesites erupted that form much of the southeastern coastal zone of Stromboli. The second phase, between 64000 and 54000 years ago, produced calc-alkaline basalts and andesites with a lower potassium content, which are exposed on the lower southeastern flanks of the volcano. The third phase, between 54000 and 35000 years ago, expelled andesites with biotite, transitional andesitic basalts and **shoshonitic basalts**. In the fourth phase, between 35000 and 25000 years ago, high potassium, calc-alkaline eruptions returned to prominence, forming the basalts and andesites lying on both the southwestern and northeastern flanks of the island. The fifth phase, marking the culmination of the old stratovolcano at Vancori, was one of the most eventful in the history of Stromboli. It began about 25000 years ago with the collapse of the summit area. The subsequent eruptions formed the Vancori complex, which is composed mainly of thick piles of lava flows, ranging from shoshonitic basalt to latite and trachyte, which have armour plated the southeastern two thirds of Stromboli and also filled the first summit hollow. Fragments were rare at first, but they erupted more frequently later, so that they often occur on the Vancori summit itself. The Vancori phase ended about 13000 years ago with a collapse that probably included the partial sinking of the northwestern

third of the island, and it may also have eliminated an upper reservoir where magmas had previously been able to evolve (Francalanci et al. 1993). The end of this phase concluded the emissions of the basic to intermediate lavas typical of the older periods in Stromboli's history.

The more recent history of Stromboli was marked by eruptions of basic non-evolved lavas that clearly must have risen more rapidly from the mantle to the surface without a significant halt in a reservoir. This period begins with the sixth, or neo-Strombolian, phase, which began when the centre of activity shifted about 300 m northwestwards and built up the Pizzo sopra la Fossa. These new vents emitted the potassic basalts that now cover the surface of most of the lowered northwestern third of Stromboli. Towards the end of this phase, satellite eruptions formed the spatter cones of the Timpone del Fuoco and the Vigna Vechia in the west of the island.

Then, about 5000 years ago, the most recent major tectonic event inaugurated the seventh and latest phase of activity on Stromboli, when a sector of the stratovolcano collapsed between the scarps now forming the Fili di Baraona and the Filo del Fuoco. Between them lies the smooth straight slope of the Sciara del Fuoco, inclined at an angle of 40° like a large blue-grey scree, 1 km wide at sea level and continuing to a depth of at least 700 m (Kokelaar & Romagnoli 1995). The terrace carrying the present active vents of Stromboli stands at its crest, about 750 m above sea level. Today's eruptions, mildly explosive in the main, are apparently a product of the formation of the Sciara del Fuoco and its consequent repercussions on the vent system of Stromboli.

The origin of the Sciara del Fuoco has long intrigued scientists. In the nineteenth century, many followed Poulett-Scrope (1825) in believing that the Sciara del Fuoco had been formed by a large explosion like the one then thought to have destroyed the summit of Monte Somma–Vesuvius in AD 79. Williams (1941) designated the Sciara del Fuoco as a downfaulted segment of the stratovolcano, which had been subsequently filled by erupted materials. Rittmann (1962) believed that it had been formed by large-scale gravity-sliding along the slippery surface of an arched sill. Rosi (1980) considered that the Sciara del Fuoco was initiated by tectonic movements at right angles to the major northeast-trending fault transecting Stromboli. The Fili di Baraona and the Filo del Fuoco marked the edges of the resultant fault trough, between which the sector had slipped seawards, aided by lubricating injected sills. But massive debris avalanche deposits, produced by several sector collapses during the past 13 000 years, extend over 10 km beyond the Sciara del Fuoco down to a depth of more than 2200 m below sea level (Kokelaar & Romagnoli 1995). At all events, the Sciara del Fuoco is one of the predominant features of the relief and personality of Stromboli. It directs the lava flows and most of the erupted fragments straight into the sea, and thereby fosters the illusion in islanders and visitors that Stromboli is a meek and mild volcano.

The Sciara del Fuoco, Stromboli, forming the smooth slope on the right.

AEOLIAN ISLANDS: LÌPARI

Lìpari is the largest of the Aeolian Islands, covering an area of 38 km², rising more than 1500 m from the floor of the Tyrrhenian Sea and 602 m above sea level. It represents only the emerged summit of a volcanic field that has seen marked shifts in the focus of its activity, and changes from andesitic to rhyolitic eruptions.

Volcanic products made Lìpari a major industrial centre from Neolithic times. Obsidian, with the sharp cutting edge given by its natural cleavage, was a valued tool throughout Neolithic and Graeco-Roman times. The Pomiciazzo flow, which erupted sometime between about 11 400 and 8600 years ago, was for instance the source of the most widely exported obsidian artefacts in the Mediterranean area during the Neolithic. Most of the abundant and easily accessible pumice deposits, such as those of Monte Pilato, erupted well after Lìpari was first colonized, and now cover 8 km² in the northeast of the island. Pumice is a foundation material of toothpaste, soap, beauty creams, abrasive and polishing

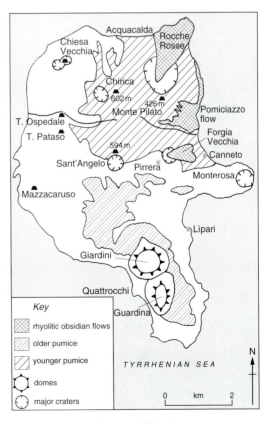

Selected volcanic features of Lìpari.

pastes (notably for computer screens), sound-proofing and insulation – and a source of silicosis for those who extract it.

The greater complexity of the development of Lìpari compared with its companions is at once evident from its much more varied skyline. There have been few repeated eruptions from a central cluster of vents. Thus, the highest point of Lìpari, Monte Chirica, rises a mere 602 m above sea level. These vents emitted chiefly ash and pumice, short, thick and viscous lava flows, and domes of even more viscous material that are found notably in the south. Explosive eruptions have expelled much rhyolite in the more recent phases of activity. These rhyolites now tend to dominate the scenery, most remarkably in the Rocche Rosse, which forms a broad twisted tongue of obsidian that juts out to sea at the northeastern corner of Lìpari.

Lìpari first erupted from the sea floor, and an island emerged over 220 000 years ago (Lanza & Zanella 1991). Lava flows, welded tuffs, and ash and cinders then formed at least a dozen low stratovolcanoes (Pichler 1980, Cortese et al. 1986). They were largely composed of andesites, especially quartz latite-andesites and quartz andesites, which are now best exposed on the northwestern coast between Acquacalda and Quattropani. They also form the low eroded stratovolcanoes of the centre and west: Timpone Carrubbo, Monte Mazzacaruso, Timpone Pataso, Timpone Ospedale and the Chiesa Vecchia.

As these stratovolcanoes were erupting, they were bevelled by a single interglacial marine terrace, which, because of contemporaneous and subsequent dislocation, is now found at different altitudes above sea level, but mainly between 18 m and 35 m. Some eruptions continued in the east and formed the twin volcanoes that make up the Monterosa peninsula, north of Lìpari town. They were formed during a period of low sea level and thus no raised marine terrace was developed around them.

There was then a long dormant interval before eruptions resumed about 60 000 years ago. This activity focused on Monte Sant'Angelo, where a long spell of explosive activity formed a stratovolcano in central Lìpari that was composed of quartz latite-andesite and quartz latite fragments and nuées ardentes. The paleosols between the beds show that the eruptions were often separated by dormant periods. During the same episode, much more effusive eruptions formed the Costa d'Agosto, about 2 km to the north, on the western flanks of the old Monte Chirica volcano. Then thick rhyodacitic lava flows were emitted from the southwestern flanks of Monte Sant'Angelo and came to rest on the west coast. Lastly, about 40 000 years ago, yet more explosive eruptions and the emission of a large quartz latite-andesitic flow were the prelude to the collapse of a caldera on the southern part of Monte Sant'Angelo, which seems to have extended, largely below sea level, from southern Lìpari to northern Vulcano.

The caldera became the main site of eruptions during the period between about 40 000 and 10 000 years ago. The change in the focus of activity in Lìpari was accompanied by a marked change in the materials expelled: rhyolitic pumice and lava domes and flows predominated during explosive eruptions that formed the southern peninsula where the scenery is dominated by Monte Guardia and Monte Giardina. A major

Rocche Rosse obsidian flow, with the pale outcrops of the Monte Pilato pumice on the right.

The dome of Monte Guardia, Lìpari, with the uppermost slopes of the Quattrocchi cliffs on the extreme right.

submarine hydrovolcanic eruption between Lìpari and Vulcano exploded the first layers of brown ashflow tuffs between 35 000 and 23 500 years ago (Crisci et al. 1983). These fine-grain layers extended over Salina, Panarea, Filicudi and the northern coast of Sicily, and they covered at least 7 km² on Lìpari itself. Their shoshonitic character suggests that they were derived from the Vulcano magma rather than that beneath Lìpari, and it is possible that they came from the still-submerged Vulcanello vent.

Eruptions resumed on Lìpari itself with the formation of the Monte Guardia sequence. Soon after andesitic lapilli from Salina had blanketed the island about 22 480 years ago, several rhyolitic domes extruded whose remains can be seen in the cliffs of Quattrocchi. They were then largely destroyed by a sub-Plinian explosion that covered most of Lìpari with an andesitic and rhyolitic breccia in a pumice matrix. Then craters near sea level exploded the bulk of the Monte Guardia sequence during a phase that occupied only short intervals in a period lasting no more than 2000 years. Eventually, the hydrovolcanic aspects decreased and explosive fragmentation gave way to viscous rhyolitic extrusions to form the domes of Monte Guardia and Monte Giardina. Between about 20 300 and 16 800 years ago, Lìpari was blanketed by another layer of brown ashflow tuffs, which no doubt came from the submerged Vulcanello vent, like their predecessors.

Thereupon, the focus of activity changed to the northeastern quadrant of Lìpari and it lasted from about 10 000 to about 1300 years ago (Cortese et al. 1986). Each eruptive episode was probably brief: each started with an explosion of rhyolitic pumice and ash, and each ended with extrusions of rhyolitic flows. The oldest eruption lies on a paleosol on the uppermost brown tuffs and formed a small breccia cone, and then emitted an obsidian flow in the Canneto–Dentro area. It was soon followed by a larger **hydrovolcanic eruption** that formed a tuff ring and the Gabellotto–Fiume Bianco beds, which cover half the island with more than a hundred thin layers of pumice and ash. They probably also covered northern Vulcano, where they were identified as the Lower Monte Pilato tephra. Between about 11 400 and 8600 years ago, activity concluded when rhyolitic lavas formed the Pomiciazzo dome and flow.

The period of repose that then ensued was marked by a paleosol, 1.5 m thick, that contains artefacts dated between 4800 and 1220 years ago. Activity resumed with the explosion of the Forgia Vecchia beds from two allied vents near Pirrera. They cover about 1 km² and are composed of many thin layers of coarse white ash and breccia. Further explosions at Monte Pilato formed a 150 m-high pumice cone, with a crater 1 km wide, that covered most of the Pomiciazzo flow. They also expelled pumice as far as Vulcano, where they form the Upper Monte Pilato layers. Near the town of Lìpari, they also cover the Roman ruins of Contrada Diana, which date from the fourth and fifth centuries AD.

After the eruption of Monte Pilato reached its climax, the Pirrera vent, north of Lìpari town, emitted the rugged lobe of the rhyolitic Forgia Vecchia lava flow, which has been dated to about 1600 years ago (Bigazzi & Bonadonna 1973). At Monte Pilato, rhyolite oozed up the vent,

The Quattrocchi on the southwestern coast of Lìpari with Vulcano in the distance.

breached the northeastern sector of the cone and spread in a bristling tongue of black obsidian, 2 km long, with a weathered red surface crust, which juts out into the Tyrrhenian Sea. This was the famous Rocche Rosse flow, which marked the latest episode of volcanic activity on Lìpari. It could have been observed by St Willibald in AD 729 (Cortese et al. 1986) or it could have taken place a century earlier and given rise to the legend that San Calogero (AD 524–62) expelled devils from the black stone of Lìpari to Vulcano. But recent archaeomagnetic research indicates two distinct dates for these eruptions of Monte Pilato – the first about AD 700, and the second about AD 1200 (Lanza & Zanella 1991, Tanguy et al. 1999, and unpublished data). At all events, the most recent two volcanic formations of Lìpari occurred after the decline of the Roman Empire. Paradoxically, therefore, the obsidian that gave Lìpari its fame in antiquity is not that which can be seen so clearly in its present flows.

AEOLIAN ISLANDS: ALICUDI

Alicudi covers an area of 5 km² and forms the summit of a large conical stratovolcano rising from a depth of 1100 m on the floor of the Tyrrhenian Sea to a height of 675 m above sea level. Its eruptions seem to have come from a central vent devoid of satellites (Villari 1980a). The details of its growth are uncertain, but the oldest eruptions formed the Galera complex, which has been dated to about 120 000 years ago (Gillot & Villari 1980). It was a mainly effusive volcano, but its activity ended with the formation of a large caldera, which was eventually filled mainly by the Dirituso cone of ash and cinders. In turn, this cone developed a summit depression (perhaps a caldera), that was covered, between 27 000 and 41 000 years ago, by the Montagna complex of domes and andesitic flows. The last eruptions formed the short thick lava flows of the Filo dell'Arpa on the southern flank of the stratovolcano. The rocks erupted range from basalts to potassium-rich andesites (Capaldi et al. 1985).

AEOLIAN ISLANDS: FILICUDI

The island of Filicudi covers an area of $9.5\,km^2$ and represents only the emerged portion of a complex volcanic structure that piled up from a depth of 1100 m on the floor of the Tyrrhenian Sea (Villari 1980b). At least six centres formed stratovolcanoes and domes composed of basalts and potassium-rich andesite. At the Fili di Sciacca, a potassium-rich andesite has given a potassium–argon date of about 210 000 years. However, $^{40}Ar/^{39}Ar$ dating has produced ages of 1.02 million years for the Zucco Grande and 0.39 million years for the Filo del Banco (Santo et al. 1995). This suggests that Filicudi, and perhaps the Aeolian Islands in general, are older than previously thought. The steep-sided andesitic pile of Fossa Felci is the chief, and youngest, stratovolcano on Filicudi. It constitutes the summit, at 774 m, and most of the rest of the island. In addition, two lava domes, the Montagnola and Capo Graziano, erupted during the last phase of activity on the island, about 40 000 years ago. Half the Capo Graziano dome has been downfaulted to reveal much of its internal structure. During the glacial period, several marine terraces were carved around the island after the eruptions had ceased.

The Fossa Felci, Filicudi, from the south.

AEOLIAN ISLANDS: SALINA

Salina is the hub of the Aeolian Islands, at $27\,km^2$ the second largest island after Lìpari, rising from a depth of 2000 m on the Tyrrhenian sea floor. The Fossa delle Felci (962 m), which is in fact the highest point in the whole archipelago, is separated by a saddle from the Monte dei Porri (860 m) in the west. Together they form the distinctive double-hump profile that gave it its Greek name of Didýme – "twin island".

The relief reflects the volcanic development of Salina. Eruptions have been concentrated on two distinctive clusters of vents beneath the humps, both of which functioned about half a million years ago, as the older island grew up. The Capo and Rivi volcanoes, in the east, and Corvo, in the west, mark the first phase of basaltic eruptions. Soon afterwards, the Fossa delle Felci grew up in the east, a more varied stratovolcano than its predecessors, with dacitic, andesitic and basaltic emissions. Glacial marine erosion has bevelled terraces into the flanks of these deeply eroded volcanoes. Potassium–argon dates of 430 000 years have been obtained from basalts in the Monti Rivi (Gillot & Villari 1980).

The western vents of Salina sprang into life again about 75 000 years ago when some of the most violent explosions in the archipelago's history built up the andesitic cone of the Monte dei Porri, and spread the grey Monte Porri tuffs all over Salina, Panarea and Lìpari. The latest volcanic event occurred in the far northwest of Salina about 13 000 years ago, when a great hydrovolcanic explosion of white pumice made the crater that cradles the village of Pollara. It is one of the most impressive craters in the whole archipelago, 1.5 km across and surrounded by a rim some 300 m high. Salina has been quiet ever since that event.

Monte dei Porri, Salina

AEOLIAN ISLANDS: SEAMOUNTS

Potassium–argon dates from lavas dredged from the Aeolian seamounts in the Tyrrhenian Sea show a range of ages (Beccaluva et al. 1985). In the west, the seamount, Sisifo, yielded samples of calc-alkaline basalts ranging from more than 1.5 million years old to 900 000 years old, as well as a rhyolite dating from 640 000 years ago. Eolo and Enarete seamounts, farther east, seem to be rather younger. Their rocks have not, so far, produced potassium–argon dates exceeding 850 000 years old; several date from 770 000 to 790 000 years ago, and one sample from Enarete is 670 000 years old. Dates of 800 000 years were obtained from the submerged slopes of Panarea. A sample from Stromboli canyon is 530 000 years old, but another sample from its submerged northern slope is only 180 000 years old. These great ages imply perhaps that most of the Aeolian seamounts are extinct, and that, instead of being under construction, they might have been carried downwards as the floor of the Tyrrhenian Sea sank. However, a recent investigation of Marsili reportedly revealed fresh morphological features that indicate renewed activity.

AEOLIAN ISLANDS: VULCANO

Vulcano is the southernmost of the Aeolian Islands, and forms an ellipse, 22 km² in area, 3 km broad and stretching 6 km from north to south. It is apparently a young island, for it bears no trace of old marine terraces around the coast (Keller 1980a). Since Vulcano was built up above sea level, the migration of activity has created a varied but low-lying island that reaches only a modest 500 m at Monte Aria, although its base lies at a depth of 2000 m on the floor of the Tyrrhenian Sea. Indeed, the Fossa cone, which has been the focus of most of the eruptions during historical times, only reaches 391 m (Frazzetta et al. 1983). However, what Vulcano lacks in height, it makes up in aesthetic quality. The volcanic landforms were often created side by side, instead of being superimposed upon each other, by a progressive and continuing northward shift of activity along a major fissure. Thus, historical eruptions in the straits between Lìpari and Vulcano might have developed or given

The Gran Cratere of the Fossa cone, with Vulcanello in the middle distance and Lìpari on the horizon.

birth to Vulcanello, which became linked to the main island only after its latest eruption in about 1550. The remains of the stratovolcano; several calderas that formed, and then filled, to create the Caldera del Piano; the Fossa cone, with its deep crater, and two other satellite craters as well as a recent obsidian lava flow on its flanks; and the separate cones of Vulcanello, not to mention a beach with hot springs and mudpots – all combined to give Vulcano its beautiful landscape.

References to activity on Vulcano go back some 2500 years, although there are gaps in the historical record, especially in the early Middle Ages. Vulcano seems to have witnessed an eruption during nearly every century for which fairly reliable information is available, with about 30 active episodes altogether (De Fiore 1922, Stothers & Rampino 1983, Frazzetta et al. 1984). The Vulcano of antiquity was intimately linked with myth, theology and geology, for it was reputed to be one of the major homes of the god of metalworking, fire and war, known to the Greeks as Héphaistos and to the Romans as Vulcanus. The ancient Greeks called it Hierá Hephaístou ("sacred to Héphaistos"); and, as Virgil described in the *Aeneid* (8: 416), the ancient Romans believed that Vulcanus worked his forges on the island. The geographer Strabo (VI: 2.10) described Vulcano at that time as

"deserted, wholly rocky, and entirely in the hands of the fire, with three erupting vents . . . the largest of which projects glowing blocks in the midst of the flames, that have already filled up much of the straits." Strabo added that "Polybius [who died about 120 BC to 125 BC] reported that one of the three craters had partly collapsed, but that the other two still remained intact. The largest of them, which was circular, measured five stadia (925 m) around its rim, but it then narrowed down to no more than 50 feet (15 m) . . . and it was one stadium (185 m) from the sea". This seems to be a quite accurate description of the Fossa cone, and all three craters could belong to it. On the other hand, one of the other craters could be Vulcanello, but the site of the third would still be a mystery.

All the lavas of Vulcano are potassic (Ellam et al. 1988). The lavas of the southern stratovolcano are mainly shoshonitic basalts and leucite-tephrites arranged in layers of lava flows and fragments. On the northwestern edges of the stratovolcano, rhyolites erupted to form the Monte Lentia volcano. Fragments of shoshonitic basalt and leucite-tephrites constitute the bulk of the materials that now fill the Caldera del Piano. The Fossa is essentially a trachytic tuff cone that lies on a base of trachyte and leucite-tephrite lava flows. It has exuded a rhyolitic obsidian lava flow in the Pietre Cotte, which resembles parts of the Monte Lentia volcano. Vulcanello, similarly, lies on a base of trachyte and leucite-tephrite lava flows, and its cones are built of tuffs of similar composition. Earthquakes are weak on Vulcano. They are centred at a depth of 2–2.5 km on the Acquacalda beach, on the fissure joining Vulcanello and the Fossa, which suggests a common magmatic source for them.

South Vulcano

South Vulcano forms two thirds of the island and is its largest and oldest volcanic accumulation. Its eruptions began at about 2000 m deep on the floor of the sea. It grew above sea level and formed an elliptical stratovolcano that erupted innumerable shoshonitic lava flows separated by relatively thin beds of cinders. One of these flows has been dated to 120 000 years ago (Gillot et al. 1980). The whole mass is also riddled by radial lava dykes and some satellite vents. Fluvial erosion has eaten deep gullies into the flanks of South Volcano, and marine erosion has cut

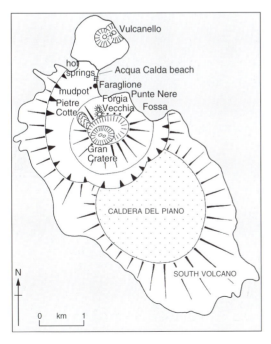

Selected volcanic features of Vulcano.

Vulcanello and the Fossa cone.

jagged cliffs, often 30 m high, around its edges.

The second main episode in the history of Vulcano began when the summit of the strato-volcano collapsed about 100 000 years ago and formed a caldera about 2.5 km across. Subsequent eruptions then filled the caldera to the brim with tuffs, leucitic lapilli, and occasional flows of shoshonitic basalts. However, even as the caldera was being filled, more collapses took place in its northern sectors. Successive calderas were formed, with each new rim biting into the previous caldera. But more eruptions compensated for each partial collapse. The infilling often included a few cinder cones, such as La Sommata, as well as the trachytic, latitic and, especially, rhyolitic domes and flows, about 15 500 years old, which form the Monte Lentia in the northwest. The infilled zone forms the Caldera del Piano, named after the piano ("the plain") about 300 m above sea level, that supports most of the rural population of Vulcano.

Caldera della Fossa

When the Caldera del Piano had been filled, the northwestern flank of South Vulcano collapsed a fourth time and formed the Caldera della Fossa, which sank below sea level. It forms a slightly elliptical hollow, about 3 km across, with a pronounced cliff, mostly 100 m high, that marks its inner perimeter except on the northeast. This sector either suffered the greatest amount of sinking or was destroyed by a blast or landslide. The Caldera della Fossa has been the theatre of all the activity on Vulcano for the past 14 000 years. New eruptions, concentrated on a fissure running from north to south, have slowly carpeted

the caldera floor, but more than a quarter of it still lies below sea level, and at least another quarter hardly rises more than 10 m above it. The stage was then set for the activity that has given Vulcano its renown – the growth of the Fossa and Vulcanello. However, in the wider perspective of the history of the island they probably represent only initial contributions to the filling of its youngest caldera (Plate 5a).

The Fossa and Vulcanello are tuff cones, and the proximity of the sea has had a distinct influence on their growth, for both have had hydro-volcanic components. The activity of the Fossa may indeed owe such distinctive features as it possesses chiefly to the interaction of sea water and an upsurging trachytic or rhyolitic magma (Frazzetta et al. 1983, Sheridan & Malin 1983, Wohletz & Sheridan 1983, Frazzetta et al. 1984, Clocchiatti et al. 1994, Montalto 1996, De Astis et al. 1997).

The Fossa cone

The Fossa is a squat cone of trachytic tuffs that forms the backcloth to the Porto di Levante. Although it rises only 391 m, its base is 1 km across and its steep slopes range between 30° and 35°. Its northern flanks are scarred by the crater of the Forgia Vecchia, created by two successive satellite eruptions and by the bristling grey lava flow of rhyolitic obsidian forming the Pietre Cotte. Two lava tongues, the trachytic Palizzi flow and the rhyolitic Commenda flow, emerge from the southern flanks of the Fossa, and the trachytic Punte Nere flow juts out seawards from its eastern base. However, these small lava flows seem only to mark late spurts of viscous emissions as the explosive phases closed. The Fossa is crowned by the Gran Cratere, which is 500 m in diameter and no less than 175 m deep. It has the typical dimensions of a tuff-cone crater and a distinct funnel shape. In particular, its splayed upper northern walls are stained yellow by beautiful sulphurous crystals from many active fumaroles. The almost flat floor of the Gran Cratere is composed of two intersecting centres, clearly linked to a couple of closely spaced vents. The Fossa is a rather shy volcano, now erupting on average about once per century. Most of its products are trachytic. The Fossa has probably been active for about 6000 years and it covers the layer of Lower Pilato rhyolitic pumice that exploded from Lìpari somewhere between 11 000 and 85 000 years ago. Its growth has been characterized by a westward migration of the vents by about 500 m, by cyclical eruptive phases, and by a gradual decrease in the volume of materials erupted. The Fossa was constructed by four or more very similar eruptive cycles, in which a short eruptive outburst was followed a long period of repose. The initial cycle probably built up the bulk of the Fossa, and eruptions during the past two centuries have expelled relatively little.

The products of the initial, or Punte Nere, cycle came from the easternmost vent of the Fossa. It is uncertain when the Punte Nere cycle started, but it ended about 5500 years ago, when trachytic lavas flowed down the cone to form the promontory of Punte Nere. The bulk of the Fossa had thus already been built up by the end of this cycle, and its old crest is nearly 300 m high on the eastern summit of the present cone.

After an interval of dormancy, the second, or Palizzi, eruptive cycle began. The active vent shifted about 300 m to the west, and the western sector of the cone was partly destroyed. Some of the earliest recorded observations of activity on the Fossa date from this time. The Palizzi cycle had two parts. First came hydrovolcanic eruptions of tuffs and then pumice, which have been dated to about 2200 years ago. Given the margin of dating error, one of these could have been the eruption noted by Thucydides, "where great flames are seen rising up at night, and, in the daytime, the place is under a cloud of smoke" (*Peloponnesian War* III: 88). This eruption probably occurred in 425 BC when the author was in Sicily. Explosive eruptions also took place between 370 BC and 350 BC, and again in about 330 BC, which might have been the activity noted

by Aristotle and Callías (De Fiore 1922, Stothers & Rampino 1983). Aristotle (*Meteora* II: 8), described how the mountain swelled up and burst with a loud noise; fire escaped, and a violent wind hurled a great quantity of ash over Lìpari and as far as Italy. Callías (book III) noted that, when Agathocles (361–289 BC) was tyrant of Syracuse, one of two craters on Vulcano could be seen shining very brightly from a long distance, and threw out glowing stones of monstrous size with a noise that could be heard 500 stadia away (92 km). According to Pliny the Elder, "an island", possibly Vulcanello, is reputed to have risen above the waves in 126 BC.

Both Vulcano and Vulcanello were probably erupting between 29 BC and 19 BC. One of these episodes could have inspired Virgil to compose the famous passage describing Vulcan's forge – the Gran Cratere of the Fossa – in the *Aeneid* (8: 416–453), where "strong blows are heard resounding on anvils and, echoing their groans, . . . ingots hiss within the caverns and the fire pants within the furnaces". The second part of the Palizzi cycle, dated about 1600 years ago, occurred when hydrovolcanic eruptions were followed by the emission of the Palizzi trachytic flow, which moulded the southern slopes of the Fossa cone. Activity was reported on Vulcano, for instance, between AD 200 and AD 250.

The third, or Commenda, cycle used virtually the same crater as its predecessor. After resting for a few centuries, an unusually violent hydrovolcanic explosion cleared the plugged vent of the Fossa and expelled a breccia that gradually changed into a **block** and ash flow. These eruptions may be tentatively matched with reported activity perhaps in either AD 526 or AD 580. This activity could correspond with a hydrovolcanic explosion on the northern flanks of the Fossa, dated to about 1400 years ago, that formed the first yellow tuff cone of the Forgia Vecchia (Frazzetta et al. 1983). Their deposits were then covered by the Upper Pilato ash from Lìpari, which is believed to have erupted in AD 729, but which recent archaeomagnetic studies suggest could be as young as about AD 1200. This marker bed was, in turn, overlain first by dry surge layers and then by the Commenda lava flow that moulds the southwestern slopes of the Fossa cone. The Commenda flow is transitional between rhyolite and trachyte. It has been dated to about AD 700, which conflicts with the stratigraphic relationships unless the margins of error in the dating are taken into account. In any case, all these events probably occurred in quick succession. Eruptions were also reported between AD 900 and AD 950, and during the thirteenth century.

After a further interlude of repose, the fourth, or Pietre Cotte, cycle began when an eruption came from a new crater about 200 m west of the Commenda crater and coated much of the Fossa cone with reddish tuff. This cycle probably started with the vigorous eruption on 5 February 1444, when "Vulcano burned with perpetual fire . . . gave off thick fumes . . . sooty clouds, while pale flames rose up from the abyss and escaped from fissures and joints." (Fazellus 1558). The Fossa was active again in 1618, 1626, 1631 and in 1646, when Bartoli observed abundant fumes emerging from the incandescent crater, and again in 1651 and 1688. In 1727, d'Orville saw a small cone within the Gran Cratere of the Fossa, which erupted glowing stones and fumes with a noise like thunder; in the same year, a small tuff cone erupted on the northern edge of its larger predecessor, the Forgia Vecchia. In 1739, after about eight years of intermittent and unusually prolonged activity, the bristling rhyolitic lava flow forming the Pietre Cotte oozed down the steep northwestern slopes of the Fossa, but it solidified even before it reached the base of the cone. The Pietre Cotte cycle trailed on with the eruptions in 1771, 1786 (two weeks in March), 1873, 1876, and came to an end, for the time being, with the activity between 1888 and 1890 (Judd 1875a, Johnston-Lavis 1888a,b, Butler 1892, De Fiore 1922, Sheridan et al. 1981b, Frazzetta et al. 1984). But, the roof of the magma reservoir still probably lies no more than 2 km deep (Clocchiatti et al. 1994, Montalto 1996).

The Fossa has remained quiet since 1890,

except for persistent fumaroles that occur most notably on radial and concentric fissures on the upper northern slopes of the Gran Cratere. They are associated with beautiful, if extremely fragile, sulphur crystals and little stalactites. A sulphur flow, 2 m long, even issued from a small vent in late September 1989. The fumaroles have varied in intensity and temperature, with an initial maximum of 615°C in 1924, then decreasing to a minimum of 110°C in 1962. A period of unrest started in 1980, which was probably caused by injections of gas from the magma below, that increased the pressures within the hydrothermal system near the surface. Temperatures rose again to nearly 700°C in 1993, but have subsequently lowered and no eruption occurred. Such activity should be followed with care, because the development of tourism on the island implies that any new eruption could cause many casualties.

The Faraglione, mudpots and the Acquacalda Beach

The fissure extending northwards from Vulcano to Lìpari did not remain inert while the Fossa was active. It gave rise to mudpots, hot springs, the Faraglione and Vulcanello, which together add much to the scientific and tourist interest of Vulcano. The oldest feature on this fissure is probably the Faraglione, which now forms a steep bulky pinnacle rising 56 m above the harbour of Porto di Levante. The Faraglione is far from being in its original state. Indeed, it has been so decayed by thermal activity, so pared by marine erosion, and so extensively depleted by quarrying, that it is hard to decipher what its

The eruption of the Fossa cone on Vulcano (3 August 1888 to 22 March 1890)

For all its long list of recorded eruptions, only the latest outburst of Vulcano was described in any detail. After a period of complete calm, the initial vigorous explosion removed the old solidified material plugging the vent and deposited a thin yellow breccia, and shifted the vent slightly southwards to the present Gran Cratere. But the main efforts of the eruptions were the powerful intermittent explosions, "like the detonations of landmines", that sent billowing clouds of gas, steam, fine ash, and lapilli several kilometres into the atmosphere. These explosions varied in both intensity and frequency. Most expelled fine pale-grey or reddish trachytic ash, but the largest also ejected big blocks and especially the "breadcrust" bombs of rhyolitic obsidian, with a surface like the

best French loaves, which were thrown as much as 1.5 km from the vent. The spasms were separated by irregular periods of rest and there were abrupt changes from febrile outbursts to mere fumarole activity, or even total calm, that sometimes lasted several days. These eruptions expelled very thin alternating layers of fine trachytic ash and lapilli that apparently have no counterpart in the history of the Fossa. However, the eruption was a relatively mild affair, full of sound and fury, but signifying little change in the Gran Cratere, and only a layer of fragments around its rim, which over a century of intense Mediterranean showers have reduced from their original thickness of 5 m to 1.7 m.

Hot thermal emissions from fissures on the Acquacalda beach, with Vulcanello on the right.

initial shape and function can have been. It may have been formed by an explosion breccia (Frazzetta et al. 1984) or it may be the remains of an early Vulcanello (Keller 1980a). It is matched by other isolated pinnacles, the Faraglioni, in the straits between Vulcano and Lìpari. All may be the relics of cones formed by prehistoric Surtseyan eruptions, which, unlike the Fossa cone and Vulcanello, did not emit enough lava to enable them to withstand marine erosion for long.

Near the Faraglione are the mudpots and hot springs that give their name to the Acquacalda beach, linking the Porto di Levante harbour with Vulcanello. The grey mudpots at the foot of the Faraglione have been made into a large bath for the benefit of tourists. The muds vary in temperature from 20°C to 60°C and are reputed to cure rheumatism and ugliness if spread in the appropriate places. To the north, the Acquacalda beach and the adjacent shallows are riddled with small

The sea-cut eastern cone of Vulcanello.

The mudpot at Vulcano.

easternmost cone of Vulcanello and provided a magnificent section of its alternating layers of reddish-brown leucite-tephrite lavas and cinders. The section also demonstrates that this cone erupted above sea level in Strombolian rather than Surtseyan eruptions. On the other hand, the second cone was formed by a Surtseyan eruption when sea water could penetrate the new vent, a little to the west of the first. It is composed of fine buff tuffs that are also scattered over the surface of the first cone. The ungullied surface of these cones confirms their recent origin. Then, two eruptions of lavas ensued, which show that water could not penetrate the vents at this time. The first emissions, of leucite-tephrite, have pahoehoe surfaces and form the broad plinth around the first two cones of Vulcanello. The much smaller second emission, of more viscous trachyte, with a more bristling surface, emerged from the northern flanks of the second cone and covers the edges of its predecessors. In places, the Upper Pilato ash layer from Lìpari lies on these flows.

The third, westernmost and latest, cone was

fissures, from which nearly-boiling water and carbon dioxide bubble out. The Acquacalda beach is itself part of the isthmus of black ash, about 500 m wide, linking Vulcano to Vulcanello.

Vulcanello

Vulcanello is a buff-coloured cone, rising to 123 m, that lies on the broad pedestal of two lava flows. In fact, it is composed of three successively formed cones, with intersecting craters, which have joined together in a single elongated mass. During its growth, the location of the vent in relation to sea level was often crucial in determining what was expelled.

The eruptions of Vulcanello began below sea level and the date of its emergence is not known exactly. Eruptions occurred on the new island perhaps in 126 BC, in 91 BC, and for some time between 29 BC and 19 BC (De Fiore 1922, Frazzetta et al. 1984), and the latest eruption occurred in about 1550 (Fazellus 1558), but it is uncertain when Vulcanello was active in between times.

The sea has eroded the eastern flanks of the

The most recent crater of Vulcanello, with the cliffs of Lìpari in the background.

The birth of a new island in the Aeolian archipelago

Strabo describes how a new island rose above the waves. He writes that "Posidonius [c. 135 BC to c. 50 BC] recounts that, in his time, at sunrise on the summer solstice, between Hiera [Vulcano] and Euonymos [Panarea, probably a mistake for nearby Lipara, i.e. Lìpari], the sea rose up to an extraordinary height and, by a sustained blast, stayed puffed up for a considerable time, before it fell back down again. The boatmen who had dared steer their boats up to it made off when they saw that the current was bringing dead fish towards them, and that some of the men had been struck down by the heat and the stench. But one of the boats that had approached more closely had lost some of its occupants. The survivors had reached Lipara only with great difficulty and had kept on falling down like epileptics, suffering from seizures of madness that alternated with sudden returns to sanity [perhaps caused by carbon dioxide emissions]. Several days afterwards, mud was seen spurting up from the sea, accompanied in many places by jets of flame, exhalations of fumes, and sooty emissions. Then, it solidified and formed rock with the consistency of millstone [tuffs?]. The governor of Sicily, Titus Flaminius, brought these facts to the notice of the [Roman] Senate, which immediately sent a deputation to offer sacrifices to the gods of the underworld, and to the divinities of the sea, both on the new island and on Lipara." (Strabo VI: 2.11)

This account forms one of the main pillars of the view that Vulcanello emerged in 126 BC. But the historical evidence is not strong. Nothing is known of Titus Flaminius, so it is assumed that Strabo mis-spelled the name of Titus Flamininus, who is known to have been Governor of Sicily in 123 BC. This view is said to offer support to a statement by Pliny the Elder [II: 88.89] that "before [his] time . . . an island rose up in the Aeolian Islands, another near Crete . . . and yet another, in the third year of the 163rd Olympiad [126 BC], in the gulf of Etruria." But only the last island is clearly and unambiguously dated in his text. Moreover, the location of the new island is imprecise and there is, in fact, no indication that it survived for long after the sacrifices were made. But the most telling argument springs from the archaeomagnetic direction of the lavas forming the plinth of Vulcanello, which shows that they cannot have erupted in 126 BC and are probably several centuries older (Tanguy, unpublished). It is, therefore, most unlikely that this passage describes the *birth* of Vulcanello, although the eruption in 126 BC may have either enlarged the island or taken place nearby.

formed, in the Middle Ages, by two short Surtseyan eruptions that came from a vent situated some 50 m west of the second crater. The third cone of yellowish trachytic tuffs rises only 100 m high, but is remarkable for its fresh-looking steep-walled crater. The first eruption probably occurred here in the thirteenth century (De Fiore 1922). It was followed by a quiescent interval during which a soil formed on the cone. The latest eruption gave off yellowish tuffs in about 1550 (Fazellus 1558). Marine currents are said to have redistributed the erupted fragments and formed the isthmus joining Vulcanello to Vulcano, but the ash making the isthmus is black and is thus quite unlike any products expelled from Vulcanello. The bulk of the accumulation must therefore have come from Vulcano to the south. No subsequent eruptions have yet taken place on Vulcanello, and fumarole activity ceased in its craters at the end of the nineteenth century.

SICILY: ETNA

Etna is the largest active volcano in Europe (Plates 2, 3, 4a). It presides over the landscape of eastern Sicily and the Straits of Messina and it forms the graceful backcloth to the Greek theatre at Taormina, as well as the constant threat on the northern skyline of Catania. The Sicilian giant rises some 3320 m above sea level, stretches 47 km from north to south and 38 km east to west, covering about 1200 km², with a volume of 350–500 km³ (Tanguy 1980, Romano et al. 1982, Chester et al. 1985, Kieffer 1985).

Etna is a complex volcano, with four large summit craters and more than 200 cinder cones, as well as many faults on its flanks. The people living on the mountain have always divided it into three parts (Tanguy & Patanè 1996). The piedmont, or cultivated region, is the lowest part and it corresponds to the edge of the old shield that forms the base of Etna. Its broad and gentle slopes are scattered with white villages, set between citrus groves, vineyards and cereal plots that flourish on the fertile weathered lavas. Up slope, between 1000 m and 2000 m above sea level, lies the second, wooded, region, with many pine forests that eventually give way to Mount Etna broom, whose brilliant yellow blooms contrast with the black and reddish

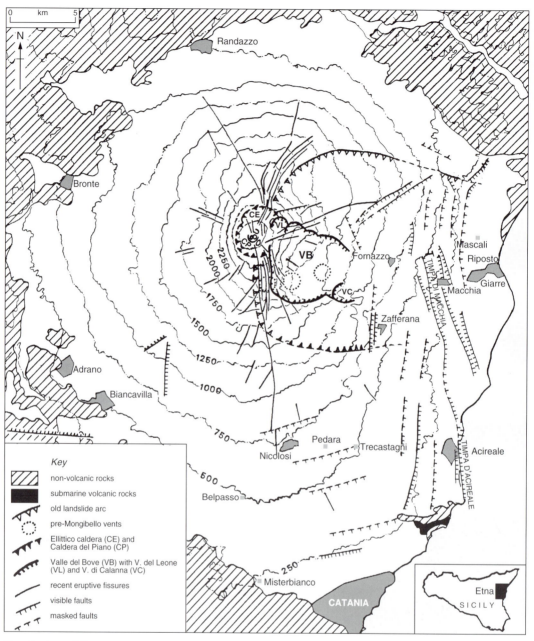

The structure of Etna.

rugged basalts where it has gained a foothold. This is where many of the cinder cones have erupted, notably in the south, southeast and northeast. Some are already wooded, others still starkly bare. Elsewhere, they would make imposing landmarks, but here they are dwarfed by the mass on which they stand. Eventually, at about 2000 m, the third, bare and sterile region begins. The slopes steepen at last to 20° to form the upper cone of Mongibello, the Sicilian name given by Earth scientists to the most recent of the several stratovolcanoes that compose the mountain. Here, the snows melt in spring to reveal the naked black lavas erupted so recently that weathering has scarcely altered their surface. This wilderness, of a vastness seen on no other European volcano outside Iceland, extends up to an infilled caldera that forms the summit plateau, the Piano del Lago, at about 3000 m. It is here that persistent activity most often changes the face of Etna. In its centre lies the summit cone, scarred by two large chasms, the Voragine and the Bocca Nuova, which joined together in 1999. Both chasms can be more than 200 m deep, but lavas well up within them from time to time, as occurred, for instance, in July 1998. The Central Cone is flanked on the northeast and southeast by two sources of sub-terminal activity. Constant eruptions from the Northeast Crater since 1911 built up cinder cones that grew even higher than the summit cone in 1978, whereupon their efforts began to wane and the baton was taken up at once by new persistent eruptions that quickly constructed the Southeast Crater.

However, in spite of all its rapid summit changes, the most remarkable single landform on Etna is perhaps the Valle del Bove, a great horseshoe trough, 5 km wide and 1000 m deep, which opens on its eastern flanks as if it had been scooped out by a giant bulldozer. Its nearly vertical walls provide almost the only major geological sections, where the history of the volcano can be deciphered, because repeated eruptions have quickly blanketed the evidence elsewhere. However, the Valle del Bove remains enigmatic,

"Snowy Etna, the Pillar of the Sky", celebrated by Pindar nearly 2500 years ago, seen from Taormina.

and its own origin, together with the full story of Etna, are still subjects of debate.

Etna provides a relatively safe and fascinating laboratory with the happy knack of always producing something new within its range of activity. Few volcanoes have been so obliging and its bibliography is now probably the largest of any European volcano. Even as early as about 430 BC, the philosopher Empedocles is reputed to have fallen into the crater during an investigative mission – an exploit that luckily has not been emulated since. But serious study reaches back more than a century, and Etna has been under intensive care by Italian, French and British research teams, especially since it took on a new lease of life in 1971. It has also been subject to practically every surveillance technique known to volcanology (e.g. Rymer et al. 1995, 1998).

Etna has displayed almost continual activity throughout historical times and it has the longest record of eruptions in the world. It has registered eruptions from its flanks or summit, ranging from the formation of cinder cones and lava flows to emissions of gas, steam or ash, which have taken place in periods stretching from days to many months, in the course of about 250 of the 500 years that have elapsed since reports became more reliable in the early sixteenth century (Tanguy 1981, Romano et al. 1982, Chester et al. 1985). Prolonged phases of persistent activity have marked the summit and, although flank eruptions have been more episodic, they have nevertheless occurred in more than 80 of the past 400 years. Etna is thus a treasure house of basaltic lava flows and has the finest collection of flank fissures and cinder cones in Europe.

The higher zones of Etna are of course most affected by eruptions, and it has been calculated that half the area above 2000 m, mainly comprising the Mongibello cone, would be covered by new lavas every 250 years, whereas it would take 525 years to cover half the area between 1000 m and 2000 m, and almost 2000 years to cover half the area below 1000 m with new lavas (Guest & Murray 1979). These eruptions represent an appreciable long-term threat to property. Moreover, since 1971, Etna has enjoyed one of its most active phases in historical times and it shows no signs of stopping, with practically every year bringing some form of activity. Etna is thus one of the most active volcanoes in the world. Its average **effusion** rate of 0.8 m³ per second is exceeded only by Kilauea in Hawaii; in fact, Etna erupts more gas, in terms of tonnes per day some

The eruption of the Monti Silvestri in 1892

The eruption of the Monti Silvestri was typical of the medium-size flank activity on Etna. It lasted from 9 July until 28 December 1892 and emitted some 150 million m^3 of magma. The eruption began on a flank fissure at a height of between 1800 m and 2000 m, where continuous explosions quickly built up three large cinder cones that have since been visited by millions of tourists. In the words of Riccò & Arcidiacono (1902–1904):

We are stupefied by the prodigious scene before us, although we are 2 km from the eruption. In the upper part, three craters are in flames and the one in the centre is giving off a double jet of fire. The cinders and glowing bombs are being violently thrown straight upwards and reach a height of about 400 m, while a broad, fiery cloud forms a splendid purple veil above this cannonade. Lower down, an immense flood of fire stretches as far as the eye can see [and] an infinity of channels, streams and cascades of lava are running in all directions across [it]. The lava on the surface, which has already cooled and solidified, is being dragged along by the liquid current, and the fragments bump into each other and make a noise like a pile of falling tiles. There is a continuous rain of cinders and small hot lapilli, and hot suffocating air gusts out from time to time. Added to this inconvenience is the intense heat of the lava that is flowing at the foot of the hill where we are standing – but we are so absorbed by the terrible beauty of the scene before us that we never even notice it. The lava front is advancing so slowly that it is hardly perceptible. We estimate that its speed is about a metre a minute, but it varies according to the slope and the form of the land. We can feel only pity for all these fine trees that are condemned to be burned alive. The leaves shrivel up and lose their colour before they burst into flames. The branches twist. The lava arrives and surrounds and suffocates the unfortunate plant in its fiery grip. The victim gives out a kind of strident cry. At length, the rapid hydrocarbon distillation creates the naked flames that bring its torment to an end. We feel both sad and horrified as the enormous, sinister black mass, with its fiery base, advances into the wood.

The Monti Silvestri, erupted on Etna in 1892.

Newly solidified lavas crossing an older weathered flow covered with Mount Etna broom. The Montagnola marks the horizon on the right.

200 000 of steam, 70 000 of carbon dioxide and 4500 of sulphur dioxide (Caltabiano et al. 1994, Allard 1997).

In spite of all this agitation, violent outbursts on Etna have been both rare and brief. Etna causes damage, not death; and it has not killed a hundred people in all its recorded history – even though a hydrovolcanic explosion killed nine tourists near the rim of the Bocca Nuova on 12 September 1979. The greatest damage has been caused by the inexorable progress of lava flows emanating from flank fissures. More than 35 eruptions during the past four centuries have damaged crops or property. Part of Fornazzo, for example, was buried in 1971, nearby Mascali was destroyed in 1928, and, most notorious of all, part of Catania was overwhelmed when the famous eruption of 1669 discharged one of the largest lava flows of historical times. In 1971, Etna was even impertinent enough to destroy the old volcano observatory and the upper parts of the Funivia cable track on the southern flank. A new observatory has now been built just below the Pizzi Deneri (2847 m) on its northern flanks.

The geological setting of Etna

Most of Etna is young. In the first half of its existence, activity was hesitant and sporadic. But during the second half – the past 250 000 years – Etna has expelled nine tenths of its volume, with about one third forming the upper cone in the past 40 000 years or so (Kieffer 1985, Kieffer & Tanguy 1993). The volcano rises in the midst of a zone of fearsome tectonic complexity. Sicily has long been the hub of repeated collisions between the African and Eurasian plates and their adjacent microplates. Associated with these collisions are the opening of the Tyrrhenian Sea, the opening and sinking of the Ionian Sea, and the subduction, bending and possible compression of the Aeolian–Calabrian arc. Then, as the subduction apparently waned, the edge of the Eurasian plate was thrust over the edge of the African plate, causing the displacement of the continental rocks of Sicily (Barberi et al. 1973, 1974, Lentini 1982, Selvaggi & Chiarabba 1995). Consequently, Sicily was shattered and transected by deep major faults. The Tyrrhenian side of Sicily is being crushed while the Ionian side is being extended. This extension keeps the faults open near the Ionian Sea (Lanzafame & Bousquet 1997, Monaco et al. 1997). Etna rises close to, but curiously not exactly upon, the intersection of three of these major faults: the Mt Kumeta–Alcantara fault, trending from east to west, that now probably marks the Afro/European boundary; the Messina–Giardini fault, running north-northeast to south-southwest that now delimits the coast north of Etna; and the conjugate Aeolian–Maltese fault, running from Vulcano to Malta. These faults probably transect the crust down to the mantle and thus direct magma to the surface with relative ease. It has been suggested recently that Etna could have originated by suction of asthenospheric material from beneath the African plate along the major Aeolian fault system (Gvirtzman & Nur 1999).

Etna also lies on a zone that has undergone uneven uplift and thus its volume is less than it seems at first sight, because, although the lava base lies at sea level in the south, it occurs at an altitude of over 1000 m on its northwestern flanks. Uplift of 6 m has taken place during the past 2000 years; and 30 cm has been recorded within the past 50 years (Kieffer 1985). The east-facing fault scarps, forming a staircase on its lower Ionian flanks, are among the main results of this displacement. The continued foundering towards the Ionian Sea has removed much support from the accumulating eastern flanks of the volcano, whereas the other flanks are bolstered by the Sicilian landmass. Thus, a broadly arcuate area, 15 km wide, including both Etna and its basement, has been gradually slipping seawards for some 100 000 years. Within this structure, the volcano was depleted further when gigantic landslips and debris avalanches formed the Valle del Bove. This zone of eastward displacement is now associated with a distinct rift zone curving from the northeast, through the South-east Crater, and then extending southwards. Here, fissures have developed and have been responsible for many of the lateral eruptions of recent times (Kieffer 1983, 1985, McGuire & Pullen 1989, Borgia et al. 1992). The most important result in human terms has been that the highly populated eastern and southeastern flanks have by far the most eruptions, and indeed earthquakes, whereas the western flank, in particular, has been relatively immune.

Although Etna erupted tholeiitic basalts at the start of its life, alkaline series have since predominated, with trachybasalts (similar to hawaiites), trachyandesites (mugearites), and some benmoreites and trachytes (Tanguy 1978, Cristofolini & Romano 1982, Métrich & Clocchiatti 1989, Tonarini et al. 1995, Tanguy et al. 1997). The petrological and geophysical evidence indicates that the magma now comes from depths of perhaps 40–100 km in the mantle and rises to about 15 km below the volcano (Hirn et al. 1997, Clocchiatti et al. 1998). Beneath the mantle/crust boundary within this zone, basaltic magma accumulates in a subcrustal or mantle reservoir (Tanguy 1980) that could, perhaps, be as large as 150 km³ or 300 km³ in volume. The primary basalts could stay there for some 1500 years and differentiate towards hawaiites (Condomines et al. 1995). The presence of this deep reservoir would thus explain the predominance of the slightly differentiated hawaiites that account for 70–80 per cent of the total volume of Etna. However, it seems that smaller shallower reservoirs must also develop from time to time, in order to account for the occasional eruptions of mugearites and trachytes, which have often been associated with caldera collapses.

The trachybasaltic lavas erupted in such quantities that they have sometimes been called "etnaites" (Rittmann 1962); however, they are more often called hawaiites. Field measurements indicate effusion temperatures of 1070–1090°C (Archambault & Tanguy 1976, Tanguy et al. 1983, Clocchiatti et al. 1999). Although it is relatively low for basaltic lavas, this temperature is in accordance with their high crystal content, which is usually 30–50 per cent. These crystals are almost always plagioclase and less abundant pyroxene, with a little olivine and titanomagnetite. It is likely that most of the crystallization takes place during slow ascent from the deep reservoir, during which plagioclase growth is enhanced by continuous degassing.

Ancient Etna

Eruptions began more than 500 000 years ago with emissions of tholeiitic lavas, which are still exposed on the southern fringes of the volcano (Gillot et al. 1994). These eruptions are often considered to belong to a pre-Etna phase of activity. They took place from many vents aligned along fissures stretching inland from a silty Sicilian bay between the Peloritani Range and the Iblean Plateau. They expelled hot and fluid tholeiitic basalts that rose quickly to the surface. On land, they cover the uppermost terrace alongside the River Simeto near Adrano. Similar basalts erupted below sea level and formed pillow lavas and breccias that now provide the plinth for the castle at Aci Castello on the Riviera dei Ciclopi (Romano 1982). The eruptions were separated by long intervals of repose and they remained so scattered and infrequent that no marked single volcano ever developed.

Imperceptibly, ancient Etna began to grow up when the scattered and sporadic eruptions became more concentrated beneath the present summit. Alkaline basalts erupted from the deep reservoir situated just below the base of the crust, but differentiation within the reservoir soon caused the eruption of hawaiites in both flows and fragments. It seems that by about 140 000 years ago, the volcanic accumulations had covered much the same area as that of present-day Etna (Condomines et al. 1982, Gillot et al. 1994). Indeed, in all, these initial eruptions encompassed more than half the total history of Etna, although they produced only a fraction of its volume. These lavas had piled up on slippery beds of clay that dipped down eastwards to the Ionian Sea. Unlike the other flanks of Etna, which were buttressed by the bulk of Sicily, the eastern flank had little support; as a result, it began to slip eastwards, forming a horseshoe depression, about 15 km across, which was of course open to the east. Thus began the long series of eastward slides towards the Ionian Sea that has persisted until the present day.

Several eruptive centres grew up within the vast caldera-shape depression of ancient Etna. One remains in the eroded piton of Monte Calanna (1325 m); another is represented by many dyke and lava-flow accumulations at Trifoglietto, about 4 km from the present summit of Etna. At Trifoglietto, two large volcanoes apparently formed in succession about 70 000 years ago. The presence of amphibole crystals means that their trachyandesitic (mugearite) lavas can be readily identified; and it seems that they erupted from a small shallow reservoir that developed above the much larger reservoir at depth. After the Trifoglietto centres had been active for several tens of thousands of years, the focus of eruptions once more shifted farther west. At the same time, the unsupported eastern flanks of the volcanic pile began to slip down towards the Ionian Sea. Each time that the magma arose to the west, the eastern flanks of Trifoglietto were destabilized a little further. As the pressure upon the magma was reduced, violent explosions and debris avalanches ensued. Thus, the enormous depression of the Valle del Bove began to form in successive stages on the eastern flanks of Etna (Kieffer 1985).

The Ellittico caldera and the origin of the Valle del Bove

After Trifoglietto had lapsed into extinction, many more eruptions took place to the west, although the exact sequence of events is far from being well established. For instance, the Vavalaci, Cuvigghiuni and Leone centres are still largely the subject of debate. However, one major centre can be identified: the Ellittico volcano. It is the direct predecessor of modern Etna and was also called Ancient Mongibello by Earth scientists. At its apogee, some 15 000 years ago, the Ellittico volcano approached a height of 3700 m. At about the same time, a shallow reservoir developed probably less than 6 km below the crest of the volcano. It erupted the lava cicirara – the "chick pea" trachybasaltic lava – which is so called because of its large white plagioclase crystals. Thereafter, mugearites, benmoreites and trachytes were expelled as the magma continued to evolve (Kieffer & Tanguy 1993).

About 14 000 years ago came the catastrophic eruption that gave off Plinian columns and nuées ardentes that cascaded down the flanks of the volcano, mainly to the northeast, northwest and southwest, where they form thick pink deposits near Biancavilla (Kieffer 1979). This eruption eventually led to the foundering of the summit and the formation of the Ellittico caldera, which was 4 km long and 3 km wide.

One of the major episodes in the history of Etna was the formation of the Valle del Bove. Its steep scalloped walls rise 1000 m above the floor to more than 2800 m above sea level at its western end. Its floor slopes down towards the east and is coated by many recent lava flows, and the walls reveal the eruptive centres of ancient Etna. The Valle del Bove thus forms a huge trough that bites deep into the flanks of Etna and opens eastwards towards the Ionian Sea. It is 5 km wide and more than 7 km long, and perhaps as much as 10 km^3 of material have been removed from the trough and deposited on the lower flanks of the volcano and in the fan-shape Chiancone conglomerate jutting out into the Ionian Sea.

The origin of the Valle del Bove is still debated.

At present, it could be five or six coalescent landslides caused by large-scale slope failure on the unsupported coastal flank of Etna, or a succession of debris avalanches generated by hydrovolcanic explosions. The landslides may have taken place in scallops – from the summit downwards – in rapid succession over a period of 2000 years, beginning between 5000 and 6000 years ago under wetter warmer conditions (Chester et al. 1985, 1987). On the other hand, the Valle del Bove may have been formed by blasts and debris avalanches, which developed from the base upwards. They were apparently generated by hydrovolcanic eruptions that occurred at intervals over a much longer period, beginning about 50 000 years ago, and ending, for the moment, about 3000 or 4000 years ago (Kieffer 1985).

Mongibello

After a period of repose, eruptions resumed within the Ellittico caldera and gradually filled it to the brim, although its rim can still be detected as a slight shoulder at Punta Lucia on the upper western slopes, at Pizzi Deneri on the upper northern slopes of Etna, and near the Torre del Filosofo in the south. As the eruptions continued, they built up the large cone of flows and fragments that has been christened Mongibello. At first, these eruptions were silicic and explosive, and many nuées ardentes swept down the wooded slopes of the volcano. The carbon-14 analysis of the burnt wood indicates dates of eruptions at intervals of about a thousand years – 8460, 7100, 6100, 5000 and 4280 years ago. Although most of these eruptions came from the central cone of Mongibello, others sprang from the last phases in the growth of the Valle del Bove.

The last explosive episode of the series caused the formation of the Caldera del Piano when the upper reaches of Mongibello collapsed. It occurred just over 2000 years ago, probably in 122 BC (Kieffer 1985, Coltelli et al. 1998). Eruptions soon began to fill the Caldera del Piano, but Etna reverted to the style that had prevailed before the explosive interlude that had marked the growth of Mongibello. Basaltic eruptions returned and violent phases were relatively few, and their products were almost invariably hawaiites. Lavas filled the caldera to the brim and formed the plain (piano) that gave it its unusual name. Then they built the present central cone. However, at the same time the eastern flanks of Etna continued to slip down towards the Ionian Sea along a whole series of rift faults, whose displacements are helped by Hyblean–Maltese tectonic movements. They now form clear fault scarps, which the Sicilians call timpa, the best known of which are the Timpa d'Acireale, and the Timpa di Santa Tecla, farther north.

The Central Cone

Sir William Hamilton and Alexandre Dumas figure among those who have written graphic accounts of their ascent of the Central Cone. But although the thrill still remains, the morphological details change with almost every passing year. At the start of the twentieth century, the Central Cone rose 300 m above its base and had a deep crater some 500 m across. Persistent eruptions completely filled this hollow and formed an undulating summit plateau. Thereupon, in 1945, its northeastern part again collapsed in a great abyss, the Voragine, which widened as its walls crumbled and gas, fumes, steam and

Empedocles and the Torre del Filosofo

The southern flanks of the volcano, rising from the city of Catania, are the most accessible sector of Etna. Most tourists take the cablecar and the bus up these slopes, and reach the foot of the central cone near the abandoned refuge of the Torre del Filosofo at an altitude of 2918 m. This name comes from the ruins of a narrow tower, which, legend has it, used to be the observatory of the Sicilian Greek philosopher, Empedocles, who lived at Agrigento. He claimed to be the equal of the gods and, therefore, wished to give the impression that he had rejoined the immortals on his death. Thus, he is said to have thrown himself into the crater of Etna, so as to leave no trace of his remains on Earth. Unfortunately, one of his sandals was recovered and gave his game away. In fact, there is no other evidence that he ever came to Etna. Some assert that he died in Messina after breaking a leg in a road accident, others that he was strangled. Aristotle, perhaps jealous of such a glamorous demise, said that Empedocles had died a natural death in Greece at the ripe old age of 60. The tower itself was only 1 m across and was much too narrow to have been anything more than a temporary refuge. It might, for instance, have been built for the visit of the Emperor Hadrian, who, of course, really was a god. Today the Torre del Filosofo is disappearing beneath the many lava flows emerging from the Southeast Crater.

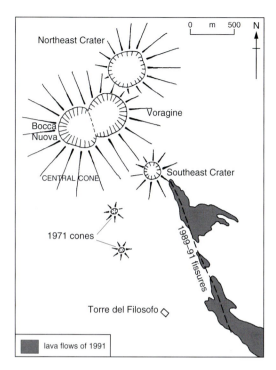

The summit of Etna.

occasional lavas welled up within it. Several smaller vents also erupted little cones of spatter and fragments on the southwestern part of the summit plateau, notably in 1955 and in 1964.

On the night of 9–10 June 1968, a new narrow abyss foundered in the western part of the summit plateau. This new vent, called Bocca Nuova, widened eventually until it joined up with its predecessor. Both are the source of occasionally upwelling lavas, almost continuous emissions of fumes and steam, and of episodic, often violent, hydrovolcanic explosions, as well as of spectacular lava fountains, as in July 1998. In September and October 1999, strong Strombolian activity almost entirely filled the Bocca Nuova with fragments, so that it once again formed a rugged platform. From it, copious lava flows fanned out for 5 km over the western flank of Etna and descended to 1700 m.

The Northeast and Southeast Craters

Some persistently spectacular eruptions have come from two subterminal vents lying at the foot of the Central Cone, which have been blessed with the banal names of Northeast and Southeast Crater. The Northeast Crater started as

a collapse pit on 27 May 1911 at 3100 m at the base of the summit cone. After emitting gas and lava flows for decades, it began to construct a typical cinder cone in 1955. Between 1966 and 1971, rather viscous hawaiite lavas also accumulated in a fan 4 km wide and 200 m thick. The cone is now horseshoe shape because upwelling lavas removed a sector in October 1974 and again in January 1978. Nevertheless, by January 1978 the cinder cone was 250 m high and had become the highest point of Etna at 3345 m, and even 3350 m in 1981. Upon this achievement, activity waned remarkably, although there was a vigorous eruption in September 1986. However, this crater was again active in 1995 and 1996, and it gave off lava fountains on 27 March 1998. During 1999, the vent stayed open and produced continuous emissions of gas and frequent explosions of cinders.

The Southeast Crater had first exploded as a pit in 1971, but it burst into magmatic activity only on 29 April 1978. Its initial eruptions were punctuated by periods of repose, but persistent activity began in 1980. It has expelled lava flows and built up a cinder cone faster than its predecessors, especially since 1998, when it rose almost as high as its neighbour, the Central Cone.

Hydrovolcanic eruptions

The summit craters are prone to hydrovolcanic eruptions, which are characterized by sudden brief explosions that usually last from a few minutes to a few hours. They form plumes of ash and steam rising more than 10 km into the air. They often disperse ash for 40 km around, damage crops, disturb travel and sometimes cause the closure of Catania airport. This type of eruption is particularly dangerous when it occurs at the Voragine. For example, the eruption on 17 July 1960 threw glowing lava clots half way down the northeastern flanks of the volcano. The eruption on 4 September 1999 not only showered bombs of up to 5 m in diameter all over the Central Cone, but ejected cinders of up to 10 cm across as far as the Ionian Sea beyond Fornazzo and Giarre. Moreover, these explosions can also occur from the Bocca Nuova, the Northeast Crater (on 24 September 1986) and the Southeast Crater, from which several metres of bombs and cinders were expelled on 4 January 1990.

These violent outbursts are caused when magma intrudes laterally into a wet zone of the

A hydrovolcanic explosion from the Bocca Nuova on 5 September 1999, with Torre del Filosofo in the foreground.

Inside the Bocca Nuova in June 1997.

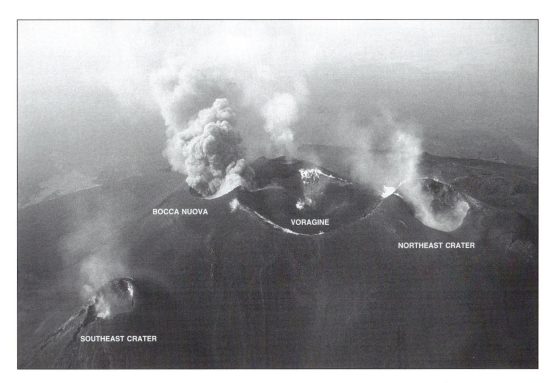

The summit craters of Etna from the air (October 1999). Two vents are erupting within the Bocca Nuova.

volcano and suddenly forms masses of steam. They also occur when damp materials fall from the walls of the vent and form a thick blockage deep in the conduit, where temperatures may approach 1000°C. Pressure builds up for several days until a vigorous explosion shatters and expels the blockage and often also widens the vent. These eruptions only expel older re-heated volcanic materials. In other instances, the damp blocking material may enter into contact with molten lavas already standing relatively high in the vent. The explosion ensues more rapidly and more violently. These hydrovolcanic eruptions are more frequent on Etna than on similar volcanoes, probably because of the permeability of the upper walls of the vents and the presence of a basement layer of impermeable clays (Kieffer 1982, 1985).

Hydrovolcanic explosions can also form part of lateral eruptions. Several steep-sided craters like maars, forming, for example, the Cisternazza of 1792 and the Padella of 1819, are punched through older volcanic materials. They occur especially often at high altitudes in much-fissured areas such as the rift zone, where water and snowmelt can readily infiltrate and be heated. Such eruptions also formed the first

phases of the growth of the Montagnola cinder cone in June 1763 and contributed largely to the unusual explosivity of the eruption in 1879.

Explosions can also occur when a tongue of lava flows over snow or a damp land surface. This is how the eruption of 1843 produced the largest number of deaths recorded on Etna. Near the summit of Etna, hydrovolcanic eruptions constitute a major element of danger, not only because they multiply the violence of its usually relatively mild activity, but also because they develop very rapidly. In addition to the nine people killed and the twenty-three injured on 12 September 1979, two people were killed at the Northeast Crater on 2 August 1929, and another two were killed at the Southeast Crater on 17 April 1987. These explosions are difficult to predict because the preliminary conditions for their development – the lateral injection of magma, water infiltration, or vent blockage – cannot often be witnessed.

Lateral eruptions

The lateral eruptions of Etna are almost invariably those that cause damage to crops and

The explosion at a lava front near Bronte in 1843

"A little after noon on 25 November . . . many people were watching the progress of the lava and were desperately trying to save everything that could possibly be saved from their cultivated land. The lava was advancing slowly and was, indeed, giving the unfortunate folks time to finish their task, when, all of a sudden, a violent explosion burst from the front of the flow. Its immeasurable power shattered the glowing lava into fragments of stones and fine ash, and lifted the very soil from the ground. A vast and dense cloud of fumes, laden with fine glowing ash, spread all around. All these materials were scattered with such force that, not only were the trees and the people nearby struck down and ravaged, but also those that were standing over 30 canne [60 m] away. In the commune of Bronte alone, 59 men were killed and 10 injured. Although many others had come to see the lava from neighbouring communes, none of them was reported missing . . . [The official gazette, *La Cerere,* published in Palermo on 6 December 1843, stated that 36 people had perished on the spot, 23 victims had died of their injuries in Bronte during the following afternoon and night, and 10 of the injured had recovered] . . .

Two men from Bronte, Pietro Foti and Vincenzo Tirenti, who were present when the explosion happened, averred that they saw the lava gradually doming up just before the explosion, and this had made them decide to run away. An uncle of Pietro Foti, a farmer in the commune, who was standing about 20 canne [40 m] from the side of the flow, declared that his arm had been drenched in hot water when he had raised it to protect himself from the explosion . . ." (Gemmellaro 1843).

Gemmellaro then demonstrates that the explosion must have been related to the generation of steam when the lava met some water, and then continues: "It only remains to prove that water was present at that spot, and all the official reports agreed that a stream used to flow down there from the Baril spring. But, opinions differed about how this water then spread out. Some said that it ran into a small pond. Others said that there had been a large ancient irrigation conduit, which the rains of the previous night and morning had filled with water. Yet others declared that there had been a big pothole, full of water, in that place. At all events, it is certain that water was present on the site". Gemmellaro then explains, with striking perspicacity, that, if such features had never been noted before, it was because such explosions require just the right amount of water to be trapped under the flow to generate enough steam to shatter the molten lavas.

property, and have sometimes destroyed villages. They are heralded by localized earthquakes as magma rises up vents branching at depth from the main conduits that generate fissures radiating from the summit zones of Etna (Patanè et al. 1984, 1996). Most of the fissures concentrate together on the rift zone, forming a band 25 km long and 2–3 km wide. Most of the lateral vents occur between 1500 m and 3000 m, but some of the most voluminous and dangerous eruptions, such as that in 1669, have occurred below 1000 m. Curiously, lateral eruptions often start in November, March or May, possibly in relation to high rainfall or snowmelt (Tanguy 1980). Since 1600, there has been an average of a dozen lateral eruptions per century, but their frequency has increased fivefold since 1971 (Tanguy & Kieffer 1993). They emit fluid hawaiite lavas that form long flows, but also construct many cinder cones and some spatter cones, because they have a relatively high gas content considering their composition. Almost two thirds of the lateral eruptions have built cinder cones, varying in height from 10 m to over 200 m, which tend to form on the upper reaches of the fissures, whereas spatter cones, **hornitos** or quiet lava effusions characterize their lower reaches.

Half the lateral eruptions continue for less than 25 days, but they can last from as little as a few hours, as on 29 April 1908 in the Valle del Bove, to a few days, as when the Monte Leone was formed from 22 to 24 March 1883, to a few weeks, such as between 2 and 20 October 1928, when Mascali was destroyed, or to several months, as from 14 December 1991 to 31 March 1993 in the Valle del Bove. The eruption lasting from 1614 to 1624 was almost as exceptional in duration as the eruption of 1669 was in intensity.

The effusion rate of lava has varied with time (Chester et al. 1985, Murray 1990, Tanguy et al. 1996). Output was apparently low from 1500 to 1610 but unusually high from 1610 until 1669. However, the great eruption of 1669 seems to have impoverished the system for a while, because output remained low for almost a century thereafter until 1763. Output was moderate from 1763 to 1971, but since then eruptive activity has never been so brisk. The lateral eruptions with slow effusions are also the most prolonged. They may start with an explosive phase that forms a cone, but soon give way to fluid lava effusions that build up a thick cover of thin flows close to the vent, chiefly because they can often last for more than a year, as in 1991–3, or even more, as in 1614–1624, when 1 km^3 erupted, or in 1651–4, when 0.5 km^3 erupted.

The total length attained by lava flows on Etna depends critically on the effusion rate (Walker

1973, Kilburn 1993, Pinkerton & Wilson 1994) Thus, the flows in 1669 had effusion rates of 50–100 m³ per second and travelled more than 17 km in less than a month, whereas the flows in 1991–3 had average effusion rates of 6 m³ per second and did not exceed 8 km in length, although the eruption lasted a long time, as did those in 1614–24. This has direct implications for volcanic risk, because careful calculation of the effusion rate could indicate the probable maximum length of the flow and, therefore, the possible threat to the areas down slope (Tanguy et al. 1996). Although such measurements cannot always be taken, simply watching the advancing lava fronts during the early stages of an eruption would offer a useful guide to what might occur.

Most flank lava flows develop **aa** surfaces, but pahoehoe surfaces have been formed on flows produced by more prolonged eruptions with lower rates of effusion (Kilburn 1981, Chester et al. 1985, Kilburn & Lopes 1988, Calvari & Pinkerton 1999), The lavas of 1614–24 are composed of piles of thin pahoehoe flows featuring tumuli, pressure ridges and lava tubes, as well as the well known large hornitos forming the Due Pizzi. Many flows have pahoehoe surfaces where they first emerge, but develop aa surfaces further down slope. Effusion rates are high enough to maintain the forward impetus of the snout while the sides congeal rapidly and restrain lateral flow. Thus, most of the lava flows retain their tongue-like outlines, even where they could easily spread sideways over gentle slopes.

Lateral eruptions are most common and most voluminous in the southeastern and southern sectors, and least common in the northern, and especially the western, sectors. Half the lateral eruptions of Etna occur on the rift zone that curves through the summit area, and most of the remainder occur in the Valle del Bove. The area most vulnerable to eruptions in the south and southeast is where the population is most concentrated on the richest agricultural land. Unfortunately, the eruptions also tend to occur lower down in this sector and to produce the greatest volumes of the most fluid lava discharged at the fastest rates. They reached their climax during historical times with the destruction of part of the city of Catania in 1669.

Archaeomagnetic dating of historical lavas

It is practically impossible to identify the products of old historical eruptions, because the documents do not locate them exactly and the dates given on geological maps are often based on doubtful judgements. Archaeomagnetic analysis overcomes this problem. When lavas solidify, they acquire the magnetization related to the Earth's magnetic field at that time. Although the Earth's magnetic field can subsequently vary by several degrees, the thermoremanent magnetization that the lavas acquired does not change and can be revealed by laboratory analysis. It can then be compared with, and plotted onto, the curve of variations established for the geomagnetic field over the past centuries. The thermoremanent magnetization thus gives the real age of the lavas. Thus, the Scorcia Vacca flow on the northeastern slopes of Etna was thought to date from 1651, whereas in reality it dates from about AD 1020. The flow that was believed to have erupted in 1595 on the southwestern flanks of Etna is really two distinct flows: one erupted in about AD 1060 and the other in about AD 1200. And the lava that was believed to have erupted in 1566 above Linguaglossa dates, in fact, from about AD 1180. The precision of the dates depends on the accuracy of the measurements, the number of the specimens taken and the speed of the secular variation of the Earth's magnetic field. The illustration shows some of the revised dates achieved by this method (Tanguy et al. 1985, 1999).

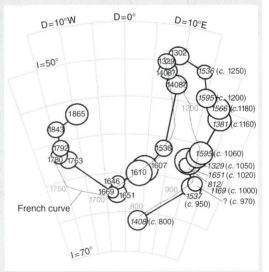

Archaeomagnetic dating of historical lavas on Etna.

Flows of erroneous age are set in italics, with the correct archaeomagnetic age in parentheses. The French reference curve was derived from c. 200 accurately dated archaeological sites.

Historical eruptions on Etna

Reports of eruptions on Etna go back more than 2500 years (Tanguy 1981, Romano & Sturiale 1982, Stothers & Rampino 1983, Chester et al. 1985, Tanguy & Patanè 1996). Eruptions on Etna were perhaps first described obliquely in the Polyphemus story of Homer (*Odyssey,* X), which was written about 750 BC (Scarth 1989). However, contrary to an oft-held opinion, Hesiod makes no reference to Sicily in his Theogony. Pindar's *Pythian odes* (I: 19–28) and then Aeschylus' *Prometheus bound* (365–74) are the oldest works in which Etna is specifically mentioned. They describe the eruption that took place at the same time as the battle of Plataea in August 479 BC, as was recorded on the Greek text know as the Arundel Tables. Another eruption, in the spring of 425 BC, was recorded by Thucydides (*Peloponnesian War*, III: 116). Diodorus Siculus (*The library of history,* XIV.59.3) described how, in 396 BC, a lava flow reached the sea and prevented the Carthaginian army from marching along the Ionian coast. The great Plinian eruption of 122 BC, which rained ash on Catania and probably formed the Caldera del Piano, was described by Lucretius *(On the nature of things,* VI: 639–46), St Augustine (*The city of God*, III.31) and Orosius (*Against the pagans,* V: 13). The large eruption of 44 BC features prominently in Virgil's *Aeneid* (III: 571–82), and its cloud of ash probably darkened the Roman sky when Julius Caesar was assassinated (Forsyth 1988, Scarth 1999, 2000). Ash erupted in AD 38 frightened Caligula, even though he was viewing Etna from the safety of Messina (Suetonius, *Caligula* 5.1). A lava flow that erupted in AD 252 was apparently arrested at the gates of Catania, when the veil of the newly martyred St Agatha was waved at it. In 1169, a widespread earthquake caused 15 000 deaths in Catania and the collapse of the eastern flank of the summit cone, but it is doubtful if an eruption ensued. In 1329, the Monte Rosso was built at an altitude of only 550 m on the southeastern slopes of Etna, and its flows threatened Acireale on the Ionian coast. The eruption in 1408 destroyed part of the village of Pedara and another perhaps happened in 1444, but no major activity then seems to have occurred for about a century.

A large fissure eruption on the southern flanks in 1536 was followed in 1537 by an effusion of lava that initiated the collapse of the summit cone. From 1607 to 1610, repeated eruptions

A contemporary picture in Catania Cathedral of the eruption of 1669 on Etna.

The eruption of 1669

The greatest eruption on Etna in modern times began on 11 March 1669 and ended 122 days later on 11 July. It was heralded from 25 February 1669 by earthquakes that severely damaged Nicolosi, a village 700 m high on the southern slopes. A gaping, glowing 2 m-wide fissure opened on 11 March. It stretched 12 km from an unusually low altitude of 850 m, just north of Nicolosi, to 2800 m at the foot of the summit cone. During the night, the vent that was to form the cone of the Monti Rossi burst into activity as the fissure spewed out lava that overwhelmed Malpasso. The hot fluid hawaiites surged forth at an exceptionally fast rate that approached 100 m^3 per second, almost ten times the average discharge of lavas on Etna. Within a few days, village after village was threatened and swamped. On 16 March, the lavas reached San Giovanni di Galermo, 6 km from the Monti Rossi. On 25 March, there was a violent hydrovolcanic explosion at the summit crater, and the summit cone collapsed in a shudder of tremors. Meanwhile, cinders and lapilli were quickly building the twin cones of the Monti Rossi and, at the same time, the lava flow divided into three arms, with its main trunk directed towards Catania. Nine villages had been destroyed by the time the lava flow reached Catania, 17 km from the vent, on 12 April 1669. During the next three days the lavas wrapped around the city walls, rose above one stretch and pushed it down. The flow reached the sea south of Catania on 23 April and by 30 April it had started to invade the western part of the city. On 6 May, with courage and scientific acumen, Diego Pappalardo and at least 50 Catanians seem to have succeeded in diverting the flow by opening its solidified sidewalls so that the molten lava in the core would turn sideways and stop its forward surge. Unfortunately, this newly created branch advanced towards Paternò, whose alarmed inhabitants drove off the Catanians before they could cause them any more trouble. This first known attempt at volcanic hazard control was thus thwarted, because the new breach solidified and the lava flow resumed its original course. No lives were lost in this eruption, because, after all, the lavas had travelled an average of only 500 m a day; but they had advanced inexorably, sustained by the rapid supply of fluid magma. It is very doubtful whether Pappalardo and his men could have saved west Catania, unless they had managed to divert the whole flow onto Paternò. When the eruption ceased on 11 July 1669, lavas between 12 m and 15 m thick had covered an area of nearly 40 km^2, and the twin Monti Rossi had reached a height of 250 m – among the highest cinder cones on the flanks of Etna. The lava and fragments had a combined volume of about 600–800 million m^3 (Tanguy & Patanè 1996 and Scarth 1999).

occurred on both the northwestern and the southwestern flanks. The ten-year period of oft-repeated eruptions between 1614 and 1624 marked the longest episode of lava-flow emissions during historical times. The lavas covered 21 km^2, and formed many lava-tube systems, the notable hornitos called the Due Pizzi, as well as mega-tumuli such as the Monti Collabasso, in a lava field composed largely of pahoehoe surfaces (Guest et al. 1984). Prolonged flank eruptions also occurred intermittently between 1634 and 1638, and again between 1651 and 1653. This seventeenth-century activity was merely preparing the terrain for the eruption of 1669, which was the largest and most famous eruption recorded during historical times on Etna.

After the eruption of 1669, the flanks of Etna saw very little activity for almost a century, although the weak emissions of gas and occasional lavas occurred intermittently from the summit zone. The chief event on Etna during this period was the great tectonic earthquakes of 9 and 11 January 1693, which severely damaged eastern Sicily and caused over 60 000 deaths, including 16 000 in Catania. The present summit cone was perhaps initiated in 1723, whereas a flank eruption in 1763 formed the Montagnola,

that is still a prominent landmark on the upper southern flanks of Etna. It was soon followed in 1764–5 by the prolonged emission of the vast pahoehoe lavas coating its upper northwestern flanks. In July 1787, the crater on the Central Cone gave a spectacular display of lava fountaining rising up to 3 km high, and an explosion at the beginning of 1792 created the Cisternazza pit near the western brink of the Valle del Bove.

Intermittent lateral eruptions and frequent persistent activity at the summit marked the first half of the nineteenth century. Two episodes in 1832 and 1843 affected the western flanks and threatened Bronte. During the eruption of 17 November 1843, the snout of the lava flow invaded a wet area near Bronte and the resulting explosion killed 59 unwary observers.

The second half of the nineteenth century saw more vigorous, rather more frequent, lateral eruptions. After those of 1852 in the Valle del Bove and 1865 on the northeastern slope, an unusually vigorous eruption to the north on 26 May 1879 created Monte Umberto–Margherita and a lava flow 9 km long. On the southern slopes, Monte Leone formed at 1100 m during a three-day eruption in March 1883, Monte Gemmellaro grew up in about a week in late May

The Valle del Bove invaded by lava flows.

1886, whereas the Monti Silvestri accumulated over five months in 1892, and both these eruptions were accompanied by extensive lava effusions. However, Etna was calm at the time of the Messina earthquake (28 December 1908), although a small eruption had occurred on 29 April in the same year.

In March 1910, very fluid lavas burst out at about 2000 m on the southern flanks of Etna and quickly reached the area around Nicolosi and Belpasso. On 27 May 1911, a pit collapse formed the Northeast Crater, which was to become the major active focus on Etna for the next 60 years, although it also experienced hydrovolcanic explosions. Meanwhile, lateral fissure eruptions took place on the northeast slopes in September 1911 and in June 1923. On 2 November 1928, a fissure opened on the northeastern flank and descended to the unusually low altitude of 1200 m within three days. Torrents of lava surged forth, reached and overwhelmed Mascali in less than 36 hours, and stopped near the coast on 20 November. During the next 15 years, effusions gradually and intermittently built up the Central Crater, but the Voragine reappeared in its northeastern parts in October 1945. In 1955 the Northeast Crater began to build up what was to become a considerable cinder cone surrounded by a thick lava apron during the next 20 years. A week of eruptions alongside the Voragine in May 1964 formed a small cone and gave off lavas that escaped by a notch from the rim of the summit cone and flowed down its flanks. On 10 June 1968, the Bocca Nuova appeared on the western side of the old crater of the Central Cone as a hole, about 4 m wide, that exhaled hot gases. Its walls foundered again in the winter of 1970 to form a chasm 100 m across, which widened intermittently until it joined up with the Voragine.

In 1971, Etna began its most active period of modern times. The lateral eruptions beginning on 5 April 1971 occurred from unusually high fissures at the southern base of the summit cone and their lavas buried the Volcano Observatory and the upper parts of the cablecar line. Lavas from other fissures on the outer crest of the Valle del Bove also travelled 7 km down to the outskirts of Fornazzo. From 30 January to 29 March 1974, what was, for modern Etna, an unusually explosive eruption formed the two Monti De Fiore cinder cones on its western flanks. It was followed by vigorous activity from the Northeast Crater, which lasted until just after its cone had become the highest point of Etna in January

1978. On 29 April 1978 the eruptions of the Southeast Crater began, which had built a cinder cone by late summer and, on 3 August 1979, another fissure sent more flows towards Fornazzo. Part of the walls of the Bocca Nuova collapsed on 2 September 1979 and the blockage was removed by the lethal hydrovolcanic explosion on 12 September.

The lateral fissure eruption on 17 March 1981 occurred in an area of the northern slopes that had not been active for at least 400 years. Effusion was very rapid and fluid lava flows soon covered 6 km^2, entered the River Alcantara and threatened, but spared, Randazzo, before they halted on 23 March. On 27 March 1983, another lateral fissure opened on the southern flanks of Etna. The thick lavas coincidentally also covered 6 km^2, but with a much slower rate of effusion, for the eruption lasted until 6 August 1983. For the first time since 1669, serious attempts were made to divert these flows. Another prolonged low-discharge eruption took place from the Southeast Crater from 28 April to 16 October 1984. It was followed by eruptions from both the Southeast Crater and the adjacent southern flanks, which lasted from 8 March 1985 until 13 July 1985, with a smaller eastern flank eruption at Christmas 1985. Yet another eruption started in the same area on 30 October 1986.

Eruptions resumed on 29 August 1989 when ash was ejected as far as 19 km from the summit vent. Two weeks of spectacular eruptions from the Southeast Crater began on 11 September 1989, which included lava fountains rising 600 m. Two fissures opened in the east: one on 25 September produced a 6 km-long lava flow towards Milo, the other formed an open rift, reaching 1 m wide, from which no lava emerged. More cinders and lava fountains were expelled from the Southeast Crater in January and February 1990, before a brief respite ensued until the end of 1991.

After the eruption of 1991–3, Etna was relatively quiet for two years. There were several phases of lava fountaining from the Northeast Crater in November and December 1995. Then, a phase of explosions and effusions between 21 July and 19 August 1996 sent lava flows cascading into the nearby Voragine in the Central Crater. All four summit craters were active throughout 1997, during which the Southeast Crater filled with lava flows. Activity increased during the winter of 1997–8, when the Southeast crater exploded vigorously and the Bocca Nuova

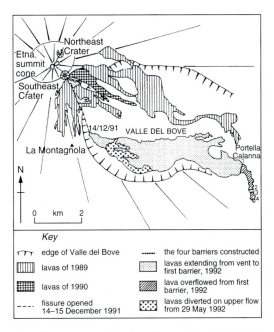

Key

⊤⊤⊤	edge of Valle del Bove	······	the four barriers constructed
▥	lavas of 1989	▦	lavas extending from vent to first barrier, 1992
▦	lavas of 1990	▨	lava overflowed from first barrier, 1992
- - -	fissure opened 14–15 December 1991	▦	lavas diverted on upper flow from 29 May 1992

The efforts to control the lava flow erupted in the Valle del Bove, 1991–3.

glowed continuously, emitting a plume of sulphur dioxide. On 27 March 1998, seismic tremors accompanied explosions from the Northeast Crater for two hours, but the Southeast Crater soon resumed its more moderate behaviour, which lasted for several months. In June 1998, the Voragine started an episode of Strombolian activity, and it was soon joined by the adjacent Bocca Nuova. These vigorous eruptions came from cones that had formed on the floors of the chasms, and they often expelled bombs and large clots of lava, notably between 12 and 13 July. On the evening of 13 July, the easternmost cone in the Voragine emitted three lava flows that carpeted the floor of the chasm. On 22 July, lava fountaining and vigorous explosions from the Voragine sent a column of ash and gas 10 km into the air, scattered ash all over the eastern flanks of Etna, and closed Catania airport. The Voragine then ejected two lava flows, one reaching the road between the north and south flanks of Etna, and the other cascaded into the Bocca Nuova. On 15 September, the Southeast Crater exploded much ash, followed on 16–19 September by Strombolian eruptions, ending with lava fountains and small flows. They were repeated at intervals of five or six days, separated by periods of almost total calm, until the end of January 1999. A vigorous eruption from the Southeast

The eruptions of 1983 and 1991–3 and the diversion of lava flows

The eruption on 28 March 1983 occurred, in the midst of tourist installations, on a fissure between 2200 m and 2400 m on the southern flanks of the volcano. The effusion rate was only about 10 m³ per second and, in spite of the claims of certain journalists, no settlements were in danger. Indeed, the flow took over a month to extend beyond Monte San Leo, which the flow that erupted in 1910 had reached in less than a day. But, this was the first eruption in this sector for 73 years and there was strong political pressure to halt this terrifying scourge.

The diversion began at the beginning of May. The aim was to use bulldozers to thin the solidified walls of the flow and then explode a hole in them in order to divert it at source and thereby expose more molten lava to the open air and make it solidify more rapidly. Enormous earth barriers were also built to stop the lava from invading land alongside it. When the explosion occurred on 14 May, molten lava flowed from the new hole, but stopped after two days when its surface solidified. A mere 50 000 m³ had been diverted out of the total of 79 million m³ that the eruption eventually gave out. On the other hand, the barriers did play a useful role in preventing the new flows from spreading sideways. The lava continued to flow until 6 August, without further human interference, but it never extended beyond the position that the front had already reached the previous April (Chester et al. 1985, Tanguy & Patanè 1996).

On 14 December 1991, lava fountains burst forth from a fissure at the foot of the Southeast Crater. During the following night, activity extended down slope and several vents opened at about 2200 m in the Valle del Bove. Lava soon spread down and invaded the rich orchards of the Valle di Calanna. In January 1992, the Italian army built a barrier of ash and cinders, 21 m high, across the exit of the Valle di Calanna to try and stop the lavas from reaching the lower valley leading directly to the town of Zafferana. Just before they reached the barrier, the lavas halted for three months. Then, suddenly at the end of March they resumed their progress and swept the barrier away on 8 April 1992. The citizens of Zafferana now feared the worst and, by the end of May, the snout was within 1 km of town. But it was a false alarm: the snout of the degassed flow was now 8 km from its source and it proved too viscous to reach the town, although molten lavas broke out near the destroyed barrier several times during the following month. Meanwhile, during April and May, concrete blocks were dropped onto the flow near its source, and explosives were used five times to stem the supply to the advancing snout (Barberi et al. 1993b). This tactic seemed to work eventually, although the most successful attempt took place on 29 May, which coincided with a marked reduction in the rate of lava discharge on 1 June. Thereafter, the flows were confined to the Valle del Bove and the eruption weakened considerably, although emissions lingered on until March 1993. (Calvari et al. 1994, Tanguy et al. 1996)

The barrier built to contain the spread of lavas in the Valle del Bove, 1992.

Before 14 000 years ago

14 000 years ago

Ellittico Caldera

10 000 to 4000–5000 years ago

Valle del Leone

Caldera del Piano, 122 BC

Mongibello in the nineteenth century

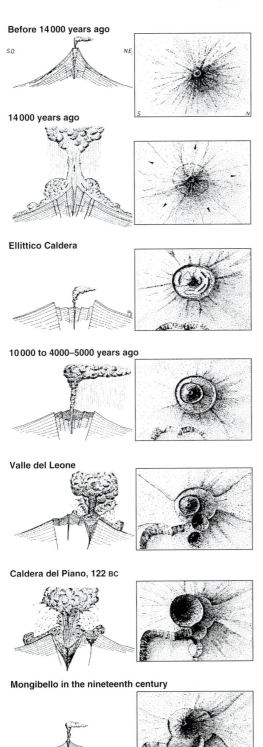

The later stages of the growth of Etna.

Crater on 4 February was followed by the opening of a fissure that almost reached the base of the cone and by the emission of quantities of lava until November 1999. A strong lava fountain shot from the Voragine on 4 September and large eruptions from the Bocca Nuova in September and October 1999 entirely changed its morphology. Then from 26 January 2000, the Southeast Crater started some of the most violent and frequent eruptions in its history, which continued until the end of June 2000. There were two further eruptions on 28 and 29 August.

To some extent, these summaries of eruptions give an unbalanced picture of Etna's behaviour, by stressing the unusual at the expense of the more mundane elements. For example, throughout the period from 1971, persistent activity – ranging from gas jets, ash and cinder explosions, to lava that formed lakes, flows and fountains – were often registered from the terminal and subterminal vents, and rapid changes have taken place on the volcano. But lateral eruptions are emphasized because of their potential danger to the people and their property. For instance, there were 14 flank eruptions between 1971 and 1996, or five times the average for the previous three centuries; and the average output of lava was almost three times that of the previous century. The apparently unprecedented scale, variety and frequency of activity have been matched by discharge rates higher than average and, since 1981, by longer eruptions. This vigour has in itself made Etna the best European volcanological laboratory during the past few decades. However, in spite of the scientific interest in all the present spectacular activity on Etna, the real danger to the area around Catania lies in the threat of an earthquake that could kill tens of thousands of people. Hence, Italian Earth scientists have concentrated their research on developing both seismological and volcanological observatories throughout the region.

SICILY: PANTELLERIA

The Italian island of Pantelleria rises in the straits between Sicily and Tunisia. In spite of its almost African climate, it is green and attractive, for its weathered volcanic soils flourish with capers and vineyards. These and the fishing boats provide the main livelihood of its few thousand inhabitants. Pantelleria is about 13 km long and 8 km broad, and it reaches a height of 836 m at Montagna Grande. It is wholly volcanic in origin and began to form over 300 000 years ago in a submerged rift, running from northwest to southeast. About half the total accumulation now rises above sea level. Many different vents have taken part in the growth of Pantelleria, and its complicated history is reflected in its mountainous and varied relief. The emerged parts of the island are largely composed of unusual hyper-alkaline rhyolites known as **pantellerites**. They form layers of ash, pumice, welded and unwelded ignimbrites, cinder cones, domes, flows and even little shields of lava. In fact, although these pantellerites had a silica content of about 70 per cent, they were emitted at quite high temperatures of between 900°C and 1020°C. Thus, they were less viscous than rhyolites and they behaved more like andesites (Mahood & Hildreth 1986). But, the island has also erupted trachytes, for instance, at Monte Gibele. It seems that a longstanding shallow reservoir containing these differentiated magmas prevented the ascent of more basic materials towards the central areas of the island, and that they have only relatively recently found a passage on its periphery. Thus, the most recent eruptions have discharged alkaline basalts and hawaiites that now occupy the far northwestern part of the island, around the capital.

It seems that, when Pantelleria first emerged from the sea, it formed one or more shields, which outcrop in part on both the southeast and southwest coasts, where a lava flow at Scauri has been dated to about 324 000 years ago. Similar shields apparently formed the Cala dell'Altura and the Cuddia Khamma. The shields also contained thick layers of welded tuffs (ignimbrites). About 114 000 years ago, a powerful eruption led to the collapse of the broad La Vecchia caldera, which covers an area of some 40 km². Further eruptions, notably from the Cuddia Attalora and the Cuddioli di Dietro Isola, filled the hollow with pumice, voluminous lava flows and at least three layers of welded tuffs.

The second major eruption on Pantelleria

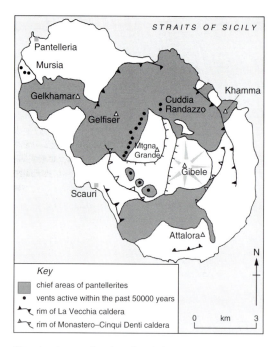

The structure and main volcanic features of Pantelleria.

took place over 50 000 years ago, when the expulsion of the green tuffs covered the whole island in a layer that reaches up to 35 m thick and effectively hides much of the earlier history of the island from geological scrutiny. The eruption, from a stratified reservoir, first expelled pantelleritic pumice, followed by thick ignimbrites, and lastly trachytic nuées ardentes. The collapse of the Cinque Denti, or Monastero caldera, brought the episode to an end. It lies within the La Vecchia Caldera, but nevertheless has a diameter of 6 km.

The eruption of the green tuffs started the first of six cycles of activity that have produced 80 eruptions during the past 50 000 years (Mahood & Hildreth 1986, Civetta et al. 1984, Civetta et al. 1988, Orsi et al. 1991). Each began with rather moderate explosions of rhyolitic fragments and ended with effusions of trachytic lavas from a shallow reservoir of stratified magma. The dispersal of the fragments was often restricted to the formation of cones of grey or yellowish pumice, ranging from 25 m to 100 m in height, the craters of which were then often filled by lava emissions. The second cycle has been dated to between about 32 000–35 000 years ago and it produced mainly trachytes, with some pantellerites, and built the cone of Monte Gibele. The third cycle occurred about 22 000 years ago. Its

Monte Gibele and southeastern Pantelleria, from the Montagna Grande.

eruptions of fragments domes and flows took place chiefly outside the caldera, and they ejected, for example, the pantellerites forming Monte Gelkhamar. The fourth cycle occurred about 20 000–15 000 years ago with eruptions around the edge of the caldera. Monte Gelfiser shows their typical development from initial explosions to concluding lava effusions. The fifth cycle, between 14 000 and 10 000 years ago, saw eruptions that were limited to the north of the island and came from fissures trending either north–south or northwest–southeast.

The sixth and latest cycle of activity began about 8000 years ago and ended about 3000 years ago. Its products outcrop in the centre of the island, especially around the Montagna Grande, and include nuées ardentes deposits, cones, domes and lava flows. These eruptions built the Cuddia Randazzo pumice cone, whereas the pumices of the Sibà Montagna and the Roncone erupted at different times from vents aligned on the same fissure. The basaltic Mursia cones and flows on the northwest coast also belong to this sixth cycle. Although submarine activity occurred in 1891 about 4 km to the northwest of the island, no historical eruptions have apparently been recorded on Pantelleria, but active fumaroles and hot springs show that the island is not extinct. Indeed, there seems to be a linear relationship between the silicic cycles, the degree of magma evolution displayed, and the volume of materials expelled. They seem to have been unleashed either by an increase in the magma pressure related to fractional crystallization or by influxes of less evolved magmas from below. As there appears to be no reason to suspect that this system has changed, the present volcanic calm of Pantelleria is probably only a temporary state.

SICILY: GRAHAM BANK AND FOERSTNER BANK

In the Mediterranean channel of the Straits of Sicily between Sicily and Tunisia, submarine banks have displayed volcanic activity during historical times, mainly in relation to the widening of the rift that gave rise to the Italian island of Pantelleria. Eruptions, for example, were recorded with varying degrees of reliability in 1632, 1707, 1831, 1845, 1846, 1863 and 1891, and possibly also in 1801 and 1832. The eruption of Graham Bank in 1831 was the largest and best documented of these events (Gemmellaro 1831, Prévost 1835, Washington 1909). It began midway between Sciacca, in Sicily, and Pantelleria, at a depth of about 250 m and it now forms a circular hump, rising 100 m from the floor. Its summit platform carries three tuff cones, one of which rises to within 8 m of sea level.

Graham Bank was first noticed in 1632, and it is possible that an eruption in 1707 was bulky enough to form a temporary island. But its largest recorded eruption announced its debut when tremors were felt in Sicily and a ship reported "unusual agitation" when it sailed over the spot on 28 June 1831. In early July, there was a fetid odour at Sciacca, and fishermen declared that the sea was "boiling and full of dead fish" about 50 km offshore. During the following days, columns of water spurted 25 m skywards at 15–30 minute intervals. On 12 July, pale trachytic pumice washed up on the shore at Sciacca, and an island 3 m high emerged on 16 July. By early August, vigorous explosions had built up a cone of hawaiitic fragments that rose 65 m above the waves and was 3700 m in circumference.

The news soon attracted the attention of the commander of the British Mediterranean Fleet, Vice Admiral Sir Henry Hotham, who sent out a boat to take charge of this strategically placed island. On 2 August 1831, Captain Humphrey Senhouse landed on the islet, raised the Union Jack, claimed it for King William IV, and named it Graham Island, in honour of the First Lord of the Admiralty. This enterprising seizure excited international alarm. King Ferdinand of Naples sent the *Etna*, whose captain also claimed the island and naturally called it Ferdinandea. Some time later, the French geologist, Constant Prévost, also landed and called it Julia, because it had first been seen in July.

But nature avoided a crisis. The eruption did not expel enough tuffs to prevent sea water from invading the vent, and the Surtseyan explosions continued unabated until the magma stopped rising. Therefore, no lava flows could be emitted that might have protected the fragile tuffs from wave attack. Fortunately for international peace, the eruption stopped and the winter Mediterranean storms destroyed the island on 28 December. The eroded crest of the tuff cone now lies some 8 m below sea level. A brief additional eruption was also reported from the Graham Bank on 12 August 1863, but no cone was built above sea level (Simkin & Siebert 1994). However, new international trouble may be brewing. In the early weeks of 2000, the waters above the shoal began bubbling and spouting up, and there were frequent tremors in the area. It is therefore possible that a further eruption could form another island.

Other submarine eruptions have also been reported from this area of the Straits of Sicily. In 1890, the people of the port of Pantelleria noticed an increase in fumarole activity, several tremors, and an uplift of 80 cm on the northeast coast of the island. They felt stronger earthquakes from 14 October 1891, which, however, decreased markedly on the morning of 17 October. That very morning, a submarine eruption began on what was later called the Foerstner Bank, about 4 km west of the little port of Pantelleria (Riccò 1891, 1892). Amid repeated rumblings, columns of smoke and steam rose into the air for about 1 km on a line trending from northeast to southwest. Close observers saw black cinders and large glowing bombs being thrown skywards in the steam. It is possible that an island about 1 km long was built about 10 m above sea level for a day or two, but it may also have been no more than a mass of floating pumice. In any case, the eruption stopped on 25 October before a cone could be built above sea level. As at the northwestern end of Pantelleria, this eruption produced alkali basalts.

BIBLIOGRAPHY

Abbruzzese, D. 1936. Sulla catastrofica esplosione dello Stromboli dell'11 settembre 1930. *Atti Accademia Gioenia di Scienze Naturali in Catania* (6th series) **1** (memoria 4), 1–13.

Albore Livadie, C. (ed.) 1986. *Tremblements de terre, éruptions volcaniques et vie des hommes dans la Campanie antique.* Naples: Institut Français de Naples, Publications du Centre Jean Bérard, 2ème série, **VII**, 1–233.

Alessi, G. 1832–5. Storia critica delle eruzioni dell'Etna. *Atti Accademia Gioenia di Scienze Naturali in Catania* (seria I) **III**, 17–75; **IV**, 23–74; **V**, 43–72; **VI**, 85–116; **VI**, 21–65; **VIII**, 99–148; **IX**, 121–206.

Alfano, B. G. & I. Friedlander 1929. *La storia del Vesuvio illustrata dai documenti coevi.* Ulm: K. Hohn.

Allard, P. 1997. Endogenous magma degassing and storage at Mount Etna. *Geophysical Research Letters* **24**, 2219–22.

Anderson, H. J. & J. A. Jackson 1987. The deep seismicity of the Tyrrhenian Sea. *Royal Astronomical Society, Geophysical Journal* **91**, 613–37.

Andronico, D., R. Cioni, P. Marianelli, R. Santacroce, A. Sbrana, R. Sulpizio 1998. Introduction to Somma–Vesuvius. *Cities on Volcanoes, International Meeting*, G. Orsi, M. Di Vito R. Isaia (eds), 14–25. Naples: Osservatorio Vesuviano.

Archambault, C. & J-C. Tanguy 1976. Comparative temperature measurements on Mount Etna lavas: problems and techniques. *Journal of Volcanology and Geothermal Research* **1**, 113–25.

Barberi, F. & L. Leoni 1980. Metamorphic carbonate ejecta from Vesuvius Plinian eruptions: evidence of the occurrence of shallow magma chambers. *Bulletin Volcanologique* **43b**(1), 107–120.

Barberi, F., P. Gasparini, F. Innocenti, L. Villari 1973. Volcanism of southern Tyrrhenian Sea and its geodynamic implications. *Journal of Geophysical Research* **78**, 5221–32.

Barberi, F., F. Innocenti, G. Ferrara, J. Keller, L. Villari 1973. Evolution of Eolian Arc volcanism (South Tyrrhenian Sea). *Earth and Planetary Science Letters* **21**, 269–76.

Barberi, F., L. Civetta, P. Gasparini, F. Innocenti, R. Scandone, L. Villari 1974. Evolution of a section of the Africa/Europe plate boundary: paleomagnetic and volcanological evidence from Sicily. *Earth and Planetary Science Letters* **22**, 123–32.

Barberi, F., H. Bizouard, R. Clocchiatti, N. Métrich, R. Santacroce, A. Sbrana 1981. The Somma–Vesuvius magma chamber: a petrological and volcanological approach. *Bulletin Volcanologique* **44**(3), 295–315.

Barberi, F., M. Rosi, R. Santacroce, M. F. Sheridan 1983. Volcanic hazard zonation: Mt Vesuvius. In *Forecasting volcanic events*, H. Tazieff & J. C. Sabroux (eds), 149–61. Amsterdam: Elsevier.

Barberi, F., G. Corrado, F. Innocenti, G. Luongo 1984a. Brief chronicle of a volcano emergency in a densely populated area. *Bulletin Volcanologique* **47**, 175–85.

Barberi, F., D. Hill, F. Innocenti, G. Luongo, M. Treuil (eds) 1984b. *The bradyseismic crisis at Phlegraean Fields, Italy.* Bulletin Volcanologique **47** [special issue], 173–412.

Barberi, F., M. Rosi, A. Sodi 1993a. Volcanic hazard assessment at Stromboli based on review of historical data. *Acta Vulcanologica* **3**, 173–87.

Barberi F., M. L. Carapezza, M. Valenza, L. Villari 1993b. The control of lava flow during the 1991–1992 eruption of Mt Etna. *Journal of Volcanology and Geothermal Research* **56**, 1–34.

Barberi, F. & M. L. Carapezza 1996. The Campi Flegrei case history. In *Monitoring and mitigation of volcano hazards*, R. Scarpa & R. I.Tilling (eds), 771–86 Berlin: Springer.

Beccaluva, L., P. L. Rossi, G. Serri 1982. Neogene to Recent volcanism of the southern Tyrrhenian–Sicilian area: implications for the geodynamic evolution of the Calabrian Arc. *Earth Evolutionary Science* **3**, 222–38.

Beccaluva, L., G. Gabbianelli, F. Lucchini, P. L. Rossi, C. Savelli 1985. Petrology and K/Ar ages of volcanics dredged from the Eolian seamounts: implications for geodynamic evolution of the southern Tyrrhenian basin. *Earth and Planetary Science Letters* **74**, 187–208.

Bertagnini, A., M. Coltelli, P. Landi, M. Pompilio, M. Rosi 1999. Violent explosions yield new insights into dynamics of Stromboli volcano. *Eos* **80**, 633–6.

Bianco, F., M. Castellano, G. Milano, G. Ventura, G. Vilardo 1998. The Somma–Vesuvius stress field induced by regional tectonics: evidences from seismological and mesostructural data. *Journal of Volcanology and Geothermal Research* **82**, 199–218.

Bigazzi, G. & F. Bonadonna 1973. Fission track dating of the obsidian of Lìpari Island (Italy). *Nature* **242**, 322–3.

Borgia, A., L. Ferrari, G. Pasquarè 1992. Importance of gravitational spreading in the tectonic and volcanic evolution of Mount Etna. *Nature* **357**, 231–5.

Bowerstock, G. S. 1980. The rediscovery of Herculaneum and Pompeii. *The American Scholar* **49**, 461–70.

Buchner, G. 1986. Eruzioni vulcaniche e fenomeni vulcano-tettonici di età preistorica e storica nell'isola d'Ischia. In *Tremblements de terre, éruptions volcaniques et vie des hommes dans la Campanie antique*, C. Albore Livadie (ed.), 145–88. Naples: Institut français de Naples.

Butler, G. W. 1892. The eruptions of Vulcano, August 3 1888 to March 22 1890. *Nature* **46**, 117–9.

Caltabiano T., R. Romano, G. Budetta 1994. SO_2 flux measurements at Mt Etna, Sicily. *Journal of Geophysical Research* **99**, 12809–819.

Calvari, S., M. Coltelli, M. Neri, M. Pompilio, V. Scribano 1994. The 1991–1993 Etna eruption: chronology and flow-field evolution. *Acta Vulcanologica* **4**, 1–14.

Calvari, S. & H. Pinkerton 1999. Lava tube morphology and evidence for lava flow emplacement mechanisms. *Journal of Volcanology and Geothermal Research* **90**, 263–80.

Capaldi, G. and 11 co-authors 1978. Stromboli and its 1975 eruption. *Bulletin Volcanologique* **41**, 260–85.

Capaldi, G., L. Civetta, P. Y. Gillot 1985. Geochronology of Plio-Pleistocene volcanic rocks from southern Italy. *Rendiconti della Societá Italiana di Mineralogia e Petrologia* **40**, 25–44.

Carta, S., R. Figari, G. Sartoris, E. Sassi, R. Scandone 1981. A statistical model for Vesuvius and its volcanological implications. *Bulletin Volcanologique* **44**(2), 129–51.

Chester, D. K., A. M. Duncan, J. E. Guest, C. R. J. Kilburn 1985. *Mount Etna: the anatomy of a volcano*. London: Chapman & Hall.

Chester, D. K., A. M. Duncan, J. E. Guest 1987. The pyroclastic deposits of Mount Etna volcano, Sicily. *Geological Journal* **22**, 225–43.

Chiarabba, C., A. Amato, P. T. Delaney 1997. Crustal structure, evolution and volcanic unrest of the Alban Hills, central Italy. *Bulletin of Volcanology* **59**, 161–70.

Chiodini, G., R. Chioni, G. Magro, L. Marini, C. Panichi, B. Raco, M. Russo 1997. Chemical and isotopic variations of Bocca Grande fumarole (Solfatara volcano, Phlegraean Fields). *Acta Vulcanologica* **8**, 228–32.

Cioni, R., L. Civetta, P. Marianelli, N. Métrich, R. Santacroce, A. Sbrana 1995. Compositional layering and syn-eruptive mixing of a periodically refilled magma chamber: the AD 79 Plinian eruption of Vesuvius. *Journal of Petrology* **36**, 739–76.

Cioni, R., R. Santacroce, A. Sbrana 1999. Pyroclastic deposits as a guide for reconstructing the multi-stage evolution of the Somma–Vesuvius caldera. *Bulletin of Volcanology* **60**, 207–222.

Civetta, L. 1998. The emergency plans of the Neapolitan area. *Cities on Volcanoes, International Meeting*, G. Orsi, M. Di Vito R. Isaia (eds), 83–5. Naples: Osservatorio Vesuviano.

Civetta, L. & R. Santacroce 1992. Steady-state magma supply in the last 3400 years of Vesuvius activity. *Acta Vulcanologica* **2**, 147–60.

Civetta, L., Y. Cornette, G. Crisci, P. Y. Gillot, G. Orsi, C. S. Requejo 1984. Geology, geochronology and chemical evolution of the island of Pantelleria. *Geological Magazine* **121**, 541–668.

Civetta, L., Y. Cornette, G. Crisci, P. Y. Gillot, G. Orsi 1988. The eruptive history of Pantelleria (Sicily Channel) in the last 50 ka. *Bulletin of Volcanology* **50**, 47–57.

Civetta, L., G. Gallo, G. Orsi 1991. Sr and Nd isotope and trace-element constraints on the chemical evolution of the magmatic system of Ischia (Italy) in the last 55 ka. *Journal of Volcanology and Geothermal Research* **47**, 213–30.

Civetta, L., G. Orsi, L. Pappalardo, R. V. Fisher, G. Heiken, M. Ort 1997. Geochemical zoning, mingling, eruptive dynamics and depositional processes – the Campanian Ignimbrite, Campi Flegrei caldera, Italy. *Journal of Volcanology and Geothermal Research* **75**, 183–219.

Civetta, L. and 20 co-authors 1998. Volcanic, deformational and magmatic history of the Campi Flegrei caldera. In *Cities on Volcanoes, International Meeting*, G. Orsi, M. Di Vito R. Isaia (eds), 26–71. Naples: Osservatorio Vesuviano.

Clocchiatti, R., A. Del Moro, A. Gioncada, J. L. Joron, M. Mosbah, L. Pinarelli, A. Sbrana 1994. Assessment of a shallow magmatic system: the 1888–90 eruption, Vulcano Island, Italy. *Bulletin of Volcanology* **56**, 466–86.

Clocchiatti, R., P. Schiano, L. Ottolini, P. Bottazzi 1998. Earlier alkaline and transitional magmatic pulsation of Mt Etna volcano. *Earth and Planetary Science Letters* **163**, 399–407.

Clocchiatti, R., C. Rivière, S. La Delfa, G. Patanè, J-C. Tanguy 1999. Lava-flow temperature measurements on Etna, Italy. *Smithsonian Institution Global Volcanism Network Bulletin* **24**(6), 6–7.

Cole, P. D. & C. Scarpati 1993. A facies interpretation of the eruption and emplacement mechanisms of the upper part of the Neapolitan Yellow Tuff, Campi Flegrei, southern Italy. *Bulletin of Volcanology* **55**, 311–26.

Coltelli, M., P. Del Carlo, L. Vezzoli 1998. Discovery of a Plinian basaltic eruption of Roman age at Etna volcano, Italy. *Geology* **26**(12), 1095–1098.

Condomines, M. & C. J. Allègre 1980. Age and magmatic evolution of Stromboli volcano from $^{230}Th/^{238}U$ disequilibrium data. *Nature* **288**, 354–7.

Condomines, M., J-C. Tanguy, G. Kieffer, C. J. Allègre 1982. Magmatic evolution of a volcano studied by $^{230}Th–^{238}U$ disequilibrium and trace elements systematics: the Etna case. *Geochimica et Cosmochimica Acta* **46**, 1397–416.

Condomines, M., J-C. Tanguy, V. Michaud 1995. Magma dynamics at Mt Etna: constraints from U–Th–Ra–Pb radioactive disequilibria and Sr isotopes in historical lavas. *Earth and Planetary Science Letters* **132**, 25–41.

Cortese, M., G. Frazzetta, L. La Volpe 1986. Volcanic history of Lìpari (Aeolian Islands, Italy) during the last 10 000 years. *Journal of Volcanology and Geothermal Research* **27**, 117–33.

Cortini, M. & R. Scandone 1982. The feeding system of Vesuvius between 1754 and 1944. *Journal of Volcanology and Geothermal Research* **12**, 393–400.

Crisci, G. M., R. De Rosa, G. Lanzafame, R. Mazzuoli, M. F. Sheridan, G. G. Zuffa 1981. Monte Guardia sequence: a Late Pleistocene eruptive cycle on Lìpari (Italy). *Bulletin Volcanologique* **44**(3), 241–55.

Crisci, G. M., G. Delibras, R. De Rosa, R. Mazzuoli, M. F. Sheridan 1983. Age and petrology of the Late Pleistocene brown tuffs on Lìpari, Italy. *Bulletin Volcanologique* **46**, 381–91.

Cristofolini, R. & R. Romano 1982. Petrologic features of the Etnean volcanic rocks. *Memorie della Società Geologica Italiana* **23**, 99–116.

De Astis, G., P. Dellino, R. De Rosa, L. La Volpe 1997. Eruptive and emplacement mechanisms of wide-spread fine-grained pyroclastic deposits on Vulcano Island (Italy). *Bulletin of Volcanology* **59**, 87–102.

De Fiore, O. 1922. *Vulcano (Isole Eolie). Zeitschrift für Vulkanologie*, Monograph 3.

De Fino, M., L. La Volpe, G. Piccarreta 1991. Role of magma mixing during the recent activity of La Fossa di Vulcano (Aeolian Islands, Italy). *Journal of Volcanology and Geothermal Research* **48**, 385–98.

De Rita, D., R. Funicello, M. Parotto 1988. *Geological map of the Colli Albani volcanic complex (Vulcano Laziale)*. Rome: Consiglio Nazionale di Ricerche.

Dellino, P., G. Frazzetta, L. La Volpe 1990. Wet surge deposits at La Fossa di Vulcano: depositional and eruptive mechanisms. *Journal of Volcanology and Geothermal Research* **43**, 215–33.

De Rosa, R. & M. F. Sheridan 1983. Evidence for magma mixing in the surge deposits of the Monte Guardia sequence, Lìpari. *Journal of Volcanology and Geothermal Research* **17**, 313–28.

Dercourt, J. and 18 co-authors 1986. Geological evolution of the Tethys Belt from the Atlantic to the Pamirs since the Lias. *Tectonophysics* **123**, 241–315.

De Vita, S., G. Orsi, L. Civetta, A. Carradente, M. D'Antonio, A. Dieno, T. di Cesare, M. A. Di Vito, R. V. Fisher, R. Isaia, E. Marotta, A. Necco, M. Ort, L. Pappalardo, M. Piochi, J. Southon 1999 The Agnano–Monte Spina eruption (4100 years BP) in the restless Campi Flegrei caldera. *Journal of Volcanology and Geothermal Research* **91**, 269–301.

Di Stefano, R., C. Chiarabba, F. Lucente, A. Amato 1999. Crustal and uppermost mantle structure in Italy from the inversion of *P*-wave arrival times: geodynamic implications. *Geophysical Journal International* **139**, 483–98.

Di Vito, M., L. Lirer, G. Mastrolorenzo, G. Rolandi 1987. The 1538 Monte Nuovo eruption (Campi Flegrei, Italy). *Bulletin of Volcanology* **49**, 608–15.

Di Vito, M. A., R. Isaia, G. Orsi, J. Southon, S. de Vita, M. D'Antonio, L. Pappalardo, M. Piochi 1999. Volcanism and deformation since 12000 years at the Campi Flegrei caldera (Italy). *Journal of Volcanology and Geothermal Research* **91**, 221–46.

Dvorak, J. J. & P. Gasparini 1991. History of earthquakes and vertical ground movement in Campi Flegrei caldera, southern Italy: comparison of precursory events to the AD 1538 eruption of Monte Nuovo and of activity since 1968. *Journal of Volcanology and Geothermal Research* **48**, 77–92.

Ellam, R. M., M. A. Menzies, C. J. Hawkesworth, W. P. Leeman, M. Rosi, G. Serri 1988. The transition from calc-alkaline to potassic orogenic magmatism in the Aeolian Islands, southern Italy. *Bulletin of Volcanology* **50**, 386–98.

Falsaperla, S., G. Lanzafame, V. Longo, S. Spampinato 1999. Regional stress field in the area of Stromboli (Italy): insights into structural data and crustal tectonic earthquakes. *Journal of Volcanology and Geothermal Research* **88**, 147–66.

Fazellus, T. 1558. *De rebus Siculis decades duae*. Panormi, Sicily: Maida & Carrara.

Ferrara, F. 1818. *Descrizione dell'Etna, con la storia delle eruzioni e il catalogo dei prodotti*. Palermo: Lorenzo Dato.

Ferrari, L. & P. Manetti 1993. Geodynamic framework of Tyrrhenian volcanism: a review. *Acta Vulcanologica* **3**, 1–9.

Fisher, R. V., G. Orsi, M. Ort, G. Heiken 1993. Mobility of a large volume pyroclastic flow: emplacement of the Campanian Ignimbrite. *Journal of Volcanology and Geothermal Research* **59**, 205–220.

Forgione, G., G. Luongo, R. Romano 1989. Mt Etna (Sicily): volcano hazard assessment. In *Volcanic hazards*, J. H. Latter (ed.), 137–150. Berlin: Springer.

Forsyth, P. Y. 1988. In the wake of Etna 44 BC. *Classical Antiquity* **7**, 49–57.

Francalanci, L., P. Manetti, A. Peccerillo 1989. Volcanological and magmatological evolution of Stromboli volcano (Aeolian Islands): the roles of fractional crystallization, magma mixing, crustal contamination and source heterogeneity. *Bulletin of Volcanology* **51**, 355–78.

Frazzetta, G. & L. Villari 1981. The feeding of the eruptive activity of Etna volcano: the regional stress field as a constraint to magma uprising and eruption. *Bulletin Volcanologique* **44**(3), 269–82.

Frazzetta, G., L. La Volpe, M. F. Sheridan 1983. Evolution of the Fossa cone, Vulcano. *Journal of Volcanology and Geothermal Research* **50**, 329–60.

Frazzetta, G. & R. Romano 1984. The 1983 Etna eruption: event chronology and morphological evolution of the lava flow. *Bulletin Volcanologique* **47**(4), 1079–1096.

Frazzetta, G., P. Y. Gillot, L. La Volpe, M. F. Sheridan 1984. Volcanic hazards at Fossa of Vulcano: data from the last 6000 years. *Bulletin Volcanologique* **47**, 105–124.

Frazzetta, G., L. La Volpe, M. F. Sheridan 1989. Interpretation of emplacement units in recent surge deposits on Lìpari, Italy. *Journal of Volcanology and Geothermal Research* **37**, 339–50.

Frepoli, A., G. Selvaggi, C. Chiarabba, A. Amato 1996. State of stress in the southern Tyrrhenian subduction zone from fault-plane solutions. *Geophysical Journal International* **125**, 879–91.

Gabbianelli, G. C. Romagnoli, P. L. Rossi, N. Calanchi 1993. Marine geology of the Panarea–Stromboli area. *Acta Vulcanologica* **3**, 11–20.

Gasparini, P. & S. Musella 1991. *Un viaggio al Vesuvio*. Naples: Liguori.

Gasparini, C., G. Iannaccone, P. Scandone, R. Scarpa 1982. Seismotectonics of the Calabrian arc. *Tectonophysics* **84**, 267–86.

Gemmellaro, C. 1831. *Relazione dei fenomeni del nuovo vulcano sorto dal mare fra la costa di Sicilia e l'isola di Pantelleria nel mese di Luglio 1831*. Catania: Pastore Carmelo.

—— 1834. Relazione dei fenomeni del nuovo vulcano sorto dal mare fra la costa di Sicilia e l'isola di Pantelleria nel mese di Luglio 1831. *Atti Accademia Gioenia di Scienze Naturali in Catania*, series 2, **8**, 271–298.

—— 1843. Sulla eruzione dell'Etna del 17 novembre 1843. *Atti Accademia Gioenia di Scienze Naturali in Catania* I(20), 225–57.

Gillot, P. Y. & L. Villari 1980. *K–Ar geochronological data on the Aeolian arc volcanism: a preliminary report*. Report 3-80, Istituto Internazionale di Vulcanologia, Catania.

Gillot, P. Y. & J. Keller 1993. Radio-chronological dating of Stromboli. *Acta Vulcanologica* **3**, 69–77.

Gillot, P. Y., G. Kieffer, R. Romano 1994. The evolution of Mount Etna in the light of potassium–argon dating. *Acta Vulcanologica* **5**, 81–7.

Guest, J. E. & R. R. Skelhorn (eds) 1973. Mount Etna and the 1971 eruption. *Royal Society, Philosophical Transactions* **A274** [whole volume].

Guest, J. E. & J. B. Murray 1979. An analysis of hazard from Mount Etna volcano. *Geological Society of London, Journal* **136**, 347–54.

Guest, J. E., R. Greeley, C. Wood 1984. Lava tubes, terraces and mega-tumuli on the 1614–24 pahoehoe lava flow field, Mount Etna, Sicily. *Bulletin Volcanologique* **47**(3), 635–48.

Guest, J. E., C. R. J. Kilburn, H. Pinkerton, A. M. Duncan 1987. The evolution of lava flow fields: observations of the 1981 and 1983 eruptions of Mount Etna, Sicily. *Bulletin of Volcanology* **49**, 527–40.

Gvirtzman, Z. & A. Nur 1999. The formation of Mount Etna as the consequence of slab rollback. *Nature* **401**, 782–5.

Hamilton, Sir W. 1772. *Observations on Mount Vesuvius, Mount Etna, and other volcanoes of the two Sicilies*. London: Cadell.

—— 1795. An account of the late eruption of Mount Vesuvius in a letter to Sir Joseph Banks. *Royal Society, Philosophical Transactions* **85**, 73–116.

—— 1776/9. *Campi Phlegraei: observations on the volcanoes of the two Sicilies*. Naples: Fabris.

Hirn, A., R. Nicolich, J. Gallart, M. Laigle, L. Cernobori, Etnaseis scientific group 1997. Roots of Etna in faults of great earthquakes. *Earth and Planetary Science Letters* **148**, 171–91.

Hornig-Kjarsgaard, I., J. Keller, U. Koberski, E. Stadlbauer, L. Francalanci, R. Lenhart 1993. Geology, stratigraphy and volcanological evolution of the island of Stromboli, Aeolian arc, Italy. *Acta Vulcanologica* **3**, 21–68.

Imbò, G. 1949. L'attività eruttiva vesuviana e relative osservazioni nel corso dell'intervallo intereruttivo 1906–1944 ed in particolare del parossismo del marzo 1944. *Annali dell'Osservatorio Vesuviano*, **5**, 185–380.

—— 1965. *Catalogue of the active volcanoes of the world including solfatara fields*, part XVIII: *Italy*. Rome: International Association of Volcanology.

Jashemski, W. F. 1979. *Pompeii and Mount Vesuvius*. In *Volcanic activity and human ecology*, P. D. Sheets & D. K. Grayson (eds), 587–622. New York: Academic Press.

Johnston-Lavis, H. J. 1888a. The islands of Vulcano and Stromboli. *Nature* **38**, 13–14.

—— 1888b. Further notes on the late eruption at Vulcano Island. *Nature* **39**, 109–111.

—— 1890. The eruption of Vulcano Island. *Nature* **42**, 78–9.

Judd, J. W. 1875a. Contributions to the study of volcanoes. The Lìpari Islands: Vulcano. *Geological Magazine* **2**, 99–105.

—— 1875b. Contributions to the study of volcanoes. The Lìpari Islands: Stromboli. *Geological Magazine* **2**, 145–52, 206–19.

Keller, J. 1970. Die historischen Eruptionen von Vulcano und Lìpari. *Zeitschrift für Deutches Geologie und Geochimie* **121**, 179–85.

—— 1980a. The island of Vulcano. *Rendiconti della Società Italiana di Mineralogia e Petrologia* **36**, 29–75.

—— 1980b. The island of Salina. *Rendiconti della Società Italiana di Mineralogia e Petrologia* **36**, 489–524.

—— 1982. Mediterranean island arcs. In *Andesites*, R. S. Thorpe (ed.), 307–325. Chichester, England: John Wiley.

Keller, J., W. B. F. Ryan, D. Ninkovitch, R. Altherr 1978. Explosive volcanic activity in the Mediterranean over the past 200 000 years as recorded in deep-sea sediments. *Geological Society of America, Bulletin* **89**, 591–604.

Kieffer, G. 1979. L'activité de l'Etna pendant les derniers 20 000 ans. *Académie des Sciences de Paris, Comptes Rendus* **D288**, 1023–1026.

—— 1982. Les explosions phréatiques et phréatomagmatiques terminales à l'Etna. *Bulletin Volcanologique* **44**, 655–60.

—— 1983. L'évolution structurale de l'Etna et les modalités du contrôle tectonique et volcano-tectonique de son activité: faits et hypothèses après les éruptions de 1978 et 1979. *Revue de Géographie Physique et Géologie Dynamique* **24**, 129–52.

—— 1985. *Evolution structurale et dynamique d'un grand volcan polygénique: stades d'édification et activité actuelle de l'Etna (Sicile)*. Thèse d'état, Université de Clermont Ferrand II, France.

Kieffer, G. & J-C. Tanguy 1993. L'Etna: évolution

structurale, magmatique et dynamique d'un volcan polygénique. In *Pleins feux sur les volcans*, R. Maury (ed.), 253–71. Mémoire 163, Société Géologique de France, Paris.

Kilburn, C. R. J. 1981. Pahoehoe and aa lavas: a discussion and continuation of the model of Peterson and Tilling. *Journal of Volcanology and Geothermal Research* **11**, 373–89.

—— 1993. Lava crusts, aa flow lengthening and the pahoehoe–aa transition. In *Active lavas: monitoring and modelling*, C. R. J. Kilburn & G. Luongo (eds), 263–280. London: UCL Press.

Kilburn, C. R. J. & R. M. C. Lopes 1988. The growth of aa flow fields on Mount Etna, Sicily. *Journal of Geophysical Research* **93**, 14759–72.

Kokelaar, P. & C. Romagnoli 1995. Sector collapse, sedimentation and clast evolution at an active island arc: Stromboli, Italy. *Bulletin of Volcanology* **57**, 240–62.

Krafft, M. & F. D. de Larouzière 1999. *Guide des volcans d'Europe et des Canaries*, 3rd edn. Lausanne: Delachaux et Niestlé.

Lacroix, A. 1908. *La Montagne Pelée après ses éruptions, avec observations sur les éruptions du Vésuve en 79 et en 1906*. Paris: Masson.

Lanza, R. & E. Zanella 1991. Palaeomagnetic directions (223–1.4 ka) recorded in the volcanites of Lìpari, Aeolian Islands. *Geophysical Journal International* **107**, 191–6.

Lanzafame, G. & J. C. Bousquet 1997. The Maltese escarpment and its extension from Mt Etna to the Aeolian Islands (Sicily): importance and evolution of a lithosphere discontinuity. *Acta Vulcanologica* **9**, 113–20.

Lentini, F. 1982. The geology of the Mt Etna basement. *Memorie della Società Geologica Italiana* **23**, 7–25.

Lo Giudice, E., G. Patane, R. Rasa, R. Romano 1982. The structural framework of Mt Etna. *Memorie della Società Geologica Italiana* **23**, 125–58.

Loddo, M., D. Patella, R. Quarto, G. Ruina, A. Tramacere, G. Zito 1989. Applications of gravity and deep dipole geoelectrics in the volcanic area of Mt Etna (Sicily). *Journal of Volcanology and Geothermal Research* **39**, 17–39.

Luongo, G. & R. Scandone (eds) 1991. Campi Flegrei. *Journal of Volcanology and Geothermal Research* (special issue) **48**, 1–227.

Luongo, G., E. Cubellis, F. Obrizzo, S. Petrazzuoli 1991. A physical model for the origin of volcanism of the Tyrrhenian margin: the case of the Neapolitan area. *Journal of Volcanology and Geothermal Research* **48**, 173–86.

Lyell, Sir C. 1858. On the structure of lavas which have consolidated on steep slopes; with remarks on the mode of origin of Mount Etna, and on the theory of "craters-of-elevation". *Royal Society, Philosophical Transactions* **148**, 703–786.

McGuire, W. J. 1982. Evolution of the Etna volcano: information from the southern wall of the Valle del Bove Caldera. *Journal of Volcanology and Geothermal Research* **13**, 241–71.

McGuire, W. J. & A. D. Pullen 1989. Location and orientation of eruptive fissures and feeder-dykes at Mount Etna: influence of gravitational and regional tectonic stress regimes. *Journal of Volcanology and Geothermal Research* **38**, 325–44.

Mahood, G. A. & W. Hildreth 1986. Geology of the peralkaline volcano at Pantelleria, Strait of Sicily. *Bulletin of Volcanology* **48**, 143–72.

Marianelli, P., N. Métrich, A. Sbrana 1999. Shallow and deep reservoirs involved in magma supply of the 1944 eruption of Vesuvius. *Bulletin of Volcanology* **61**, 48–63.

Mattson, P. H. & W. Alvarez 1973. Base surge deposits in Pleistocene volcanic ash near Rome. *Bulletin Volcanologique* **37**, 553–72.

Mazzarella, S. 1984. *Dell'isola Ferdinandea e di altri cose*. Palermo.

Mercalli, G. 1907. *I vulcani attivi della terra*. Milan: Ulrico Hoepli.

Mercalli, G., M. Baratta, B. Friedlander, A. Aguilar, O. Scarpa 1907. *Il Vesuvio e la grande eruzione dell'Aprile 1906*. Napoli: Colavecchia, Colombai.

Métrich, N. 1985. *Mécanismes d'évolution à l'origine des roches volcaniques potassiques d'Italie centrale et méridionale: exemples du Mt Somma–Vésuve, des Champs Phlégréens et de l'île de Ventotene*. Thèse d'état, Université de Paris-Sud, Paris.

Métrich, N. & R. Clocchiatti 1989. Melt inclusions investigation of volatiles behavior in historic alkaline magmas of Etna. *Bulletin of Volcanology* **51**, 185–98.

Monaco, C., P. Tapponnier, L. Tortorici, P. Y. Gillot 1997. Late Quaternary slip rates on the Acireale–Piedimonte normal faults and tectonic origin of Mt Etna (Sicily). *Earth and Planetary Science Letters* **147**, 125–39.

Montalto, A. 1996. Signs of potential renewal of eruptive activity at La Fossa (Vulcano, Aeolian Islands). *Bulletin of Volcanology* **57**, 483–92.

Murray, J. B. 1990. High-level magma transport at Mount Etna volcano, as deduced from ground deformation measurements. In *Magma transport and storage*, M. P. Ryan (ed.), 357–83. Chichester, England: John Wiley.

Murray, J. B. & J. E. Guest 1982. Vertical ground deformation on Mt Etna, 1975–1982. *Geological Society of America, Bulletin* **93**, 1166–75.

Nazzaro, A. 1997. *Il Vesuvio: storia eruttiva e teorie vulcanologiche*. Naples: Liguori.

Ntepe, N. & J. Dorel 1990. Observations of seismic volcanic signals at Stromboli volcano (Italy). *Journal of Volcanology and Geothermal Research* **43**, 235–51.

Orsi, G., L. Ruvo, C. Scarpati 1991. The recent explosive volcanism at Pantelleria. *Geologische Rundschau* **80**, 187–200.

Orsi, G., S. de Vita, M. Di Vito 1996a. The restless, resurgent Campi Flegrei nested caldera: constraints on its evaluation and configuration. *Journal of Volcanology and Geothermal Research* **74**, 179–214.

Orsi, G., M. Piochi, L. Campajola, A. D'Onofrio, L. Gianella, F. Terrasi 1996b. ^{14}C geochronological constraints for the volcanic history of the island of Ischia (Italy) over the last 5000 years. *Journal of Volcanology and Geothermal Research* **71**, 249–57.

Orsi, G., S. de Vita, M. Piochi 1998. The volcanic island of Ischia. In *Field excursion guide: Cities on volcanoes, International Meeting*, G. Orsi, M. Di Vito, R. Isaia (eds), 72-8. Naples: Osservatorio Vesuviano.

Orsi, G., L. Civetta, C. Del Gaudio, S de Vita, M. A. Di Vito, R. Isaia, S. M. Petraluzzuoli, G. P. Ricciardi, C. Ricco 1999. Short-term ground deformations and seismicity in the resurgent Campi Flegrei caldera (Italy): an example of active block resurgence in a densely-populated area. *Journal of Volcanology and Geothermal Research* **91**, 415–51.

Owen, R. 2000. British isle rises off Sicily coast. London: *The Times* (5 February).

Palmieri, L. 1872. *Incendio Vesuviano del 26 Aprile 1872*. Torino: Fratelli Bocca.

Parascandola. A. 1946 Il Monte Nuovo ed il Lago Lucrino. *Bollettino della Società dei Naturalisti in Napoli* **55**, 152–264.

Pasquarè, G., L. Francalanci, V. H. Garduno, A. Tibaldi 1993. Structure and geological evolution of the Stromboli volcano, Aeolian Islands, Italy. *Acta Vulcanologica* **3**, 79–89.

Patanè, G., S. Gresta, S. Imposa 1984. Seismic activity preceding the 1983 eruption of Mt Etna. *Bulletin of Volcanology* **47**, 941–52.

Patanè, G., A. Montalto, S. Vinciguerra, J-C. Tanguy 1996. A model of the onset of the 1991–1993 eruption of Mt Etna, Italy. *Physics of the Earth and Planetary Interiors* **97**, 231–45.

Paterne, M., F. Guichard, J. Labeyrie 1988. Explosive activity of the south Italian volcanoes during the past 80000 years as determined by marine tephrochronology. *Journal of Volcanology and Geothermal Research* **34**, 153–72.

Perret, F. A. 1916. The lava eruption of Stromboli summer–autumn, 1915. *American Journal of Science* **42**, 443–63.

—— 1924. *The Vesuvius eruption of 1906: study of a volcanic cycle*. Publication 339, Carnegie Institution, Washington DC.

Pichler, H. 1980. The island of Lìpari. *Rendiconti della Società Italiana di Mineralogia e Petrologia* **36**, 415–40.

Pinkerton, H. & R. S. J. Sparks 1976. The 1975 subterminal lavas, Mount Etna: a case history of the formation of a compound lava field. *Journal of Volcanology and Geothermal Research* **1**, 167–82.

Pinkerton, H. & L. Wilson 1994. Factors controlling the lengths of channel-fed lava flows. *Bulletin of Volcanology* **56**, 108–20.

Pliny [Gaius Plinius Secundus] 1969. Letters and Panegyricus, vol. I: letters 16 and 20 [translated by B. Radice]. London: Heinemann.

Poli, S., S. Chiesa, P. Y. Gillot, F. Guichard 1987. Chemistry versus time in the volcanic complex of Ischia (Gulf of Naples, Italy): evidence of successive magmatic cycles. *Contributions to Mineralogy and Petrology* **95**, 322–35.

Poli, S., S. Chiesa, P. Y. Gillot, F. Guichard, L. Vezzoli 1989. Time dimension in the geochemical approach and hazard estimates of a volcanic area: the Isle of Ischia case (Italy). *Journal of Volcanology and Geothermal Research* **36**, 327–35.

Prévost, C. 1835. Notes sur l'île de Julia, pour servir à l'histoire de la formation des montagnes volcaniques. *Mémoires de la Société Géologique de France* **2**(5), 91–124.

Principe, C. 1998. The 1631 eruption of Vesuvius: volcanological concepts in Italy at the beginning of the XVIIth century. In *Volcanoes and history*, 525–42. *Proceedings of the 20th International Commission on the History of the Geological Sciences Symposium Napoli–Eolie–Catania (Italy)*, N. Morello (ed.) Genoa: Brigati.

Principe, C., A. Paiotti, J-C. Tanguy, M. Le Goff 1998. Archéomagnétisme et chronologie des éruptions du Vésuve. *Bulletin de la Section de Volcanologie de la Société Géologique de France* **44**, 25.

Recupero, G. 1815. *Storia naturale e generale dell'Etna* (2 vols). Catania: Stamperia della Regia Università.

Riccò, A. 1891. Tremblements de terre, soulèvement et éruption sous-marine à Pantelleria. *Académie des Sciences de Paris, Comptes Rendus* **113**, 733–5.

—— 1892. Terremoti, sollevamento ed eruzione sottomarina a Pantelleria nell seconda metà dell'Octobre 1891. *Bolletino della Società Geographica Italiana* **29**, 130–56.

Riccò, A. & S. Arcidiacono 1902–1904. L'eruzione etnea del 1892. *Atti Accademia Gioenia di Scienze Naturali in Catania* (series 4) **15–17**.

Rittmann, A. 1930. Geologie der Insel Ischia. *Zeitschrift für Vulkanologie* **6**, 1–265.

—— 1931. Der Ausbruch des Stromboli am 11 Sept. 1930. *Zeitschrift für Vulkanologie* **14**, 47–77.

—— 1962. *Volcanoes and their activity*. New York: John Wiley.

Romano, R. (ed.) 1982. *Mount Etna volcano: a review of the recent Earth science studies*. Memorie 23, Società Geologica Italiana,.

Romano, R. & C. Sturiale 1982. The historical eruptions of Mt Etna (volcanological data). *Memorie della Società Geologica Italiana* **23**, 75–97.

Rosi, M. 1980. The island of Stromboli. *Rendiconti della Società Italiana di Mineralogia e Petrologia* **36**, 345–68.

Rosi, M. & R. Santacroce 1983. The AD 472 "Pollena"

eruption: volcanological and petrological data for this poorly known Plinian-type event at Vesuvius. *Journal of Volcanology and Geothermal Research* **17**, 249–71.

—— 1984. Volcanic hazard assessment in the Phlegraean Fields: a contribution based on stratigraphic and historical data. *Bulletin Volcanologique* **47**, 359–70.

Rosi, M. & A. Sabrana (eds) 1987. *Phlegraean Fields*. Volume 114, Quaderni de La Ricerca Scientifica, Consiglio Nazionale delle Ricerche, Rome.

Rosi, M., A. Sbrana, C. Principe 1983. The Phlegraean Fields: structural evolution volcanic history and eruptive mechanisms. *Journal of Volcanology and Geothermal Research* **17**, 273–88.

Rosi, M., C. Principe, R. Vecci 1993. The 1631 Vesuvius eruption: a reconstruction based on historical and stratigraphical data. *Journal of Volcanology and Geothermal Research* **58**, 151–82.

Rosi, M., L. Vezzoli, P. Aleotti, M. De Censi 1996. Interaction between caldera collapse and eruptive dynamics during the Campanian Ignimbrite eruption, Phlegraean Fields, Italy. *Bulletin of Volcanology* **57**, 541–54.

Rymer, H., J. Cassidy, C. A. Locke, J. B. Murray 1995. Magma movements in Etna volcano associated with the major 1991–1993 lava eruption: evidence from gravity and deformation. *Bulletin of Volcanology* **57**, 451–61.

Rymer, H., F. Ferrucci, C. A. Locke 1998. Mount Etna: monitoring in the past, present and future. In *Lyell: the past is the key to the present*, D. J. Blundell & A. C. Scott (eds), 335–47. Special Publication 143, Geological Society, London.

Santacroce, R. 1983. A general model for the behaviour of the Somma–Vesuvius volcanic complex. *Journal of Volcanology and Geothermal Research* **17**, 237–48.

—— (ed.) 1987. Somma–Vesuvius. *Quaderni della Ricerca Scientifica* **114**(8), 1–220.

—— 1996. Preparing Naples for Vesuvius. *International Association of Volcanology and Chemistry of the Earth's Interior News* **1–2**, 5–7.

Santo, A. P., Y. Chen, A. H. Clark, E. Farrar, A. Tsegaye 1995. $^{40}Ar/^{39}Ar$ ages of the Filicudi Island volcanics: implications for the volcanological history of the Aeolian Arc, Italy. *Acta Vulcanologica* **7**, 13–18.

Scandone, R., F. Bellucci, L. Lirer, G. Rolandi 1991. The structure of the Campanian plain and the activity of the Neapolitan volcanoes. *Journal of Volcanology and Geothermal Research* **48**, 1–31.

Scandone, R., L. Giacomelli, P. Gasparini 1993. Mount Vesuvius: 2000 years of volcanological observations. *Journal of Volcanology and Geothermal Research* **58**, 5–25.

Scarpati, C., P. Cole, A. Perrotta 1993. The Neapolitan Yellow Tuff: a large volume multiphase eruption from Campi Flegrei, southern Italy. *Bulletin of Volcanology* **55**, 343–56.

Scarth, A. 1989. Volcanic origins of the Polyphemus story in the Odyssey: a non-classicist's interpretation. *Classical World* **83**(2), 89–96.

—— 1994. *Volcanoes*. London: UCL Press.

—— 1999. *Vulcan's fury: man against the volcano*. London: Yale University Press.

—— 2000. The volcanic inspiration of some images in the Aeneid. *Classical World* **93**, 591–605.

Scrope, G. J. P. [later Poulett-Scrope] 1825. *Considerations on volcanoes, the probable causes of their phenomena and their connection with the present state and past history of the globe; leading to the establishment of a new theory of the Earth*. London: W. Phillips.

Selvaggi, G. & C. Chiarabba 1995. Seismicity and *P*-wave velocity image of the southern Tyrrhenian subduction zone. *Geophysical Journal International* **121**, 818–26.

Sharp, A. D. L., P. M. Davis, F. Gray 1980. A low velocity zone beneath Mt Etna and magma storage. *Nature* **287**, 587–91.

Sharp, A. D. L., G. Lombardo, P. M. Davis 1981. Correlations between eruptions of Mount Etna, Sicily, and regional earthquakes as seen in historical records from AD 1582. *Royal Astronomical Society, Geophysical Journal* **65**, 507–523.

Sheridan, M. F., F. Barberi, M. Rosi, R. Santacroce 1981a. A model for Plinian eruptions of Vesuvius. *Nature* **289**, 282–5.

Sheridan, M. F., T. C. Moyer, K. H. Wohletz 1981b. Preliminary report on the pyroclastic products of Vulcano. *Società Astronomica Italiana, Memorie* **52**, 523–7.

Sheridan, M. F. & M. C. Malin 1983. Applications of computer-assisted mapping to volcanic hazard evaluation of surge eruptions: Vulcano, Lìpari and Vesuvius. *Journal of Volcanology and Geothermal Research* **17**, 187–202.

Sigurdsson, H. S., S. Cashdollar, R. S. J. Sparks 1982. The eruption of Vesuvius in AD 79: reconstruction from historical and volcanological evidence. *American Journal of Archaeology* **86**, 39–51.

Sigurdsson, H. S., S. N. Carey, W. Cornell, T. Pescatore 1985. The eruption of Vesuvius in AD 79. *National Geographic Research* **1**(3), 332–87.

Simkin, T. & L. Siebert 1994. *Volcanoes of the world*, 2nd edn. Tucson: Geoscience Press.

Spallanzani, L. 1792. *Viaggio alle Due Sicilie e in alcune parti dell'Appennino* (vol. 2). Pavia: Stamperia Baldassare Comini.

Stothers, R. B. & M. R. Rampino 1983. Volcanic eruptions in the Mediterranean before AD 630 from written and archaeological sources. *Journal of Geophysical Research* **88**, 6357–71.

Tanguy, J-C. 1978. Tholeiitic basalt magmatism of Mount Etna and its relations with the alkaline series. *Contributions to Mineralogy and Petrology* **66**, 51–67.

—— 1980. *L'Etna: étude pétrologique et paléomagnétique; implications volcanologiques*. Thèse d'état, Université de Paris VI, France.

—— 1981. Les éruptions historiques de l'Etna: chronologie et localisation. *Bulletin Volcanologique* **44**, 586–640.

Tanguy, J-C., I. Bucur, J. F. C. Thompson 1985. Geomagnetic secular variation in Sicily and revised ages of historic lavas from Mount Etna. *Nature* **318**, 453–5.

Tanguy, J-C. & G. Kieffer 1993. Les éruptions de l'Etna et leurs mécanismes. In *Pleins feux sur les volcans*, R. Maury (ed.), 239–52. Mémoire 163, Société Géologique de France, Paris. 239–52.

Tanguy, J-C. & G. Patanè 1996. *L'Etna et le monde des volcans*. Paris: Diderot Editeur.

Tanguy, J-C., G. Kieffer, G. Patanè 1996. Dynamics, lava volume and effusion rate during the 1991–1993 eruption of Mount Etna. *Journal of Volcanology and Geothermal Research* **71**, 259–65.

Tanguy, J-C., M. Condomines, G. Kieffer 1997. Evolution of the Mount Etna magma: constraints on the present feeding system and eruptive mechanism. *Journal of Volcanology and Geothermal Research* **75**, 221–50.

Tanguy, J-C., M. Le Goff, V. Chillemi, A. Paiotti, C. Principe, S. La Delfa, G. Patanè 1999. Variation séculaire de la direction du champ géomagnétique enregistrée par les laves de l'Etna et du Vésuve pendant les deux derniers millénaires. *Académie des Sciences de Paris, Comptes Rendus* **329**, 557–64.

Tonarini, S., P. Armienti, M. D'Orazio, F. Innocenti, M. Pompilio, R. Petrini 1995. Geochemical and isotopic monitoring of Mt Etna 1989–1993 activity: bearing on the shallow feeding system. *Journal of Volcanology and Geothermal Research* **64**, 95–115.

Ventura, G. 1994. Tectonics, structural evolution, and caldera formation on Vulcano Island (Aeolian archipelago, southern Tyrrhenian Sea). *Journal of Volcanology and Geothermal Research* **60**, 207–224.

Vezzoli, L. 1988. *Island of Ischia*. Volume 114, Quaderni de La Ricerca Scientifica, Rome.

Villari, L. 1980a. The island of Alicudi. *Rendiconti della Società Italiana di Mineralogia e Petrologia* **36**, 441–6.

—— 1980b. The island of Filicudi. *Rendiconti della Società Italiana di Mineralogia e Petrologia* **36**, 467–88.

—— (ed.) 1994. The 1991–1993 Etna eruption. *Acta Vulcanologica* **4**, 1–177.

Vogel, J. S., W. Cornell, D. E. Nelson, J. R. Southon 1990. Vesuvius–Avellino, one possible source of seventeenth-century BC climatic disturbances. *Nature* **344**, 534–7.

Walker, G. P. L. 1973. Lengths of lava flows. *Royal Society, Philosophical Transactions* **A274**, 107–118.

—— 1981. Plinian eruptions and their products. *Bulletin Volcanologique* **44**(2), 223–40.

Washington, H. S. 1909 The submarine eruptions of 1831 and 1891 near Pantelleria. *American Journal of Science* **27**, 131–50.

Williams, H. 1941. *Calderas and their origin*. Publications in Geological Science 25 (pp. 239–346), University of California, Berkeley.

Wilson, L., R. S. J. Sparks, G. P. L. Walker 1980. Explosive volcanic eruptions, IV: the control of magma properties and conduit geometry on eruption column behaviour. *Royal Astronomical Society, Geophysical Journal* **63**, 117–48.

Wohletz, K. H. & M. F. Sheridan 1983. Hydrovolcanic explosions II: evolution of basaltic tuff rings and tuff cones. *American Journal of Science* **283**, 385–413.

Wohletz, K., L. Civetta, G. Orsi 1999. Thermal evolution of the Phlegraean magmatic system. *Journal of Volcanology and Geothermal Research* **91**, 381–414.

Zollo, A. and 13 others 1998. An image of Mt Vesuvius obtained by 2-D seismic tomography. *Journal of Volcanology and Geothermal Research* **82**, 161–73.

3 Greece

The Hellenic volcanic arc is the best-developed island arc in Europe. It swings for 500 km across the southern Aegean Sea from the Isthmus of Corinth to Bodrum in Turkey. It comprises the smaller volcanic centres of Aegina, Póros and Méthana, and the larger accumulations of Mílos, Antíparos, Santoríni, Nísyros, Yalí and southwestern Kós. But only Méthana, Nísyros and Santoríni have recorded eruptions of more than fumaroles during historical times. However, the Minoan eruption of Santoríni in the Bronze Age was not only one of the greatest volcanic events in Europe during the past 10 000 years but it also made an most impressive contribution to one of its most spectacular landscapes.

The volcanic arc was generated by subduction of an oceanic part of the African plate beneath the small Aegean plate on the southern edge of the Eurasian plate. The volcanic arc rises some 220 km north of the Hellenic trench, where the upper surface of subducted slab has reached a

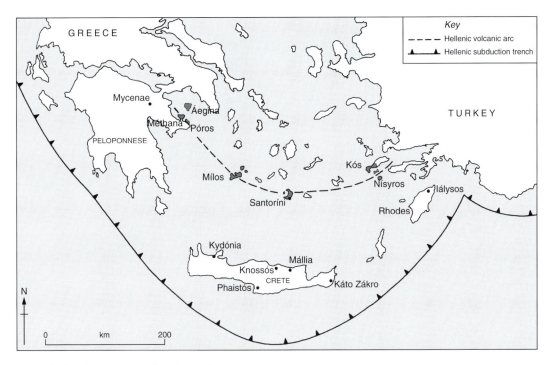

The Hellenic volcanic arc and the Minoan Aegean.

depth of 130–150 km (McKenzie 1972, Papadópoulos 1984, Huijsmans et al. 1988, Kalogerópoulos & Parítsis 1990). The oldest emerged rocks, on Mílos, are 3.5 million years old. During historical times, the Kaméni Islands in the Santoríni caldera have marked the most voluminous output from any centre in Greece. The Hellenic volcanoes have generally erupted calc-alkaline, andesitic, dacitic and rhyolitic rocks, and, among them, andesites and **dacites** each constitute about a third of the volcanic products ejected. Only Santoríni has expelled significant quantities of basalt.

In the west, the eruptions have been smaller and the lavas have usually been viscous. Thus, domes and thick flows of basaltic andesite and dacite, erupted about 1–2 million years ago, make up the hills in the central and southern two thirds of Aegina and the volcanic southern peninsula of Póros; and viscous flows also accumulated as low shields in western Mílos. However, greater magmatic differentiation, in relatively shallow reservoirs, in the centre and east of the arc occasionally generated more explosive eruptions of rhyodacitic and rhyolitic ash and pumice, and the collapse of calderas notably off southwestern Kós, in Nísyros and, of course, in Santoríni (Keller et al. 1990, Druitt & Francaviglia 1992, Druitt et al. 1999). Since the latest eruption in the arc took place in the Kaméni Islands in 1950, activity has been limited to mild fumaroles and hot springs that are manifest in all the volcanic centres, especially in Méthana, the Kaméni Islands and Nísyros.

SANTORÍNI

From a distance, Santoríni appears as a round island that rises to a broad central focus, with scattered white villages surrounded by vines and tomatoes in terraced fields of buff pumice. In the south, the rugged Cycladic limestones of Mount Profítis Elías form its highest point at 565 m. Santoríni is, in fact, a group of islands that acquired this name during the Venetian occupation, but, in classical times, it was called Théra, which is also often used today. However, it seems clearer to use Théra for the principal island and retain Santoríni for the whole group.

The glory of Santoríni is its sea-flooded caldera, about 85 km² in area, whose rim marks the formidable inner edge of three islands (Plate 6). On the main island, Théra, the northern part of the rim makes a rampart of red, brown and black layers of lava and cinders, 400 m high, but its southern part forms a less forbidding wall of buff pumice rising no more than 200 m. A similar wall faces it on the white pumice islet of Aspronísi (white island) to the west. Therasía, which has a rampart like northern Théra, then completes the circle. In the midst of the caldera, intermittent activity since 197 BC has formed the Kaméni (Burnt) Islands. They have been the site of all but

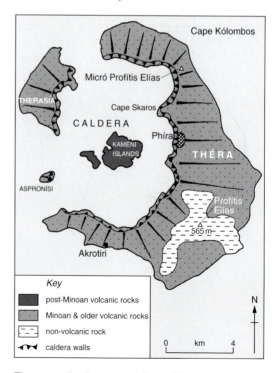

The generalized geology of Santoríni.

one of the eruptions of historical times in Santoríni, and they make a low, stark agglomeration of recent lava flows, domes and cones that offer a splendid views of the whole caldera.

The event that gave Santoríni its notoriety – and much of its spectacular landscape – was the Minoan eruption in the Bronze Age. The pumice ejected buried a city, excavated at Akrotíri, which had formed part of the Minoan civilization that was named after the legendary King Mínos, who reigned at the Palace of Knossós in Crete (Evans 1921–36). The Minoan eruption has also been blamed for the sudden end of this civilization (Marinátos 1939, Luce 1969, Doúmas 1978).

The growth of Santoríni

Before the present Hellenic volcanic arc ever existed, Mount Profítis Elías marked the summit of an island of limestones and schists that formed the most southerly of the Cycládes. It was at a depth of about 1000 m on the adjacent Aegean sea floor that eruptions began, perhaps as much as 3 million years ago. Santoríni apparently never formed a single stratovolcano, but separate vents erupted a complex volcanic field over a long period. The oldest volcanic rocks are exposed in the Akrotíri Peninsula on the southern arm of Théra, where, between 550 000 and 650 000 years ago or earlier, a dozen or so vents erupted dacitic domes, andesitic lava flows, spatter cones and cinder cones. Volcanic activity then shifted northeastwards and constructed the Peristeria volcano over 430 000 years ago, while emissions of andesitic and dacitic lavas built the *bases* of four shield volcanoes to a height of some 200 m: Megálo Vounó, Micró Profítis Elías, Therasía and Skáros (Pichler & Kussmaul 1980, Heiken & McCoy 1984, Huijsmans et al. 1988, Huijsmans & Barton 1990, Druitt et al. 1999). Events then occurred in two long cycles. The first cycle began 360 000 years ago and culminated 180 000 years ago in the powerful eruptions that expelled the rhyodacitic Lower Pumices. The second cycle lasted from 180 000 years ago until the rhyodacitic Plinian Minoan eruption in the Bronze Age. Large **silicic magma** reservoirs developed during the last 50 000 years of each cycle.

Santoríni had dozens of violent eruptions of rhyodacitic pumice during the second cycle. Many of these eruptions took place along two parallel fissure or fault systems, trending from northeast to southwest, one passing through

Cape Kolómbos in the north; the other passing through the Kaméni Islands in the centre (Druitt & Francaviglia 1990, 1992).

The caldera of Santoríni was formed not all at once but seems to have resulted from four main episodes of collapse. After the Lower Pumices erupted about 180 000 years ago, a caldera collapsed and was probably flooded by the sea. Mainly andesitic eruptions then supervened and probably filled the caldera with both lava shields and layers of exploded fragments, such as the distinctive Middle Pumice and the Upper Scoria (Druitt et al. 1989, Huijsmans & Barton 1990). A second caldera then collapsed, chiefly in the north, about 70 000 years ago. Again, eruptions of basalts, andesites and rhyodacites filled the hollow and extended well beyond its walls. They developed the Skáros shield and the Therasía shield or dome complex.

About 21 000 years ago, both these lava piles were severely damaged when a rhyodacitic Plinian eruption brought about the collapse of the third caldera, centred on Cape Riva in northern Therasía (McClelland & Druitt 1989, Druitt et al. 1999). Subsequent eruptions then seem to have formed a volcanic island in the centre of this caldera, similar to the present Kaméni Islands.

Several millennia of quiescence allowed at least two (and perhaps many more) Minoan settlements to grow up on the island. One, near the southern tip of Therasía, was exposed when pumice was being quarried to make the dykes alongside the Suez Canal in 1869, but it has not yet been systematically excavated (Fouqué 1879). The main Minoan settlement yet discovered is being excavated at Akrotíri on the southern arm of Théra. It is possible that the fine naval-flotilla fresco unearthed there could depict the Minoan landscape of Santoríni, although its lack of perspective, not to mention its artistic licence, makes it hard to fit the details into a coherent picture. But, if the fresco is at all accurate, then a large body of water occupied central Santoríni, and the chief Minoan settlement on the island must have stood near the present southern tip of Therasía, and not at Akrotíri itself.

The Minoan eruption of Santoríni

The Minoan eruption was the largest on any European volcano in recent millennia and it destroyed the Minoan Bronze Age settlements on the island. As the magma pressure rose, the inhabitants were warned by earthquakes that were powerful enough to damage sturdy two- and three-storey buildings in Akrotíri. These were being repaired when a hydrovolcanic eruption, perhaps lasting for several weeks, deposited a thin layer of pale lithic ash over southern Théra. The Akrotírians abandoned their town and no human bones have been found there. But it must be very doubtful whether any inhabitants of Santoríni survived the ensuing cataclysm (Doúmas 1974, Heiken & McCoy 1990, Sigurdsson et al. 1990).

The Minoan eruption took place in a shallow flooded caldera during four phases that occurred without respite and lasted for a total of about four days. The first phase was marked by a continuous Plinian eruption, forming a column 36 km high that spread up to 6 m of coarse roseate rhyodacitic pumice. It was followed by a thin layer of fine white ash and then by a layer of pumice ejected by further Plinian eruptions. The parent vent was probably situated near the present position of the Kaméni Islands. The pumice was strongly concentrated in southeastern Théra, but it formed only a thin layer on Therasía, which indicates that the Plinian column was winnowed by northwesterly winds, possibly when the meltemi was at its height in the summer. The first phase might have lasted up to about eight hours.

The second phase was marked by a gradual change to distinctive hydrovolcanic characteristics as erupting vents opened in the sea. Its fine ash and pumice attain a maximum thickness of 12 m. Sometimes, nuées ardentes knocked down any walls in Akrotíri still emerging from the previous blanket of pumice. This second phase probably lasted about a day.

The third phase of the Minoan eruption was also hydrovolcanic and came from a vent filled with a slurry of ash, pumice, steam and water. It produced more than half the Minoan materials on Santoríni. They comprise up to 55 m of chaotic layers of white pumice and ash, with many lithic fragments, including breadcrust bombs. They are thickest in the quarries south of Phíra, near the probable vent. The Minoan caldera may have started to collapse at this time. As more water reached the vent, ashflows merging into mudflows were generated, which radiated over most of Théra and Therasía. Most of the lithic fragments that erupted during the third phase seem to be derived from the dacitic lavas of the Therasía complex, as if they came from new vents situated to the west of those in operation during the first two phases. The third phase might have lasted a day.

The fourth and final phase might have lasted for several hours. It expelled ashflows, mudflows and ignimbrites, which mantle the outer flanks of Théra and Therasía with up to 40 m of fine ochre pumice. The close of the fourth phase saw the elaboration of the Minoan caldera, whose floor now lies some 380 m below sea level in the north. It enlarged the previous calderas considerably, and submarine slumping also extended two submerged arms on either side of Therasía. Estimates of the volume of magma ejected have varied from 19 km^3 to 39 km^3, depending on views about the nature of the landscape before the eruption and on different interpretations of the extent and thickness of Minoan deposits, but every estimate indicates that this was a massive outburst.

The effects of the Minoan eruption extended physically beyond the limits of Santoríni and scientifically well beyond the confines of the Earth sciences. The eruption presented two interrelated problems: when did the eruption take place and what were its effects on Minoan civilization in Crete, 125 km to the south? Excavations at Akrotíri showed that the eruption had buried the town in the period named the Late Minoan IA, traditionally assigned by archaeologists to about 1550 BC to 1500 BC. Further research cast doubt on this date, but generated as many problems as solutions. Radiocarbon dates obtained from shortlived shrubs killed by the eruption ranged between 1675 BC and 1525 BC, with a preference towards 1645 BC (Betancourt 1987, Hammer et al. 1987), although the data pointed to about 1629–1622 BC when tested statistically (Manning 1988, 1989). If anything, the radiocarbon dates indicated that Akrotíri was buried a century before the dates suggested by the traditional archaeological record.

Cores taken from the Greenland ice cap showed aerosol acidity peaks about 1645 BC (Hammer et al. 1980, 1987). They were linked to Santoríni because it was believed to be the only large eruption that could have caused such a concentration at about that time. Attempts to prove or disprove this link have generated much dispute, and this problem is far from being solved. For example, analysis of the same fragments in the ice cores led different authors to conclude that they did (Manning 1998), or probably did not (Zielinski & Germani 1998a,b) come from Santoríni.

The sheer walls of the Santoríni caldera rise 400m at Phira. The mule track zig-zags up the piles of lava flows.

Dendrochronological studies (whereby absolute dates could be established by the analysis of tree rings) were also brought into controversy. Bristlecone pines in the southwestern USA registered severe frost damage in about 1627 BC, which was attributed to an unusually cold spell (LaMarche & Hirschboeck 1984). Ancient oaks preserved in bogs in Northern Ireland also showed poor growth of tree rings, which was linked to a cold spell in about 1628 BC (Baillie & Munro 1988). Exceptional growth in a tree ring in a 1503-year chronology established in Anatolia was also assigned, perhaps rather arbitrarily, to 1628 BC (Kuniholm et al. 1996). The eruption of Santoríni was invoked as the source of the volcanic aerosol or dust veil that would account for these biological anomalies. But, in each case, the link to Santoríni was only circumstantial (Pyle 1989, Scarth 1994, Buckland et al. 1997). Many other explosive volcanoes, including dozens scarcely studied, would have been just as likely as Santoríni to supply the requisite stratospheric veil and cause a cold spell. Moreover, not all large eruptions cause frost damage to vegetation, and not all severe cold spells are caused by volcanic eruptions. Indeed, during the better-documented centuries after 550 AD, acidity peaks in the ice cores and restricted tree-ring growth coincided only seven times.

If the Minoan eruption caused acidity peaks in the Greenland ice cap and the anomalous tree-ring growths, then it probably occurred in about 1628 BC. But it is over a century older than the dates traditionally adopted by archaeologists, and its acceptance would apparently require the addition of 130 years to the history of Egypt, and a revision of the Bronze Age chronology throughout the eastern Mediterranean area (Bietak 1996). Thus, these still-conflicting strands of evidence mean that the exact date of the Minoan eruption remains to be established. The research involves the confrontation between new and traditional techniques and attitudes in several academic disciplines. This is truly the stuff of a prolonged academic polemic, and a consensus should not be awaited with bated breath, as the outcome may be long in coming.

Eruptions of Santoríni during historical times

Eruptions of Santoríni have been recorded for more than 2000 years and they seem to have happened with increasing frequency in recent centuries. All but one of the historical eruptions have taken place from vents in the centre of the present caldera, where they have formed the Kaméni Islands, a strikingly barren landscape of black and grey aa lava flows, and several domes and explosion pits that still give off fumaroles. The Kaméni Islands rise a total of 520 m from a depth of 380 m on the caldera floor and they represent an accumulation of 2.5 km^3 during the past 2100 years (Heiken & McCoy 1984).

In 197 BC, a new island called Hierá erupted in the centre of the caldera between Théra and Therasía. "Flames came bursting up from the sea for four days, causing the water to seethe and flare up. Gradually an island emerged and was built up, as though it had been forged by implements out of a red-hot mass" (Strabo I: 57). This island formed a tuff cone that was quickly destroyed by the waves, and it probably now remains as the Bánkos shoal, to the northeast of the present islands. In AD 46, lava eruptions formed the islet of Thía, which marked the inception of Palaeá Kaméni (Fouqué 1879). In AD 726, an emission of lavas enlarged Palaeá Kaméni and the pumice that exploded from a vent on its east coast covered the surrounding sea and floated off as far away as Turkey. The next active episode, which took place in 1570 and seems to have lasted for three years, formed the islet of Mikrá ("small") Kaméni. It was 500 m long, 320 m wide and 71 m high, but it foundered a little to 66 m in 1707 (Georgalás 1962).

In 1650, the only activity in historical times outside the caldera occurred on a fissure on the sea floor, 6.5 km northeast of Cape Kolómbos. Violent earthquakes shook Santoríni for several days in early March 1650 and again in mid-September. Many homes were damaged, the ground roared, the sea turned green, and rock crashed from the cliffs around the caldera. On 27 September the most severe earthquake rocked the houses as if they were cradles. That day, a Surtseyan eruption began offshore when lava emissions were followed by explosions of snow-white pumice and ash. In the midst of deafening explosions that were heard 400 km away in the Dardanelles, a small tuff cone formed an islet. This was the most damaging of all the historical eruptions, because it generated a small **tsunami** on 29 September that caused much destruction on the nearby coasts, knocking down orchards and five churches in Santoríni, and throwing a Turkish naval vessel onto the shore on the island

The collapse of the Minoan civilization

The Minoan civilization was the most sophisticated in the European Bronze Age. It was named after the legendary King Minos, whose palace at Knossós was the largest of several that have been excavated in Crete. Akrotíri was one of at least two Minoan settlements on Santoríni. Minoan civilization collapsed when these Cretan palaces seem to have been suddenly destroyed in the Late Minoan IB period, which has been traditionally assigned to between about 1500 BC and 1450 BC. Only Knossós survived, under the occupation of Greek Mycenaeans. They were certainly in charge at Knossós just after the catastrophe, because their clay inventory tablets were written in the Mycenaean Greek Linear B script, whereas earlier documents had been written in the different language of the Linear A script (Ventris & Chadwick 1973).

The collapse of Minoan civilization has long puzzled the experts. It has been attributed, *inter alia*, to a military takeover by Mycenaeans, to extensive earthquakes, to volcanic ash from Santoríni, and to civil war in Crete (e.g. Hutchinson 1962). However, the most controversial solution emerged when Marinátos proposed that Minoan civilization had been annihilated when the eruption of Santoríni had generated a tsunami that had destroyed the coastal settlements of northern Crete, while associated earthquakes had severely damaged those inland (Marinátos 1939). This view proved to be a red herring, but it stimulated much research and academic dispute about the role of the eruption of Santoríni, which are reviewed in more detail elsewhere (e.g. Scarth 1994, Manning 1996, Buckland et al. 1997, Rehak & Younger 1998).

Marinátos's own excavations soon showed that the eruption had buried Akrotíri in the Late Minoan IA period, assigned traditionally to between about 1550 BC and 1500 BC. Thus, this eruption could not have destroyed the Cretan palaces at the end of the Late Minoan IB. To overcome this crucial point, it was suggested the eruption had taken place in two episodes separated by an interlude of about 50 years. But, the geological consensus now indicates that this outburst took place within days, not decades, during the Late Minoan IA (Sigurdsson et al. 1990).

Neither could the eruption have caused tectonic earthquakes over 125 km away in Crete. Volcanotectonic earthquakes would have affected only Santoríni and its immediate vicinity. And even at Akrotíri, the damage caused by these precursory earthquakes was being repaired when the great eruption was unleashed. Moreover, previous Cretan earthquakes had not annihilated Minoan civilization; any palaces that were destroyed were simply rebuilt to levels of even greater splendour.

Deluges of pumice and ash over Crete might have killed crops and herds, and caused civil disruption, famine and death. But they could not have destroyed the substantially built Cretan palaces. And, for instance, a pumice layer 10 cm thick at the Minoan colony of Iálysos in Rhodes does not even seem to have broken the continuity of settlement (Doúmas & Papázoglou 1980). It also lies beneath the Late Minoan IB pottery layer, which again indicates that the eruption occurred during the Late Minoan IA. In addition, most of the ash and pumice was blown over Turkey, and indeed over the Black Sea, thus leaving Crete relatively unaffected on the very fringes of the area of deposition.

The tsunami from Santoríni would certainly have been strong enough to demolish the palaces near the north coast of Crete; and Knossós, 5 km inland, could also have been badly damaged (Marinátos 1939, Luce 1969). But this tsunami could not have touched Phaestós and Hághia Triáda, standing 100 m high near the south coast. The tsunami might not even have travelled mainly towards Crete, for the Santoríni caldera is breached in the west and northwest, which would direct it primarily towards Mycénae and the Peloponnese. The Mycenaeans would, therefore, scarcely have been in a position to profit from any weakness in Crete. And, of course, the tsunami would have developed when the caldera collapsed: it would have reached Crete in 25 minutes, in the Late Minoan IA and not about 50 years later, at the end of Late Minoan IB.

Santoríni seems irrelevant to a purely political Mycénaean takeover or to internecine Cretan struggles. But new evidence indicates that the destruction of the palaces was selective and that administrative buildings seem to have been special targets (Rehak & Younger 1998). Moreover, recent research has also indicated that, in fact, the Cretan palaces might not have all been destroyed at the same time during the Late Minoan IB. If these interpretations of events are substantiated, war or sporadic earthquakes would seem to be the chief suspects, but the eruption of Santoríni could be entirely eliminated from the enquiry. Thus, it now seems impossible to blame Santoríni for the fall of Minoan Crete. But the Minoans were, no doubt, greatly impressed by the Plinian column rising 36 km above Santoríni, the shock waves, the sea waves, and the loudest noise in the Mediterranean Bronze Age, which would have reverberated all over Europe.

of Kéos. Pumice floated all over the Aegean. The vent also gave off noxious fumes that blew over Théra, where many people were blinded for several days and about 50 inhabitants, and more than 1000 farm animals, were asphyxiated. On 2 October, a boat coming from the island of Amorgós sailed too close to the vent and was enveloped in gas. The next day, the people on the island of Íos saw it wandering aimlessly about in full sail, and went to investigate. They discovered the corpses of all nine members of the crew, with their eyes inflamed, their heads grossly swollen, and their tongues hanging from their mouths. After this macabre climax, the vigour of

Cape Skaros and the northern walls of the caldera.

the eruption diminished. However, on 4 November an explosion covered the northern part of Théra in black smoke, and about 20 agricultural workers lost consciousness. They did not recover their senses until that evening, just as they were about to be buried. The eruption then waned and stopped altogether on 6 December. A short time afterwards, the sea destroyed the tuff cone and left behind a shoal, the Kolómbos Bank, which is about 18 m deep (Fouqué 1879).

Effusive and explosive eruptions between Palaeá Kaméni and Mikrá Kaméni took place from 23 May 1707 to 11 September 1711 and led to the formation of Neá (New) Kaméni.

The next eruptions lasted from 26 January 1866 until 15 October 1870. Each started with effusions and ended with explosions. Submarine eruptions off the southern coast of the original Neá Kaméni led to the emergence of the dacitic Geórgios dome that eventually rose to

The recently formed domes and craters of Neá Kaméni.

The eruption of the Kaméni Islands, 1707–1711

The Jesuit priest, Father Goree, described the eruption of 1707–1711 for the Royal Society of London (Goree 1711). For many weeks before the eruption "fishermen had perceived an ill smell every time that they passed by that place". There was a small earthquake on 18 May and another in the night of 22/23 May. The following morning, some fishermen discovered the new island, "which they imagined to be some vessel that had suffered ship-wreck". They rushed to claim the prize, found the new island, but then "grew afraid", and hurried away to report the news. But the eruption calmed a few days later and several men, "not imagining any danger", landed on the new island. They discovered a "white stone, which cuts like bread, and resembles it so well in form, colour and consistency that, were it not for its taste, anyone would take it for real bread. But what pleased them more was a great number of fresh oysters." As they busied them-selves collecting the oysters, the island started to shake under their feet, and they fled. During the next two weeks, the island rose up, probably by calm emissions of lava, because Father Goree makes no mention of explosions there, although he noted occasional eruptions on Mikrá Kaméni. At the same time, the sea changed to a "yellow-ish colour – with a stink" that spread all over Santoríni. Smoke first appeared on 16 July "from a new ridge of black stones" that gave off molten lava on 20 July, and became the main centre of the increasing activity on the expanding new island. "It was then that the inhabitants of Santoríni . . . began to be in good earnest afraid", fearing that their homes would be blown up, if ever "the sub-terranean vitriol and sulphur caught fire". The Turks, who were on the island collecting their annual tribute, "being amazed to see fire break out of the sea where it was so deep . . . entreated the Christians to pray to God, and especially to make their children cry out *Kyrie Eleison*" (Father Goree claimed that the Turks believed that the children, being without sin, would have greater influence on the Almighty). But new vents opened, and made a noise like thunder and cannons; the sea frothed, bubbled, and turned oily; the new island expanded greatly; the fumes tarnished silver and killed the grapes upon which most of the islanders depended for a livelihood; dead fish were thrown upon the shore; and the volcanic bombs sank a small vessel that had unwisely sailed too close. On 24 August, for instance, as many as 60 vents were reput-edly active, and the explosions often resembled fire-works; one vent even whistled. Earthquakes on 18 September 1707 and 10 February 1708 gave the erup-tions even greater vigour. There was a terrifying eruption on Easter Sunday, 15 April 1708, that "confirmed several ignorant Greeks in the ridiculous opinion . . . that the new island was one of the mouths of Hell". The eruptions diminished in vigour as the summer progressed, which encouraged Father Goree and the Latin Bishop of San-toríni to try and visit the new island. But both the island and the surrounding sea proved to be too hot, and they had to content themselves with a view of the spectacle from Palaeá Kaméni, which had not erupted for nearly a thousand years. After the abortive visit of Father Goree and the bishop, the eruption continued, with more or less the same intensity, long after Father Goree left Santoríni on 15 August 1708. Activity eventually drew to a close on 11 September 1711. (Goree 1711)

Néa Kameni: the most recently erupted lavas on the right and Phíra on the horizon crowning the walls of the caldera.

131 m above sea level and was flanked by several lava flows. It still constitutes the highest point of the Kaméni Islands. Two adjacent vents also erupted, but most of their lavas were eventually covered by those expelled from Geórgios. A small ash ring also formed around this dome, which then gave off only fumes from 15 October 1870. By the end of this episode, Neá Kaméni had trebled in size.

The period between 1925 and 1950 was the most active in the history of the Kaméni Islands (Papadópoulos 1990). On 11 August 1925, submarine eruptions marked the beginning of a new episode, and new lavas finally joined Mikrá Kaméni to Neá Kaméni on 12 August. Then, the Dáfni dome arose, a 96 m-high tuff ring exploded around it, and three lava flows formed a rugged apron almost surrounding the old islet of Mikrá Kaméni. This eruption ceased in January 1926, but another, lasting from 23 January to 17 March 1928, formed the small 15 m dome of Naftílos, southeast of Dáfni.

Longer eruptions began on 20 August 1939 when submarine effusions west of Neá Kaméni formed the Tríton dome, which rose to 13 m above sea level during the first week. On 26 August 1939, the pit exploded that was to become the main centre of activity for the rest of the episode. On 26 September 1939, the Ktenás dome extruded to a height of 12 m near the crest of Geórgios. Ktenás was soon surrounded by a horseshoe cinder rampart and also gave off lava flows that spread into the sea. On 13 November 1939, the large Fouqué dome extruded from the explosion pit. It eventually rose 122 m above sea level and its lava flows covered most of the Ktenás lavas, and parts of the flows erupted in 1866–70. This dome was also linked to the violent explosions that took place between 12 July 1940 and early July 1941. Many explosive phases were soon followed by dome extrusions and lava-flow eruptions. Thus, the small Smith A and Smith B domes formed from 15 July to 8 September 1940; Reck dome extruded nearby from 17 July until 17 October 1940 and it was surrounded by a cinder rampart. But an explosion shattered the dome and thus only the rampart survives. On 24 November 1940, another explosion formed the Níki dome, which eventually rose 125 m above sea level. When these eruptions ended on 5 July 1941, Nea Kaméni had been consolidated into a round mass, 2 km across.

The latest eruptions, of explosions and lava effusions, lasted from 10 January to 3 February

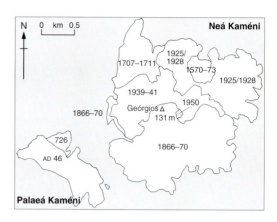

Growth of the Kaméni Islands.

1950, and formed only the small 9 m-high dome of Liátsikas and two lava flows. Since then, the emanations have been limited to solfataras and fumes at about 70–86°C.

All the historic activity of the Kaméni Islands has been concentrated along a band trending northeast, 600 m wide and 4.5 km long, following the dominant trends of fissures throughout Santoríni (Heiken & McCoy 1984). The dacitic lavas came from a reservoir about 3 km deep at a temperature of 950–1000°C. Each episode began with extrusions of viscous dacites and was followed by the formation of cinder cones and lava flows. But, in view of the marine location of the Kaméni vents, the general absence of Surtseyan eruptions is remarkable. It has been attributed to high effusion rates, which prevented water from entering the vents. On the other hand, the main explosive episodes, in 726, 1650 and 1925–8, were probably related to the highest rates of discharge, which enable the magma to retain its gases more easily (Barton & Huijsmans 1986).

The kind of activity of the past 500 years will probably continue in the Kaméni Islands. Earthquakes usually occur at intermediate depth about 15 years before activity resumes, but only one (1956) has occurred since the last eruption in 1950. However, on current trends new eruptions could take place within the next 30 years.

The depth of the recently erupted lavas on Néa Kaméni is clear against the scale of the yacht's mast.

Craters and domes formed on Néa Kaméni, with Phira on the horizon.

MÍLOS

Mílos is the southwesternmost island of the Cyclades, where it forms a small group with Kímolos, Polýegos and Antímilos. It covers an area of 160 km² and is composed of ochre plains and bare tawny hills, mantled with pumice and ash. Mílos and its neighbours are almost entirely composed of volcanic rocks, among which are those dated at about 3.5 million years as the oldest yet determined in the arc.

The activity occurred first in the southwest, and then in the northeast, before finally switching to the centre. Many eruptions sprang from fissures and created a volcanic field that never developed a central focus. Dacitic domes, lava flows and small nuées ardentes built up the western peninsula of Mílos and the islands of Kímolos and Polýegos. The eruptions that built the undulating northeastern arm of Mílos began about 1.8 million years ago when obsidian-rich rhyolite domes were expelled, as well as layers of ash and pumice. This phase ended about 900 000 years ago, when two dacitic domes extruded: Halepá dome forms a **malpaís** of rough lava, whereas the Plákes dome forms the plinth of the village of Pláka.

After a dormant interval, hydrovolcanic eruptions began some 380 000 years ago and ended with effusions of rhyolitic lavas that now compose the promontory jutting northwards from the Plákes dome. At about the same time, andesitic, dacitic and rhyolitic eruptions formed Antímilos island. In the south, hydrovolcanic eruptions began 140 000 years ago, followed by effusions of rhyolitic lavas, and they ended 90 000 years ago when more hydrovolcanic eruptions formed a tuff ring 1.5 km across. These eruptions joined the eastern and western islands together, leaving Mílos Bay between them. Less than 200 000 years ago, more hydrovolcanic eruptions formed a dozen small, often intersecting, maars on the eastern peninsula, which brought activity to a close. However, fumaroles still persist, such as those on the coast near Adámas.

KÓS

On Kós, first dacitic and then rhyolitic domes extruded in the southwestern Kéfalos peninsula between about 3.4 and 1.6 million years ago (Keller et al. 1990). Then an eruption about 500 000 years ago expelled vast quantities of pale yellow pumice and ash from a centre in the nearby Bay of Kamári. Another major eruption, about 160 000 years ago, ejected the Kós Plateau tuff, a blanket of rhyolitic pumice and ash, 30 m thick, that covers much of southern Kós (Dalabakis 1987, Keller et al. 1989, Smith et al. 1995). Its total volume probably reached 100 km³, and it spread across the sea as far as Kálymnos to the north and Tílos 40 km away to the south. Its source was a vent near the island of Yalí, and the eruption was accompanied by the collapse of a caldera, between 5 km and 10 km across, that now lies below sea level. Other eruptions occurred on the island of Yalí about 31 000 and 24 000 years ago.

Pumice with large lithic boulder on the Kos Plateau.

MÉTHANA

The isthmus linking the Méthana Peninsula to the Peloponnese is so low that it looks like an island from most directions. It forms a triangle, 10 km across, made up of an array of rugged domes and thick, viscous lava flows, rising to a hilly and inaccessible central area, where several peaks exceed 500 m. Méthana is younger than its Sarónic neighbours, the oldest rocks being only 900 000 years old (Georgalás 1962, Pé 1974).

The first three eruptive episodes all began with small effusions of basalts and basaltic andesites that were succeeded by larger and more violent eruptions of dacites and rhyodacites. However, the latest phase began with extrusions of dacitic domes that now form the thickest accumulations in the centre of the island. The latest eruption extruded the Kaméni Vounó (Burnt Hill) dome in northwestern Méthana. As it extruded, the western sector collapsed and a rugged brown latite–andesite flow, up to 50 m thick, stretched 1.5 km northwards and extended out to sea. A smaller flow, with a similar aa surface, turned southwestwards, where it now dominates the village of Kaméni Xorió. These eruptions were most probably those in about 250 BC that were noted by Strabo (I: 3–18). "In

Méthana a mountain, seven stadia [1300 m] in height, was raised up as a result of a fiery eruption. It was unapproachable in daytime because of the heat and smell of sulphur, and at night it shone for a great distance. It was so hot that the sea boiled for five stadia [925 m] and was turbid for as much as twenty stadia away [3.7 km]". The peninsula still has several active hydrothermal areas, especially those forming large hot springs and the baths at Loutrá Méthana in the southeast.

Lava flow issuing from the displaced dome near Kaméni Xorió, Méthana.

The eruptions on Nísyros in 1871 and 1873

The only well documented eruptions on Nísyros occurred in 1871 and 1873 (Gorceix 1873a,b, 1874). After an earth-quake had shaken the island, explosions like thunderclaps followed in late November 1871; red and yellow flames rose into the air, and rock fragments whistled over the highest peaks and rained down on the coast. Two craters opened at the foot of the Hághios Elías dome, spread white ash over the floor of the caldera, and covered the whole island in a curious mist. But calm returned and soon the vents were giving off no more than fumes.

On 2 or 3 June 1873, several earthquakes rocked Nísyros, a fissure opened on the floor of the caldera, and another crater exploded near those that had formed in 1871. Salty water gushed out first and then combined with the mud, ash and rock fragments that erupted during the next three days to form a mudflow that was 500 m long and 3 m thick. When the water evaporated, it left behind a salty crust, like frost, on the ground and trees. During the summer, tremors were felt, and hydrogen sulphide and steam emerged regularly from the vents. On 11 September, an earthquake opened a submarine fissure between the island of Yalí and the capital, Mandráki, where buildings were damaged and the sea turned milky. On 26 September, the new craters in the caldera once again threw out salty water, mud and rocks, and continued to gush out steam and hot water from time to time during the ensuing months. At one stage, even the prospect of abandoning the island was mooted, because several people had been injured and almost everyone had been forced to leave home and camp out in the open air. But the eruption stopped before the decisive steps had to be taken. There was a postscript to these eruptions in September 1887, when a small pit nearby briefly ejected mud, rocks and steam.

These hydrovolcanic and hydrothermal eruptions expelled buff fragments around steep craters, or explosion pits, the largest of which reaches 180 m across. These fragile features have already suffered much erosion, and six other older craters, probably formed several centuries ago, have been so degraded that they can now be identified only from air photographs. Stéfanos, the crater that had developed "a long time before", was certainly active in 1873, when a small vent also opened on its southwestern edge. Stéfanos ("the crown") is a circular hollow, 240 m across and 25 m deep, with nearly vertical sides. It is also the main source of the contemporary hydrothermal activity. It may have resulted from a hydro-volcanic explosion, but its wide, flat floor also suggests that it may have originated partly by piston-like collapse. The floor of Stéfanos has to be crossed with care, for it contains many fumaroles (at temperatures approaching 100°C), mudpots, and solfataras with fine sulphur aureoles. Other fumaroles rise from a fissure stretching from the caldera walls below Nikiá to the lower flanks of Hághios Elías dome, where they have stained the dacites yellow, and separate vents on the north coast have also produced the hot springs at the little spa of Páli.

NÍSYROS

Nísyros is the largest volcanic island in the Dodecanese. It constitutes the emerged parts of a stratovolcano, 8 km across and 42 km^2 in area, which rises to a height of 698 m. Its smooth trun-cated cone, often clothed with almond and olive groves, seems all the more luxuriant beside the bare rugged limestones of its neighbours. But the main eruptive features of Nísyros are hidden away in the 300 m-deep summit caldera. It is here that the fumaroles in the circular pit of Stéfanos provide the most evident source of contem-porary activity on the island, although reddish-grey domes have recently filled the western half of the caldera, and several yellowish cones have erupted at their base (Plate 5b).

Nísyros is a young stratovolcano, for some of the oldest exposed rocks erupted only 66 000 years ago. Its landforms are often clearly much younger: cliffs formed by marine erosion are rare; marine terraces are completely absent, soils are still thin, the domes have retained their bristling surfaces, weathering has scarcely affected many lava flows, and even the friable pumice layers have developed only shallow gullies. However, records of activity on Nísyros throughout histor-ical times are sparse and uncertain, and the volcano could possibly have been quiescent for much of that period (Gorceix 1873a,b, 1874, Di Paolo 1974, Scarth 1983, Keller et al. 1989, Stamatelópolou-Seymour & Vlassópoulos 1989, Limburg & Varekamp 1991, Marini et al. 1993).

The eruptions on Nísyros ranged from basic andesites to dacites, rhyodacites and rhyolites, and they occur both in thick lava flows and as layers of ash and pumice. Many came from a res-ervoir less than 8 km deep, although some of the more recent activity, which formed rhyodacitic domes within the caldera, probably originated in a small reservoir that was less than 2 km deep.

Submarine basaltic andesite eruptions piled up fragments and pillow lavas from a depth of 1000 m on the Aegean sea floor. As the volcano emerged, mainly effusive eruptions of similar composition formed much of the emergent stratovolcano, where one of the oldest flows has been dated to 66 000 years old. Subsequently,

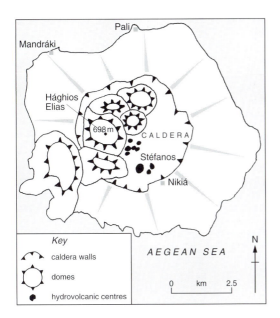

The main volcanic features of Nísyros.

and almost completely filled its western half. The domes of Nífios and Hághios Geórgios are unusually large, and Hághios Elías, which is almost 600 m high, is probably one of the largest domes in Europe. Their rocky surfaces, still bristling with their original pinnacles, suggest that they were formed only recently. At about the same time, dacitic domes also extruded on the southwestern flanks of the stratovolcano. The two southern domes distended, flowed down slope and solidified as two rocky promontories projecting out to sea. Neither weathering nor marine erosion has yet been able to alter them significantly, and they seem to be no more than a few thousand years old (Scarth 1983).

however, more andesitic or dacitic materials erupted and explosions of fragments became more common. For instance, they form a blanket about 250 m thick in both northeastern and southern Nísyros. Plinian outbursts expelled dacitic and rhyodacitic "lower pumices" that may have been linked to an early caldera collapse (Stamatelópolou-Seymour & Lalonde 1991). At about the same time, Mount Hághios Ioánnis, near Nikiá, expelled thick viscous rhyodacitic lava flows that were also voluminous enough to reach the southeast coast. Slightly later, several Plinian eruptions expelled about 10 km³ of ash and pumice that now form a yellow blanket, 100 m thick, on the northern slopes.

These eruptions probably led to the collapse of the caldera less than 24 000 years ago – which is the age of the latest dated rocks of the stratovolcano proper. But the caldera must have been in existence about 5000 years ago, because Neolithic artefacts have been found on its floor. The original caldera was circular and 3 km in diameter; its floor stood only 100 m above sea level; and it was enclosed by walls ranging from 150 m to 400 m high. When the eruptions resumed, the chemical nature of the magmas hardly changed, but fewer fragments were expelled. Instead, several domes of reddish-grey dacites extruded from fissures extending from inside the caldera southwestwards onto the flanks of the stratovolcano. Five domes extruded within the caldera

BIBLIOGRAPHY

Allen, S. R. & R. A. F. Cas 1998. Lateral variations within coarse co-ignimbrite lithic breccias of the Kos Plateau Tuff, Greece. *Bulletin of Volcanology* **59**, 356–77.

Baillie, M. G. L. 1990. Irish tree rings and an event in 1628 BC. See Hardy (1990: 160–66).
—— 1998. Bronze Age myths expose archaeological shortcomings? A reply to Buckland et al. 1997. *Antiquity* **72**, 425–7.
Baillie, M. G. L. & M. A. R. Munro 1988. Irish tree rings, Santoríni and volcanic dust veils. *Nature* **332**, 344–6.
Barton, M. & J. P. P. Huijsmans 1986. Post-caldera dacites from the Santoríni volcanic complex, Aegean Sea, Greece: an example of the eruption of lavas of near-constant composition over a 2200 year period. *Contributions to Mineralogy and Petrology* **94**, 472–95.
Betancourt, P. P. 1987. Dating the Aegean Late Bronze Age with radiocarbon. *Archaeometry* **29**, 45–9.
Bietak, M. 1996. *Avaris: the capital of the Hiksos*. London: British Museum Press.
Buckland, P. C., A. J. Dugmore, K. J. Edwards 1997. Bronze Age Myths? Volcanic activity and human response in the Mediterranean and North Atlantic regions. *Antiquity* **273**, 581–93.

Cadogan, G. 1988. Dating of the Santoríni eruption. *Nature* **332**, 401–402.

Dalabakis, P. 1987. *Le volcanisme récent de l'Ile de Kos (Grèce)*. Thèse d'état, Université de Paris-Sud, Orsay, France.
Di Paolo, G. M. 1974. Volcanology and petrology of Nísyros Island (Dodecanese, Greece). *Bulletin Volcanologique* **38**(4), 944–87.
Doúmas, C. 1974. The Minoan eruption of the Santoríni volcano. *Antiquity* **48**, 110–15.
—— (ed.) 1978. *Thera and the Aegean world* I [proceedings of the First International Scientific Congress on the Volcano Thera, Santoríni, Greece]. London: Thera Foundation.
—— (ed.) 1980. *Thera and the Aegean world* II [proceedings of the Second International Scientific Congress on the Volcano Thera, Santoríni, Greece]. London: Thera Foundation.
—— 1983. *Thera: Pompeii of the ancient Aegean*. London: Thames & Hudson.
Doúmas, C. & L. Papazoglou 1980. Santoríni tephra from Rhodes. *Nature* **287**, 322–4.
Druitt, T. H., R. M. Mellors, D. M. Pyle, R. S. J. Sparks 1989. Explosive volcanism on Santoríni, Greece. *Geological Magazine* **126**, 95–126.
Druitt, T. H. & V. Francaviglia 1990. An ancient cliff line at Phira and its significance for the topography and geology of Pre-Minoan Santoríni. See Hardy (1990: 362–9).

—— 1992. Caldera formation on Santoríni and the physiography of the islands in the late Bronze Age. *Bulletin of Volcanology* **54**, 484–93.
Druitt, T. H. and 7 co-authors 1999. *Santoríni volcano*. Memoir 19, Geological Society, London.

Evans, A. J. 1921–36. *The palace of Minos at Knossos* [4 vols]. London: Macmillan.

Fouqué, F. 1879. *Santorin et ses éruptions*. Paris: Masson.
Francalanci, L., J. C. Varekamp, G. Vougioukalakis, M. J. Defant, F. Innocenti, P. Manetti 1995. Crystal retention, fractionation and crustal assimilation in a convecting magma chamber, Nísyros volcano, Greece. *Bulletin of Volcanology* **56**, 601–620.
Furumark, A. 1941. *The chronology of Mycenaean pottery*. Stockholm: Kunglig Vitterhets Historie och Antikvitets Akademien.

Georgalás, G. C. 1962. *Catalogue of the active volcanoes of the world including solfatara fields*, part XIII: *Greece*. Rome: International Association of Volcanology.
Gorceix, H. 1873a. Sur l'état du volcan de Nísyros au mois de mars 1873. *Académie des Sciences à Paris, Comptes Rendus* **77**, 597–601.
—— 1873b. Sur la récente éruption de Nísyros. *Académie des Sciences à Paris, Comptes Rendus* **77**, 1039.
—— 1874. Phénomènes volcaniques de Nísyros. *Académie des Sciences à Paris, Comptes Rendus* **78**, 444–6.
Goree, Father 1711. A relation of a new island which was raised from the bottom of the sea on the 23rd of May 1707 in the Bay of Santoríni, in the archipelago. Written by Father Goree (a Jesuit) an eye witness. *Royal Society of London, Philosophical Transactions* **2**(332), 353–74.
Guichard, F., S. Carey, M. A. Arthur, H. S. Sigurdsson, M. Arnold 1993. Tephra from the Minoan eruption of Santoríni in sediments in the Black Sea. *Nature* **363**, 610–12.

Hammer, C. U., H. B. Calusen, W. Dansgaard 1980. Greenland ice-sheet evidence of postglacial volcanism and its climatic impact. *Nature* **288**, 230–35.
Hammer, C. U., H. B. Clausen, W. L. Friedrich, H. Tauber 1987. The Minoan eruption of Santoríni in Greece dated to 1645 BC? *Nature* **332**, 517–19.
Hardy, D. A. (ed.) 1990. *Thera and the Aegean world* III [3 vols]. London: Thera Foundation.
Heiken, G. & F. McCoy 1984. Caldera development during the Minoan eruption, Thira, Cyclades, Greece. *Journal of Geophysical Research* **89**, 8441–62.
—— 1990. Precursor activity to the Minoan eruption, Thera, Greece. See Hardy (1990: 79–87).
Heiken, G., F. McCoy, M. Sheridan 1990. Palaeotopographic and palaeogeologic reconstruction of

Minoan Thera. See Hardy (1990: 370–76).

Huijsmans, J. P. P., M. Barton, V. J. M. Salters 1988. Geochemistry and evolution of the calc-alkaline volcanic complex of Santoríni, Aegean Sea, Greece. *Journal of Volcanology and Geothermal Research* **34**, 283–306.

Huijsmans, J. P. P. & M. Barton 1990. New stratigraphic and geochemical data for the Megalo Vouno complex: a dominating volcanic landform in Minoan Times. See Hardy (1990: 433–41).

Hutchinson, R. W. 1962. *Prehistoric Crete*. Harmondsworth, England: Penguin.

Kalagerópoulos, S. & S. Parítsis 1990. Geological and geochemical evolution of the Santoríni volcano. See Hardy (1990: 164–71).

Keller, J., T. Rehren, E. Stadlbauer 1990. Explosive volcanism in the Hellenic arc: a summary and review. See Hardy (1990: 13–26).

Kitchen, K. A. 1996. The historical chronology of ancient Egypt, a current assessment. *Acta Archaeologica* **67**, 1–13.

Kuniholm, P. I., B. Kromer, S. W. Manning, M. Newton, C. E. Latini, M. J. Bruce 1996. Anatolian tree rings and the absolute chronology of the eastern Mediterranean, 2220–718 BC. *Nature* **381**, 780–3.

La Marche, V. C. & K. K. Hirschboeck 1984. Frost rings in trees as records of major volcanic eruptions. *Nature* **307**, 121–6.

Limburg, E. & J. C. Varekamp 1991. Young pumice deposits on Nísyros, Greece. *Bulletin of Volcanology* **54**, 68–77.

Luce, J. V. 1969. *The end of Atlantis*. London: Thames & Hudson.

Manning, S. W. 1988. Dating of the Santoríni eruption. *Nature* **332**, 401.

—— 1989. A new age for Minoan Crete. *New Scientist* (11 February), 60–63.

—— 1996. Dating the Bronze Age: without, with, and beyond, radio-carbon. *Acta Archaeologica* **67**, 15–37.

—— 1998. Correction. New GISP2 ice-core evidence supports 17th century BC date for the Santoríni (Minoan) eruption: response to Zielinski & Germani (1998). *Journal of Archaeological Science* **25**, 1039–1042.

Marinátos, S. 1939. The volcanic destruction of Minoan Crete. *Antiquity* **13**, 425–39.

Marini, L., C. Principe, G. Chiodini, R. Chioni, M. Fytikas, G. Marinelli 1993. Hydrothermal eruptions of Nisiros (Dodecanese, Greece). Past events and present hazard. *Journal of Volcanology and Geothermal Research* **56**, 71–94.

McClelland, E. A. & T. H. Druitt 1989. Paleomagnetic estimates of emplacement temperatures of pyroclastic deposits on Santoríni, Greece. *Bulletin of Volcanology* **51**, 16–27.

McKenzie, D. P. 1972. Active tectonics of the Mediterranean region. *Royal Astronomical Society, Geophysical Journal* **55**, 217–54.

Ninkovich, D. & B. C. Heezen 1965. *Santoríni tephra*. Colston Research Society Paper 17, University of Bristol, Bristol.

Page, D. 1970. *The Santoríni volcano and the destruction of Minoan Crete*. Paper 12, Society for the Promotion of Hellenic Studies, London.

Papadópoulos, G. A. 1990. Deterministic and stochastic models of the seismic and volcanic events in the Santoríni volcano. See Hardy (1990: 151–9).

Pé, G. G. 1974. Volcanic rocks of Methana, South Aegean arc, Greece. *Bulletin Volcanologique* **38**, 270–90.

Pernier, L. & L. Banti 1951. *Il Palazzo di Festos II*. Rome: Libreria del Stato.

Pichler, H. & W. L. Friedrich 1980. Mechanism of the Minoan eruption of Santoríni. See Doumas (1980: 15–29).

Pichler, H. & S. Kussmaul 1980. Comments on the geological map of the Santoríni islands. See Doúmas (1980: 413–26).

Pyle, D. M. 1989. Ice-core acidity peaks, retarded tree growth and putative eruptions. *Archaeometry* **31**, 88–91.

—— 1990. New estimates for the volume of the Minoan eruption. See Hardy (1990: 113–21).

Reck, H. (ed.) 1936. *Santorin: der Werdegang eines Inselvulkans und sein Ausbruch 1925–1928* [3 vols]. Berlin: Reimer.

Rehak, P. & J. G. Younger 1998. Review of Aegean Prehistory VII: Neopalatial, Final Palatial, and Postpalatial Crete. *American Journal of Archaeology* **102**, 91–173.

Scarth, A. 1983. Nísyros volcano. *Geography* **68**(2), 133–9.

—— 1994. *Volcanoes*. London: UCL Press.

Sigurdsson, H. S., S. Carey, J. D. Devine 1990. Assessment of mass, dynamics and environmental effects of the Minoan eruption of Santoríni volcano. See Hardy (1990: 100–112).

Smith, P. E., D. York, Y. Chen, N. M. Evensen 1995. The timing of an ancient Greek paroxysm on the island of Kos: towards a more precise calibration of the Mediterranean deep-sea record. *American Geophysical Union meeting, December 1995*. Abstracts, F 712.

Sparks, R. S. J., H. S. Sigurdsson, N. D. Watkins 1978. The Thera eruption and Late Minoan Ib destruction on Crete. *Nature* **271**, 91.

Sparks, R. S. J. & C. J. N. Wilson 1990. The Minoan deposits: a review of their characteristics and interpretation. See Hardy (1990: 89–99).

Stamatelópoulou-Seymour, K. & D. Vlassópoulos 1989. The potential for future explosive volcanism associated with dome growth at Nísyros, Aegean volcanic arc, Greece. *Journal of Volcanology and Geothermal Research* **37**, 351–64.

Stamatelópoulou-Seymour, K. & A. E. Lalonde 1991. Monitoring oxygen fugacity conditions in pre-, syn- and post-caldera magma chamber of Nísyros volcano, Aegean island arc, Greece. *Journal of Volcanology and Geothermal Research* **46**, 231–40.

Stothers, R. B. & M. R. Rampino 1983. Volcanic eruptions in the Mediterranean before AD 630 from written and archaeological sources. *Journal of Geophysical Research* **88**, 6357–71.

Sullivan, D. E. 1988. The discovery of Santoríni Minoan tephra in western Turkey. *Nature* **333**, 552–4.

Ventris, M. & J. Chadwick 1973. *Documents in Mycenean Greek*, 2nd edn. Cambridge: Cambridge University Press.

Warren, P. M. 1987. Absolute dating of the Aegean Late Bronze Age. *Archaeometry* **29**, 205–211.

—— 1988. The Thera eruption, III: further arguments against an early date. *Archaeometry* **30**, 176–9.

Zielinski, G. A. and 8 co-authors 1994. Record of volcanism since 7000 BC from the GISP2 Greenland ice core and implications for the volcano–climate system. *Science* **264**, 948–52.

Zielinski, G. A. & M. S. Germani 1998a. New ice-core evidence challenges the 1620s BC age for the Santoríni (Minoan) eruption. *Journal of Archaeological Science* **25,** 279–89.

—— 1998b. Reply to: Correction. New GISP2 ice-core evidence supports 17th-century BC date for the Santoríni (Minoan) eruption. *Journal of Archaeological Science* **25** 1043–1045.

PART 3 THE ATLANTIC

4 Spain: Canary Islands

The Canary Islands form Spanish provinces, situated off the Atlantic coast of Africa, between 1200 km and 1750 km southwest of Cádiz. The archipelago consists of seven main islands. Tenerife, covering 2058 km^2, is the largest, and El Hierro, only 275 km^2 in area, is the smallest, and Fuerteventura lies only 115 km from Africa. All the islands belong to the African plate. They fall into two groups. Fuerteventura and Lanzarote form the eastern group and belong to the same volcanotectonic unit that trends parallel to the African coast. Here the climate is arid, the vegetation is steppe like, and fissures have dominated the distribution of eruptions to such an extent that neither island has a marked scenic focus. In contrast, the central and western group of islands – El Hierro, La Palma, La Gomera, Tenerife, and Gran Canaria – are more mountainous. Thus, even little El Hierro reaches 1051 m and Tenerife rises to a majestic climax of 3715 m in the Pico de Teide. They lie in the path of the northeast trade winds, which bring cloud, humidity and a luxuriant vegetation to their northern windward shores, although their leeward southern slopes are often arid. Their lower parts enjoy an equable and mild climate throughout the year, and several have

become some of Europe's major tourist attractions. Thus, in any season, the volcanologist is unlikely to be alone on the summit of Teide and may even encounter tourist-laden camels in Lanzarote. On the other hand, few areas beyond the shadow of Vesuvius have done more for popular appreciation of the impact of volcanism on mankind and the environment.

The Canary Islands contain a great variety of volcanic forms that range from plateau basalts, which are the remains of large basal shields, to recent cinder cones and rugged lava flows that have often joined into malpaís; and their strato-volcanoes have sometimes been decapitated by large calderas. The islands have been considered as the type-locality of the caldera ever since Von Buch published his controversial work on their landforms in 1825. Indeed, the Caldera de las Cañadas in Tenerife is, with Santoríni, the most spectacular in all Europe. Fortunately, this well populated archipelago has not undergone a caldera-forming eruption in historical times.

The Canary Islands used to be inhabited by the Guancho peoples, who passed on some references to eruptions before they were exterminated by Spanish settlers. Lanzarote, Fuerteventura, La Gomera and El Hierro were settled from 1402, Gran Canaria from 1483, Tenerife from 1491, and La Palma finally from 1493. Thus, historical times in the archipelago amount to less than 600 years, during which activity has been dominated by basaltic eruptions along fissures, which occurred in Lanzarote, La Palma and Tenerife. In addition, parts of Fuerteventura, Gran Canaria and El Hierro all have cones and flows of such remarkable freshness that they scarcely seem to be more than a thousand years old.

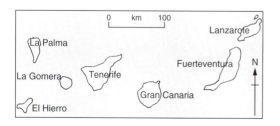

The Canary Islands.

Indeed, all the islands except La Gomera have had activity within the past 10 000 years. The recorded eruptions in the Canary Islands have occurred at average intervals of 30–35 years. The latest took place in La Palma in 1971, and by far the most prolonged and extensive historical eruption took place in Lanzarote from 1730 to 1736.

The volcanic activity in the archipelago is perhaps the most difficult to explain in all Europe. The islands lie on the passive continental margin of northwestern Africa, on one of the oldest parts of the Atlantic Ocean floor, where the basaltic oceanic crust ranges from 180 to 155 million years old from east to west (Araña & Carrecedo 1979). A basal complex outcrops on Fuerteventura, La Gomera and La Palma, and probably lies hidden beneath the remaining islands. It contains some sediments, some plutonic intrusions and some submarine basalts, but chiefly assemblies of dykes. However, most of the Canary Islands are much younger, and many parts of the surface are probably no more than a few thousand years old. The islands were born separately from different sources of magma and did not all develop in the same way. Thus, for instance, the oldest dated volcanic rocks occur in Fuerteventura, where they are about 20.6 million years old, but they are no more than 2.0 million years old in La Palma and about 1.12 million years old in El Hierro. The rocks themselves range from basalts to hawaiites, mugearites, phonolites and rhyolites. Basalts, chiefly emitted from fissures, account for the longest and most prolific eruptive episodes; and the more evolved rocks developed largely beneath the stratovolcanoes, probably in relatively shallow magma reservoirs (Valentin et al. 1990).

Tectonic movements might have played a role in the growth of the Canary Islands by opening up the oceanward prolongation of the South Atlas fault of North Africa (Anguita & Hernán 1975, Araña & Carracedo 1978). Nevertheless, the archipelago probably owes most of its growth to a hotspot, which has given rise to a broad westward development of activity in the islands (Carracedo 1994). It seems probable that masses from the rising magma have formed individual basal shields, and, at times, domed up their surface until major three-arm rift systems separated by angles of 120° have developed. The fissure systems on these rifts then enabled yet more magma to reach the surface; and they have, for example, become the zones of the most marked concentrations of recent emission centres in the islands. Thus, the rifts have been built up into large high ridges, which reach spectacular proportions in Tenerife, where they form a Y-shape pattern centred on Teide. These rift ridges grew up so quickly that they sometimes became unstable. Consequently, in El Hierro, La Palma and Tenerife, parts of these ridges have collapsed into the Atlantic Ocean in major landslides. Most of the eruptions in the Canary Islands during historical times have occurred on the fissures developed along these ridges. During these eruptions, fragments explode and form cones on the upper parts of the fissures, while fluid basalts, and perhaps spatter, emerge lower down. However, they have produced only tiny volumes of broadly alkaline basalts, and formed but small cones and thin lava flows.

TENERIFE

Tenerife is the largest of the Canary Islands, covering an area of 2058 km², and stretching 84 km from northeast to southwest and 50 km from north to south. Tenerife is the central island, with the greatest volume of erupted materials in the archipelago: more than 15 000 km³, of which some 2000 km³ lie above sea level. It rises from a depth of more than 3000 m on the Atlantic Ocean floor, where eruptions began perhaps as much as 11.6 million years ago. The island is Y-shape, with three main ridges, extending northwestwards, northeastwards and southwards, which have been formed primarily by basaltic fissure eruptions along rifts. The crowning glory of Tenerife, the Pico de Teide, rises where these ridges intersect and soars from the spectacular Caldera de las Cañadas to a height of 3715 m, the highest peak in Spain and in the whole of the Atlantic Ocean, and the highest volcano in Europe outside the Caucasus. It dwarfs its companion, the Pico Viejo, which, at 3134 m, is itself almost as high as Etna. But they rise so high because they formed on the floor of the Caldera de las Cañadas, which itself stands 2100 m above sea level. Teide has been an unmistakable landmark for six centuries of mariners, ever since the northeast trade winds led the first sailing ships

to the Canary Islands. The trade winds also brought their equable climate and orographic rainfall to the northern slopes of Tenerife, which gave rise to rich vegetation and agriculture, in contrast with the aridity of the rain shadow in the south, which is fit only for tourist resorts. Teide itself rises well above the trade winds and enjoys a dry regime of clear skies and westerly breezes.

Teide and Pico Viejo grew up less than 200 000 years ago inside the Caldera de las Cañadas, which has been famous ever since it was described by Humboldt and Von Buch in the early nineteenth century. It formed at the crest of a huge volcanic pile, now called the Cañadas volcano, which had, itself, grown up on top of basal shields that basaltic eruptions had built up from the floor of the Atlantic Ocean. Teide is crowned by a small cone, El Pitón, with a white crater emitting fumes. It is unlikely that the Pico de Teide itself has seen any activity since Columbus, en route for the Americas, witnessed an eruption in August 1492. However, several eruptions have occurred in Tenerife since the Spanish settlement, and the latest formed the cinder cone of Chinyero in 1909. These are but the most recent of a whole series of eruptions that have formed dozens of cones and lava flows in the past few thousand years on the long high ridges that form the Y-shape backbone of the island.

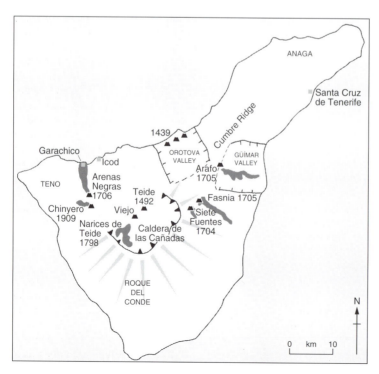

Historical eruptions of Tenerife, with the Orotova and Güímar valleys, the Caldera de las Cañadas, and the remains of the three basal shields forming the extremities of the island.

As in the rest of the Canary Islands, the lavas of Tenerife are predominantly alkaline and basaltic, and they emerged chiefly from fissures in fluid flows. But occasional evolution in shallow reservoirs produced eruptions of trachybasalts (hawaiites and mugearites), trachytes and phonolites. The trachytic and phonolitic eruptions expelled extensive ignimbrites and pumice or ash layers, with a total volume of some $70 km^3$ (Wolff 1985), but also produced lava flows such as those decorating the southern slopes of Teide, as well as some domes in the Caldera de las Cañadas. Several recent lava flows of both phonolite and basalt in the caldera also have conspicuous blocky obsidian surfaces. Teide itself has a very shallow magma chamber and its eruptions have been basaltic, trachytic and phonolitic (Araña et al. 1989, Albert-Beltran et al. 1990, Valentin et al. 1990).

The basal shields

Eruptions concentrated at first in three separate zones of the ocean floor, where basaltic flows issuing from fissures had built three island shields by about 3.28 million years ago. They now form massifs on the three extremities of Tenerife: the Anaga peninsula in the northeast, the Teno peninsula in the northwest, and the Roque del Conde in the south, near Adeje (Ridley 1970a,b, Carracedo 1994). The basal complex probably occurs below these basalts, because fragments have been thrown up as xenoliths during historical eruptions.

The Anaga Massif is an accumulation of some 1000 m of tabular flows of **alkali basalt** and mugearite, interspersed by cinder or ash layers and transected by innumerable dykes, and even occasional domes. It was formed 6.5–3.28 million years ago by three similar cycles of activity separated by intervals of repose. The Teno Massif is also composed of tabular basaltic flows, with some explosive breccias, mudflows and cinders in the lower levels and some trachytes in the upper. The eruptions began about 6.7 million ago and ended when the Roque Blanco phonolitic dome erupted about 4.5 million years ago. Near Adeje, in southwestern Tenerife, about 1000 m of basalts are exposed in and around the Roque del Conde. The youngest eruptions formed the trachytic dome of the Roque Vento about 3.8 million years ago.

A largely dormant period followed 3.28–1.9 million years ago (Carracedo 1979). **Barrancos** cut deep into the shields, and the basaltic plateaux were reduced to ridges, knife-edge cuchillos, or pinnacles, such as the Roque Imoque near Adeje. At the same time, marine erosion trimmed the faulted seaward flanks of the shields.

The Cañadas volcano

When activity resumed on Tenerife, it changed in both location and eruptive style. The new concentrated vents built two successive stratovolcanoes where the regional rift fissures of Tenerife intersect. The first stratovolcano, now called Cañadas I, erupted for nearly a million years. It grew up from vents on the west of the central area and its deposits now lie in western Tenerife. Its initial basaltic eruptions, 1.89–1.82 million years ago, were followed about 1.5 million years ago by more basalts, mugearites, trachytes and phonolites. Explosions of nuées ardentes and ignimbrites about 1.24–1.05 million years ago were probably related to caldera collapse, which brought the main eruptions of Cañadas I to an end (Ancochea et al. 1990, Martí et al. 1994).

Eruptions from close-knit vents to the northeast immediately started to build Cañadas II stratovolcano. Its deposits now lie in eastern Tenerife and are well exposed, for instance, in the eastern wall of the Caldera de las Cañadas. They form many layers of ash and lava flows, ranging from basalts to trachytes and phonolites, which are about 650 000 years old. However, effusive phases were interspersed with both explosive episodes and dormant periods when old soils formed. About 600 000 years ago, Cañadas II underwent a phase of collapse revealed by many layers of fragments exposed in the Cañada de Diego Hernández section of the caldera rim. Nevertheless, the eruptions of Cañadas II continued unabated until they had built up a large and rather unstable stratovolcano, which almost certainly rose well above 3000 m, for its highest remaining flank still culminates at 2717 m at the Montaña Guajara.

The Cañadas volcanoes together represented a considerable eruptive output. A million years of eruptions probably gave the Cañadas I stratovolcano a volume of 350–400 km^3. The Cañadas II stratovolcano probably reached 150 km^3 to 200 km^3 in volume during the ensuing 800 000 years (Ancochea et al. 1990). In comparison, however, the Teide–Pico Viejo volcanoes have

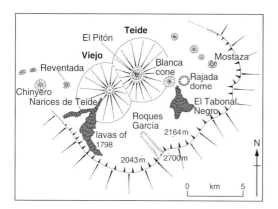

Caldera de las Cañadas, with some of the sites of the eruptions initiating its volcanic infilling.

erupted twice as fast, for together they are already approaching a volume of 150 km^3 after less than 170 000 years of activity.

The Caldera de las Cañadas

The formation of the Caldera de las Cañadas was the last and most spectacular event in the evolution of the Cañadas II stratovolcano. This majestic caldera is elliptical, double and asymmetrical, and its main axis trends some 17 km from northeast to southwest. The Caldera de las Cañadas harbours some of the most striking volcanic landforms in Europe, where the crystal-clear atmosphere, the brilliant sunshine and the absence of any continuous vegetation combine to emphasize the varied colours of the rocks, displayed in almost pristine splendour like a painted desert. Here steel-blue, brown, or black and glassy lava flows, and grey or yellow pumice piles and dark red cinder cones decorate the base of the grey Pico de Teide and its black companion, the Pico Viejo, which produced the latest eruption within the caldera in 1798. And, in the south, the rim of the Caldera de las Cañadas protects them all with a cliff that reaches 700 m high and curves for more than 20 km from El Portillo in the northeast to beyond the Boca de Tauce in the southwest. The Roques de García jut out from this cliff and divide the caldera itself into a larger eastern hollow and a smaller, lower, western hollow. But there is no sign of the northern rim of the caldera, and opinions differ about what might have happened to it.

The Caldera de las Cañadas is a classic among European calderas, and the many views about its formation illustrate the changing evaluations of the tectonic, erosional, explosive, collapse and mass-movement processes that have been proposed to explain calderas during the past 200 years. It was first considered a crater of elevation (Von Buch 1825); then an erosional amphitheatre like its supposed counterpart, the Caldera de Taburiente in La Palma (Lyell 1855); and, even later, as the headwaters of two river systems, whose exit northwards had been later covered by Teide and Pico Viejo (Fritsch & Reiss 1868). Later still, it was thought to have resulted from a massive explosion, like that which was then believed to have destroyed Krakatau in 1883 (Gagel 1910). But it became accepted that it had formed when the summit of the volcano had collapsed after massive eruptions of magma, either in one single or several repeated catastrophes (Williams 1941, Hausen 1956, Fúster et al. 1968b, Ridley 1970a, 1971a, Borley 1974, Wolff 1985, Valentine et al. 1990, Martí et al. 1994). The intricate network of galleries built to extract underground water shows that the northern rim of the caldera is entirely missing. Thus, if collapse formed the northern rim, landslides might have removed it later (Martí et al. 1994). In a simpler view, the Caldera de las Cañadas has been seen as entirely the result of a gigantic landslide, whose uppermost arcuate scar forms the majestic southern rim of the caldera (Navarro & Coello 1989, Carracedo 1996, Carracedo et al. 1999a). Thus, the upper 100 km^3 of the unstable Cañadas II stratovolcano apparently suddenly slipped northwestwards into the Atlantic Ocean, where it seems to be related to vast debris-avalanche deposits, on the submarine flank of northern Tenerife. A seaward-sloping layer of chaotic breccia, 100 m thick, could represent the sliding plane. Although the landslide was not caused by direct volcanic action, it did, however, unleash nuées ardentes. Their deposits still crown the summit of the southeastern rim of the caldera at the Cañada de Diego Hernández, and have been dated to 130 000–170 000 years ago (Martí et al. 1989, Ancochea et al. 1990). They indicate the oldest possible date for the birth of the Caldera de las Cañadas.

The Roques de García, separating the two parts of the caldera, form a ridge that has been dissected into a number of individual pinnacles. The most famous of these is the much photographed Roque Cinchado ("tightened-belt rock"), which is also known as the Arból de Piedra ("tree of stone", shown in Plate 7a). The ridge is about

The 500 m-high southern wall of the Caldera de las Cañadas, with the lower reaches of the lavas erupted from the Narices de Teide in 1798 on the extreme right.

200 m high and 2.5 km long, and is composed of an array of brightly coloured, altered volcanic rocks, notably at Los Azulejos. They form an intrusive complex that is apparently older than its surroundings. The Roques de García jut out from a distinct jog in the southern rim of the caldera and may represent some of the original vents of the Cañadas stratovolcanoes, which have survived in the landscape because they were plugged with agglomerates that offered greater mechanical resistance as the caldera developed.

Soon after the Caldera de las Cañadas formed, the Pico Viejo and the Pico de Teide grew up as a pair of large stratovolcanoes in their own right, along a major fissure trending from northeast to southwest from the Cumbre Ridge to the Sandiago area. They erupted from separate vents that issued from a common shallow reservoir, where the three rift-fissure zones of Tenerife intersect. The magma evolved towards a trachytic and phonolitic composition, but explosive eruptions of ash and pumice were punctuated by frequent emissions of viscous lava flows that form the armour plating on both volcanoes. Pico Viejo and Teide were active during the same period and their products are often interbedded, but they are dissimilar twins, and their different eruptive styles created very different crests. Pico Viejo has a wide crater, whereas the crater of Teide, La Rambleta, has been completely filled by the summit cone of El Pitón. Both volcanoes have remained active during the past millennium.

Pico Viejo

Pico Viejo is broader, lower and more gently sloping than Teide, but still forms a considerable volcano, rising 1034 m above the floor of the caldera to a height of 3134 m. It erupted many aa lava flows, but many recent flows, like those emitted in 1798, have pahoehoe surfaces. However, some of the youngest phonolitic flows were so viscous that they solidified in thick elongated masses within 300 m of their parent vents, even on slopes of 20°.

Pico Viejo is crowned by a circular crater 750 m across and 150 m deep, which is rimmed by a ragged scarp. It could have been formed by a large hydrovolcanic explosion, but it seems more likely to be a small caldera formed by recent collapse along ring fractures, because few exploded fragments occur on its flanks. The latest events on the summit occurred when hydrovolcanic explosions blasted out two pits,

Pico Viejo, clothed by recent lava flows and crowned by its jagged-rim crater. The southwestern rim of the Caldera de las Cañadas can be discerned on the horizon on the left.

one 140 m and the other 75 m deep, into the floor of the crater.

The flanks of Pico Viejo have also witnessed eruptions recently. The oldest occurred on a fissure radiating northwestwards from the stratovolcano and formed the phonolitic dome of the Roques Blancos, which sagged down on its lower, northwestern side and formed a stubby lava flow. The fissure later extended farther down slope and expelled a black obsidian flow that eventually reached 3 km in length. The second eruption formed Los Gemelos ("the twins") along the saddle linking Pico Viejo to Teide. The twin craters were produced by a brief hydro-volcanic explosion that expelled pumice, ash and breadcrust bombs. Their sharply defined outlines indicate their youth. Most recently, in 1798, the only eruption recorded within the Caldera de las Cañadas took place 1 km farther down the western flanks of Pico Viejo and formed the Chahorra vents, or the Narices ("nostrils") de Teide.

Pico de Teide

The Pico de Teide is a majestic volcano in a splendid setting (Plate 7a). It is a large symmetrical cone, streaked with dark grey lava flows spilling like paint from its summit. Eruptions of hawaiite and especially of phonolite have built Teide within the past 170 000–130 000 years. Its

lower northern flanks are clothed with distinctive reddish-brown or bluish-grey aa and blocky flows of fairly viscous phonolites, which form tongues, about 12 m to 15 m thick, with lobate ridges, lava moraines and steep rubble-covered margins. On the steepest slopes in the east, lava masses have sometimes detached themselves from the snouts of the flows and have rolled forwards, like snowballs, as accretionary balls. In contrast, the flows reaching the floor of the Caldera de las Cañadas sometimes spread out in wide lobes with crescent-shape pressure ridges that run parallel to their margins. Most of these flows have suffered such limited atmospheric alteration that they cannot be many thousands of years old.

Like Pico Viejo, Teide probably had an open summit crater for a long time, the rim of which can still be discerned as the shoulder at La Rambleta. The cone of El Pitón now fills this crater and crowns Teide like a giant sandcastle. It is itself a small stratovolcano, 150 m high, which encloses a crater, La Caldereta, that is 70 m in diameter and 40 m deep. Apart from the contemporary fumaroles that have changed its crater walls to yellow and brilliant white, El Pitón has been as quiet as its parent, Teide, for the past 500 years. The latest magmatic eruption gave off fresh black, glassy, phonolitic aa lavas that spilt from El Pitón and flowed in ridged tongues down the flanks of Teide as far as the Caldera de las

Cañadas. This emission was then followed by the throat-clearing explosion that gave El Pitón its present funnel-shape crater.

Seven satellite vents surround Teide. Five of these vents lie on a circle, 3 km from Teide, which probably represents a ring fracture in the Caldera de las Cañadas. The eruptions probably occurred where this ring fracture intersected fissures radiating from Teide.

The Pico Cabras, the Montañas de las Lajes and the Montaña Abejera, in the northeastern sector of Teide, have much in common. Each eruption began with the extrusion of a brown phonolitic dome, which distended and breached on its northward downslope side, and delivered a more fluid phonolitic flow extending towards the northern shore of Tenerife. In contrast, the Montaña Majua and the Montaña Mareta, which erupted on the gentle slopes beyond the southern base of Teide, produced only lobes of unusually fluid black phonolitic lava that are 10 m thick and over 2 km long and wide. The Montaña Majua lavas eventually accumulated thickly above the vent, but the Montaña Mareta remained as only a low mound. It is very similar to El Tabonal Negro ("the black table"), which

The Chahorra or Narices de Teide lavas, which erupted in 1798 on the flanks of Pico Viejo.

issued from a vent low on the southeastern flanks of Teide and emitted thick blocky black phonolitic flows, decorated by arcuate ridges, that spread over 3 km across the caldera floor.

The two remaining flank eruptions of Teide took place about the same time. They formed the Montaña Blanca and the Montaña Rajada, which both had more complicated histories that are reflected in their more varied relief. The first eruptions of Montaña Blanca came from five

The summit of Montaña Rajada, with the pumice cone of the Montaña Blanca behind, with a lava flow issuing from it on the left, and Teide on the horizon.

vents along a fissure radiating from Teide and gave off flows of brown phonolite (Ablay et al. 1995). They were followed by a sub-Plinian explosion of frothy phonolites, which was the most violent eruption on Tenerife since the formation of the Caldera de las Cañadas. It formed the Montaña Blanca, a mound about 300 m high and 1 km across, and composed of loose yellow pumice, with lumps attaining up to 30 cm in diameter. The eruption ended with a return to calmer conditions and the emission of small domes and flows. All the flows from the Montaña Blanca are notably rugged, with rough angular blocks on their aa surfaces, and their margins are invariably very steep and rubble strewn. These eruptions took place about 2000 years ago.

The Montaña Rajada, 1 km east of Montaña Blanca, erupted bulky phonolitic lava flows. They travelled 5 km, wrapped around the older Montaña Mostaza cinder cone, and formed the remarkable chaos of blocky and shining phonolites of the Valles de las Piedras Arrancadas ("valley of the uprooted stones"). Succeeding emissions were even more viscous and travelled scarcely more than 200 m from the vent. The very latest lava emissions from this vent formed the Montaña Rajada itself ("the split mountain"), which is about 250 m high and 1 km in diameter. It forms a rugged dome of reddish-brown phonolites, whose summit was burst open by an explosion that formed a ragged central hollow into which a similar but smaller dome then intruded. The Montaña Rajada also has a scattering of fine pumice on its surface, which could be derived from the Montaña Blanca nearby. The northwestern flank of the Montaña Rajada also burst open and extruded a very viscous phonolitic flow that congealed before it reached the foot of the 20° slope.

Rift-fissure eruptions

As the Cañadas stratovolcanoes were growing up, Tenerife also resumed, or continued, its old pattern of rift fissure eruptions that have given the island its characteristic Y shape, and they have persisted into historical times. The emissions seem to have come from blade-like insertions of magmas that could rise fairly quickly from a deep source and which formed basaltic cones about 100 m high and flows up to 10 km long.

Northwest of the Caldera de las Cañadas, for example, a set of rift fissures trending from northwest to southeast gave rise to the Sandiago volcanic field of recent cones and basaltic flows. Where these fissures extend into the caldera, they formed the Montaña de Samara, the Volcán Botija, the Montaña de la Cruz de Tea, and the Montaña Reventada ("breached"), which has been dated to AD 1000–1300 (Soler & Carracedo 1986). This set of fissures also produced eruptions in 1706 and 1909. At the northwestern end of these rift fissures, the large Taco cone erupted near the sea at Buenavista. Its unusual size – it is almost 1 km across and 200 m high – and its large, deep crater, as well as its fine tuff layers, all indicate that it was formed by Surtseyan eruptions in shallow water, during a recent period of higher sea level.

A companion set of fissures erupted aligned cinder cones and fluid basaltic flows that built the backbone of northeastern Tenerife, the Cumbre Ridge. It is 25 km long, 18 km wide and 1600 m high, and probably represents one of the fastest volcanic accumulations experienced on the island. Such rapid construction helped make its flanks unstable and eventually led to the landslides by which the Orotava Valley in the north and the Güímar Valley in the south collapsed between 830 000 and 560 000 years ago (Ancochea et al. 1990).

Both the Orotava and Güímar valleys are 10 km wide, with smooth floors sloping gently seawards and bounded by crisply outlined inward-facing straight cliffs extending at right angles to the Cumbre Ridge. They look like two piano keys pressed down on either side of the Ridge. They have been recently interpreted as great landslides displacing more than 100 km^3 of lavas with a total thickness of 150–600 m (Navarro & Coello 1989, Carracedo 1994). They now lie upon a layer of chaotic breccia and sandy clay that was apparently crushed when the volcanic materials slid over them.

One of the most notable fissures recently erupted Montaña Colmenar and the Siete Fuentes cones on the Cumbre Ridge. It extends into the eastern part of the Caldera de las Cañadas, where it has given rise to the Montañas Negra, Los Tomillos, Los Corrales and the highest, Mostaza, which rises over 100 m. Montaña Mostaza was formed before the Montaña Rajada, whose lavas are wrapped around its base.

Historical eruptions on Tenerife

Most of the activity on Tenerife during historical times took place along rift fissures. The early references to eruptions on the island are so vague that it is uncertain when, or even if, they occurred. However, there can be little doubt that, before the Spanish conquest, the local Guanche peoples must have witnessed eruptions on the island, notably on Teide. For the Guanches, Teide was Echeide, the Inferno. As the Italian engineer Torriani wrote in 1592, "because of the terrible fire, noise and tremors that come from it, they consider it to be the home of demons".

Andalusian and Basque sailors might have seen an eruption on Teide in 1393 or 1399. Humbolt reported that the Guanches had told their Spanish conquerors of an eruption in about 1430, which might have formed the cinder cones of Las Arenas, Los Frailes and Gañañías, in the lower Orotava valley. The Venetian mariner, Cadamasto, approached Tenerife in 1455 and later declared that "the mountain that rose above the clouds [Teide] . . . was glowing incessantly." Alonso de Palencia, chronicler of the conquest of Gran Canaria, wrote that, somewhere between 1478 and 1480, "fire surged continually from an infernal mouth in the centre of the highest mountain [Teide] . . . Small chips of stone were carried on the wind to the very edge of the sea" (Romero 1991). Christopher Columbus called at the Canary Islands on his first voyage to the Indies. On 24 August 1492, his logbook recorded that, while La Pinta was anchored off La Gomera, "they saw very large flames coming from the mountain, which filled the crew with wonder. [Columbus] explained the cause of such fire, saying that it was just like Etna." It is possible that one of these fifteenth-century accounts could refer to the eruption that gave off the fresh-looking lavas that stream down from El Pitón. On the other hand, it is strange that no record survives, of what must have been a spectacular display, from the Spaniards who were established on nearby La Gomera from 1402.

Then, for over a century after the Spanish conquest of Tenerife in 1493, the only signs of volcanic activity were the fumes that issued from the crater of El Pitón. But the first decade of the eighteenth century saw four basaltic eruptions and the destruction of half a town. The crisis began on Christmas Eve in 1704. Twenty-nine earthquakes had been counted before dawn on Christmas Day. During the next few days, more

and more earthquakes and rumblings were felt throughout the island. Calm returned on 29 and 30 December. Then, on 31 December 1704, a fissure opened on the Cumbre Ridge and began the little eruption that built the Siete Fuentes cone. A string of vents expelled lava fountains, ash and cinders, which constructed three cones – the largest of which reached only 22 m high – and sent an olivine basalt lava flow, 1 km long, into the neighbouring valley. Siete Fuentes had probably stopped erupting when the next episode began about 900 m to the northeast.

At 08.00 on Monday 5 January 1705, a new set of earthquakes started and, that afternoon, a fissure opened and the Fasnia volcano began to form. Over a length of about 1400 m, 30 initial vents formed an array of cones, explosion pits, hornitos, lava fountains and lava flows. The number of active vents fell to eight on 7 January, but they still "made as much noise as the artillery", and fumes and ash descended on the nearby towns of La Orotava and Güímar. When the eruption ended, on 16 January 1705, the largest cone was 550 m long but only 37 m high, and half of it had been destroyed as basalts had flowed out into the valley nearby. This aa flow has some fine lava moraines and flow channels. No more lava emerged during the rest of January, but the earthquakes continued, the ground rumbled, the surface cracked and fumes escaped. It is said that one shock threw down 70 houses and killed 16 people in Güímar (Romero 1991). If this tale is to be believed, this volcanoseismic event was the most lethal of all the recorded eruptions in the Canary Islands.

The next eruption arose on the same fissure, about 7 km northeast of Fasnia volcano. The Volcán de Arafo began to erupt between 16.00 and 17.00 on Monday 2 February 1705, at the very head of the Güímar valley, whose upper walls tower 400 m above it. Fountains of lava spurted 30 m into the air, and ash and cinders rose much higher. Arafo produced higher columns of fumes, more fragments, more lava, and more noise than its three predecessors – and it lasted at least until 27 February. Thus, Arafo reaches a height of 102 m. One of its two main craters is breached by an aa flow that developed distinct flow channels and many accretionary balls, and almost reached the sea, 8 km away.

These three eruptions were brief and small scale, even by the modest standards of their type. All gave out olivine basalts, and covered a total area of only 12 km^2. However, earthquakes

The destruction of Garachico, 1706

The eruption that began at 03.00 on 5 May 1706 was the most destructive that has been clearly recorded on Tenerife. There was only one preliminary earthquake before fluid basalt started gushing from a fissure on the ridge rising behind the flourishing port of Garachico, on the northwestern coast. Activity tended to migrate from northwest to southeast along this fissure. In the northwest, a spatter rampart, 5 m high and about 400 m long, formed only along its southern edge, because continuous streams of lava prevented it from developing on the downslope side to the north. On the southeastern part of the fissure, on the other hand, a mixture of explosions and effusions formed a cinder cone. But there again, persistent lava emissions stopped its northern sector from forming. This cone has been called the Montaña Negra, or the Volcán de las Arenas Negras ("the black sands"), but it is often known as the Volcán de Garachico, because of the damage that its eruption caused in the little town.

Founded in 1505, Garachico had soon become the main port in northern Tenerife. But, in 1645, floodwaters from a burst dam had badly damaged part of the town and, in 1697, fire had ravaged some of the best buildings along the shore. The new volcano was to bring yet another threat.

At 03.00 in the morning of 5 May 1706, the basalt flowed into the hamlet of Tanque, burned the church and several houses, and destroyed the vineyards nearby. At about 21.00 that same evening, a larger flow swept across the plateau and poured in seven separate cascades down the steep scarp dominating Garachico. It entered the town and filled the harbour to such an extent that little more than a creek remained. At 08.00 on 13 May, an even stronger flow rushed down the scarp, burned watermills and windmills, buried orchards and blocked springs. It reduced the San Francisco monastery and the Santa Clara convent to ashes, overwhelmed the finest quarter of Garachico, set fire to many houses and warehouses along the shore, and then spread in a fuming lobe out to sea. It is said that the basalts glowed for 40 days. But the citizens had time to flee with their goods towards the neighbouring town of Icod, often accompanied by – now equally homeless – members of the religious orders, who sang psalms to maintain morale. Many people in both the town and the surrounding countryside lost their livelihoods, but nobody was killed. The eruption probably ended on 28 May 1706, when the cinder cone had reached a height of 80 m and its flows had spread 8 km from their vent. During the following decades, Garachico was rebuilt, and more or less the same town plan was retraced over the surface of the new lava flows.

continued to shake Tenerife for more than a year afterwards, hinting that another eruption was yet to come. But it came on the northwestern rift fissure and severely damaged Garachico in 1706.

The only eruption that is certain to have occurred within the Caldera de las Cañadas in historical times took place in the summer of 1798. This time, at least, it happened well away from any settlements, 2300–2800 m high on the southwestern flanks of the Pico Viejo. However, it is rather confusingly called the Chahorra, or Narices, de Teide eruption. It took place on a fissure, 850 m long, and it was unusual for such an eruption on a stratovolcano because the activity started at the lower end of the fissure and progressed up slope. If it did indeed last for 99 days, as some have supposed, it could have been one of the longest eruptions in the Canary Islands during historical times. Earthquakes had been felt as early as 15 June 1795, but they did not reach their climax until late April 1797. However, the Chahorra magma did not make its appearance until just after 09.00 on 9 June 1798. Fifteen vents opened in the next two days as the fissure spread up the slope, as if Pico Viejo were being unzipped. By 13 June, eruptions had concentrated on three vents: the top vent giving off billowing fumes, the middle vent ejecting lava and fragments in hydrovolcanic and Strombolian activity, and the lowest delivering only lava flows. On 14 June, a large explosion joined the two upper vents together. The new vent gave off a column of snow-white steam before resuming the habits of the old middle vent, with noisy explosions and spurting lavas. It eventually built a cinder cone with two craters. The lower parts of the fissure produced a continuous wall of spatter, but the chief eruptive effort of the lower vent was to disgorge copious lava flows. These have both aa and pahoehoe surfaces that vary in

Garachico, rebuilt on the lava delta formed by the eruption of 1706.

aspect from strikingly rugged to smoothly shining. These basanitic lavas broadened out as they reached the floor of the Caldera de las Cañadas and came to a halt and solidified, with a slabby pahoehoe surface, in a big lobe at the foot of its great boundary scarp, some 8 km from their source. On 16 June, one observer saw such a flow reach the floor of the caldera in less than a day. Another observer later declared that the eruption ended in mid-September, but little information about the course of events after June has apparently survived, and it is thus not certain when the activity stopped.

The eruption of Chinyero was the latest to occur in Tenerife. It was a typical of the Canaries and lasted for only ten days. It started between 13.00 and 15.00 on 18 November 1909 in the Abeque plateau, not far, in fact, from the Volcán de Garachico. Chinyero caused nothing like the same amount of damage. Although many small earthquakes had been recorded during the previous year, they became more intense and frequent in the autumn of 1909. In the week before the eruption began, they were even accompanied by underground noises, and the ground became warm where the lava was eventually to reach the surface. The eruption started with explosions of ash from three or more vents on a fissure some 650 m long that opened on the flanks of an older cinder cone also called Chinyero. Activity migrated towards the northwestern end of the fissure as the new cone grew. Windblown ash fell at least 25 km from the vent, at La Orotava, for instance, before the first lava emerged late on 18 November. Next day, four craters were operating, bubbling out frothing lava fountains 50 m high and throwing columns of ash 500 m into the air, often to the accompaniment of loud explosions. For several days, each crater seemed to take up the main role in turn, but from 25 November activity began to decline, and from noon on 27 November 1909 only fumes issued from the vents. The new cone of Chinyero was of crescent shape, about 50 m high, and its olivine basalt flow had wrapped itself half way around the base of the cinder cone of the Montaña de Bilma, which had no doubt grown up in a similar fashion a few centuries or millennia before.

All the historical eruptions of Tenerife covered an area of only about 25.3 km², which represents a low rate of production. On average, another eruption on one of the rift fissures might be expected to occur within the next few decades, and add a little more to this modest total.

Bombs and ash erupted from Chinyero in 1909.

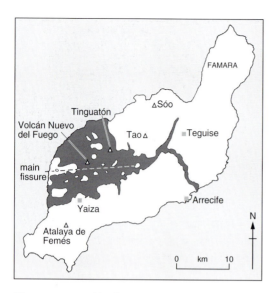

The area covered by the eruptions of 1730–36 in Lanzarote and the volcanoes that erupted in 1824.

LANZAROTE

Lanzarote is the northeasternmost of the Canary Islands, covering an area of about 795 km². It is 56 km long, has a maximum width of 21 km and reaches a height of 671 m. The higher parts of Lanzarote form the Famara Plateau in the north and the Los Ajaches Plateau in the south, both of which are composed of piles of basaltic lavas and bordered by steep straight cliffs. Between these plateaux lies the main axis of Lanzarote, an area often below 300 m, where lava flows form aprons around cinder cones that have erupted along fissures. The products of the older eruptions are covered with pale ochre **caliche**, but, in the northwest, black or reddish basalts form a grim wilderness of cones and flows that erupted between 1730 and 1736, and briefly again in 1824.

The volcanic forms of Lanzarote are all the more striking because the rainfall averages less than 200 mm per year and the natural vegetation is often limited to xerophytic plants ranging from sempervivum to prickly pear. There are thus no permanent streams on Lanzarote and, although many older cones are ribbed by gullying, there are very few barrancos. However, much of the centre of Lanzarote is blanketed with layers of black ash and lapilli, or **picón**, in which the farmers have planted vines or vegetables.

Practically all the volcanic activity on Lanzarote sprang from fissures that generally follow the overall trend of the island. Eruptions from similar fissures also no doubt built up the whole mass of Lanzarote from a depth of at least 2700 m on the floor of the Atlantic Ocean. The first eruptions above sea level in Lanzarote have been dated to about 15.5 million years ago, but three quarters of the surface area of the island erupted from more or less parallel fissures probably less than 500 000 years ago – including one quarter that erupted between 1730 and 1736 (Carracedo 1994). Apart from a minute outcrop of trachytes in the extreme south, all the lavas of Lanzarote are basaltic. The older basalts were typically alkaline, but the eighteenth- and nineteenth-century eruptions expelled basalts with a tholeiitic tendency (Araña & Carracedo 1979).

These oldest basalts form the Famara and Los Ajaches plateaux, which are similar in age, origin and nature, and are covered in caliche (Hausen 1959, Driscoll et al. 1965). Their olivine basalts are at least 670 m thick and they came from fissures that first produced thin fluid flows, then cinder cones, and then yet more thin flows. The emissions took place between 15.5 million and 5 million years ago, but were often separated by long intervals when thick fossil soils developed. The plateaux were then tilted about 15° down to the east, so that fault scarps or marine cliffs formed on their western edges. The cliffs reach a height of 500 m on the west coast of Famara, but have been widely masked by later eruptions on the west of Los Ajaches.

After over 3 million years of rest, basaltic emissions resumed 1–2 million years ago. Some 25 vents, aligned on east-northeast to west-southwest fissures, formed aa lava flows and cones that now have eroded craters and flanks.

Less than about 10 000 years ago, more basaltic eruptions took place on two main fissures,

Tomato plants growing in picón in central Lanzarote.

one following the north coast and the other along the south coast. In all, more than a hundred vents can be identified and their lava flows are the most extensive in Lanzarote at present. These eruptions began with an explosive phase, forming cinder cones and ash layers, but later gave off abundant lava flows, which have now weathered enough to be extensively cultivated. The northern fissures include the Sóo volcanoes, near the north coast, which show a sequence of Surtseyan eruptions from aligned vents that were so close together that each cone was partly destroyed when its successor erupted. At the same time, several fissures inland produced, for example, the Pico del Cuchillo and the Caldera Blanca. The latter is not a true caldera but one of the largest cones on the island, rising 175 m, whose vast lava-breached crater is 150 m deep and 1200 m across. Although it erupted inland, it has the typical dimensions of a Surtseyan cone and owes its name to the pale caliche covering its surface. The beautiful shallow ribbing on its outer flanks displays the typical dissection of the cones of this period.

The southern bunch of closely parallel fissures produced essentially similar forms. Its cones include Tahiche, Zonzomas and the Montaña Blanca, and two notable cones in the southwest: the Montaña Roja (red) and the Atalaya de

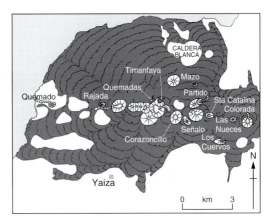

The cones and malpaís formed in 1730–36 in Lanzarote (Carrecedo et al. 1992).

Femés ("lookout"). Both erupted on the same fissure. Montaña Roja is the older of the pair and it dominates the Rubicón plain in the far west of Lanzarote. As its name indicates, it forms a reddish cinder cone, 130 m high, that is thickly mantled with caliche and many thin reddish basaltic flows covering about 7 km² on the western Rubicón plain. At 608 m, the Atalaya de Femés lives up to its name as one of the highest points in Lanzarote, forming a cone 100 m high. It grew up in three stages: the smaller northern

Spatter cones (the larger features) and hornitos (their smaller counterparts) near the Islote de Hilario.

113

Islote de Hilario: an artificially induced geyser created by pouring a bucket of water into a hole; a vigorous column of steam erupts noisily a few seconds later, generated by the subterranean heat.

crater probably formed before the larger southern crater, and a minor cone then erupted on the rim between them. Its blocky lava flow reaches the sea and covers about 20 km² of the eastern Rubicón plain. This flow is quite young because it covers a recent beach now raised about 5 m above the present shore.

At about the same time, the seaward ends of these fissures gave rise to notable Surtseyan cones. At El Golfo, in the far west, the sea has gouged out the deep crater to form the gulf that gave the volcano its name, and exposed innumerable multicoloured layers of tuffs in the cliffs enclosing the green lagoon. Originally, El Golfo might have resembled the five large cones composing the islet of Graciosa that lie at the opposite end this northern group of fissures. Each cone is almost 1 km across, with a typical deep Surtseyan crater that has been protected from marine erosion because the eruptions took place some 5 m or so above sea level.

The latest eruptions before the settlement occurred after the latest raised beaches had formed. They include the six cones aligned on a single fissure that crosses the Famara Plateau from northeast to southwest. They probably date from just before the Spanish settlement, because their forms are remarkably fresh, just as quemado (burnt) describes their appearance. Their cones include Quemado de Orzola, the Montaña de los Helechos, La Quemada, and especially the Montaña Corona. These vents have erupted a malpaís of olivine basalts that extends mainly eastwards

to the coast and covers about 50 km². The main contributor was the Montaña Corona, which was probably the youngest cone in the series, 609 m high and with a crater 418 m deep, which was breached on the downslope side by a lava flow that built a broad bulge out into the ocean. This flow contains some well developed **lava tunnels**, here called jameos. One exceeds 6 km in length and is 35 m high in places and wide enough to have been converted into a nightclub.

The eruptions in 1730–36

The historical eruptions of Lanzarote took place from 1730 to 1736, after at least three centuries of calm. Between 1 September 1730 and 16 April 1736, the island was the scene of one of the longest eruptions in Europe during historical times. One quarter of Lanzarote was given an entirely new landscape of cones and lava flows, and farms and villages in the west and centre were buried completely. Access to part of this area is restricted to protect the landscape in an educational and ecological volcanic reserve in the Timanfaya National Park. But the volcanic forms are equally fine in the rest of the area, where access to this spectacular tourist attraction is free.

During the five and a half years, the focus of activity switched from place to place, and eruptions may not, in fact, have been incessant. However, the magma probably came from the same deep source, and the olivine basalts with a tholeiitic tendency erupted from fissures forming a belt, about 18 km long and some 4 km broad, running east-northeast to west-southwest. The eruptions produced an extensive malpaís decorated with a variety of cones, often with evocative names, such as Calderas Quemadas (burnt), Montaña Rajada (split), Pico Partido (cloven), M. Colorada ((red)-coloured), M. Roja del Fuego (fire-red), Caldera Fuencaliente (hot spring) and Volcán Negro (black). They make the starkest and most brutal landscape in the Canary Islands, where scarcely a bush has yet taken root after more than two centuries (Carracedo et al. 1992).

During the first 16 months of the eruption, Father Andrés-Lorenzo Curbelo, parish priest of the village of Yaiza, kept a diary of events. The original is lost, but the German scientist Leopold von Buch published a summary, albeit spiced with his own interpretations of events. Other accounts have emerged recently in the Spanish National Archives at Simancas, which describe some of the first efforts ever made to manage a volcanic crisis (von Buch 1825, Carracedo & Rodríguez Badiola 1991, Scarth 1994, 1999).

It is uncertain whether any earthquakes preceded the eruption on 1 September 1730. It began modestly with the formation of the Los Cuervos cone over the next 18 days. Calm then returned until 10 October, when the Santa Catalina and Pico Partido cones and flows began to erupt on top of the villages of Santa Catalina and Mazo respectively. The eruption could now be seen from Gran Canaria. An ad hoc committee was set up in Lanzarote to deal with the crisis, but it could do little to alleviate the ensuing distress. Santa Catalina, and Pico Partido stopped erupting on 30 October, but Pico Partido resumed with even greater violence on 10 November, and probably continued for the rest of the winter. By 20 March 1731, a new series of vents had begun more than three months of eruptions that built the Montañas del Señalo. In June, the focus of activity suddenly switched to the sea off the west coast, where Surtseyan eruptions occurred for a time before the fissures extended onto the land nearby. For the rest of the year, successive vents exploded the cones of El Quemado, in June, Montaña Rajada, in July, and the four Montañas Quemadas from October 1731 to January 1732.

The construction of Timanfaya, a little to the east, seems to have occupied most of 1732 and probably the following year, for this is one of the largest in the area. Eruptions may have been less vigorous in 1734 and 1735, but in 1736 they found a new site, in the east, in March, when the Las Nueces cone was formed; and they reached their last abode in early April when the Colorada cone erupted. The whole episode finished on 16 April 1736.

These eruptions gave rise to some impressive and varied landforms. Timanfaya is a complex accumulation of reddish-brown lapilli and cinders forming a sharp crescent-shape ridge, 190 m high, surrounding a crater about 80 m deep. The chief lava emissions came from small vents on its western flanks and made a major contribution to an extensive malpaís, decorated with a sinuous lava tunnel that has partly collapsed. East of Timanfaya, Corazoncillo is one of the most striking cones, attracting attention as much by its colour as by its form. The squat cone, 65 m high, forms a circular rim 500 m across, enclosing a deep funnel-shape crater. Its steep and perfectly smooth slopes are clothed entirely in small pink tuffs. Corazoncillo bears the hallmark of a Surtseyan eruption, although it is some distance from any visible water bodies. In direct contrast,

Spatter cones on the horizon and malpaís with a typically rugged aa surface formed on Lanzarote in 1730–36.

The early months of the eruption seen from Yaiza

"On the first of September 1730 between nine and ten in the evening, the earth suddenly opened up near the village of Timanfaya, two leagues [in fact, 8 km] from Yaiza. During the first night an enormous mountain [Los Cuervos] rose up from the bosom of the Earth and it gave out flames from its summit for 19 days. A few days later, a fissure opened up . . . and a lava flow quickly reached the villages of Timanfaya, Rodeo and part of Mancha Blanca. This first eruption took place east of the Montaña del Fuego, half way between that mountain and Sobaco. The lava flowed northwards over the villages, at first as fast as running water, then it slowed down until it was flowing no faster than honey. A large rock arose from the bosom of the Earth on 7 September with a noise like thunder, and it diverted the lava flow from the north towards the northwest. In a trice, the great volume of lava destroyed the villages of Maretas and Santa Catalina lying in the valley. On 11 September, the eruption started again with renewed violence. The lavas began to flow again, setting Mazo on fire and then overwhelmed it before continuing on its way to the sea. There, large quantities of dead fish soon floated to the surface of the sea or came to die on the shore. The lavas kept flowing for six days altogether, forming huge cataracts and making a terrifying din. Then everything calmed down for a while, as if the eruption had stopped altogether. But on 18 [in fact, 10] October, three new openings formed just above Santa Catalina, which was still burning, and gave off great quantities of sand and cinders that spread all around, as well as thick masses of smoke that belched forth from these orifices [Santa Catalina and Pico Partido] and covered the whole island. More than once, the people of Yaiza and neighbouring villages were obliged to flee for a while from the ash and cinders and the drops of water that rained down, and the thunder and explosions that the eruptions provoked, as well as the darkness produced by the volumes of ash and smoke that enveloped the island. On 28 October, the livestock all over the area nearby suddenly dropped dead, suffocated by an emission of noxious gases [carbon dioxide?] that had condensed and rained down in fine droplets over the whole district. Calm returned on 30 October.

Ash and smoke started to be seen again on 1 November 1730 and they erupted continually until 10 November, when a new lava flow appeared, but it covered only those areas that had already been buried by previous flows. On 27 November, another lava flow [from Pico Partido] rushed down to the coast at an incredible speed. It formed a small islet that was soon surrounded by masses of dead fish. On 16 December, the lavas changed direction and reached Chupadero, which was soon transformed into what was no more than an enormous fire. These lavas then ravaged the fertile croplands of the Vega de Ugo [1 km east of Yaiza]. On 17 January 1731, new eruptions [from Pico Partido] completely altered the features formed before. Incandescent flows and thick smoke were often traversed by bright blue or red flashes of lightning, followed by thunder as if it were a storm. On 10 January 1731, we saw an immense mountain rise up, which then foundered with a fearsome racket into its own crater the self-same day, covering the island in ash and stones. Burning lava flows descended like streams across the malpaís as far as the sea. This eruption ended on 27 January. On 3 February, a new cone grew up [Montaña Rodeo] and burned the village of Rodeo. The lavas from this cone . . . reached the sea. On 7 March, still further cones were formed, and their lavas completely destroyed the village of Tíngafa. New cones with craters arose on 20 March [Montañas del Señalo] and continued to erupt until 31 March. On 6 April, they started up again with even greater violence and ejected a glowing current that extended obliquely across a previously formed lava field near Yaiza. On 13 April, two [of the Montañas del Señalo] collapsed with a terrible noise. On 1 May, the eruption seemed to have ceased, but on 2 May a quarter of a league farther away, a new hill arose and a lava flow threatened Yaiza. This activity ended on 6 May and, for the rest of the month, this immense eruption seemed to have stopped completely. But, on 4 June, three openings occurred at the same time, accompanied by violent earthquakes and flames that poured forth with a terrifying noise, and once again plunged the inhabitants of the island into great consternation. The orifices soon joined up into a single cone of great height, from which exited a lava flow that rushed down as far as the sea. On 18 June, a new cone was built up between those that already masked the ruins of the villages of Mazo, Santa Catalina and Timanfaya. A crater opened up on the flanks of this new cone, which started to flash and expel ash. The cone that had formed over the village of Mazo then gave off a white gas, the like of which nobody had ever seen before. More lava flows also reached the sea. Then, about the end of June 1731, the whole west coast was covered by enormous quantities of dead fish of all kinds, including some that had never been seen before. These eruptions took place under the sea. A great mass of smoke and flames, which could be seen from Yaiza, burst out with violent detonations from many places in the sea off the whole west coast. In October and December, further eruptions [of the Montañas Quemadas] renewed the anguish of the people. On Christmas Day 1731 the whole island was affected by the most violent of all the earthquakes felt during the previous two [sic] years of disasters. On 28 December, a new cone [in the Montañas Quemadas] was formed and a lava flow was expelled from it southwards towards Jaretas. That village was burned and the Chapel of San Juan Bautista near Yaiza was destroyed."

Many of the panic-stricken villagers of Yaiza then decided to take refuge in Gran Canaria.
(Andrés-Lorenzo Curbelo, in von Buch 1825)

The smooth cone of Corazoncillo in the background and neighbouring malpaís, with a hornito in the foreground.

there is no evidence of the slightest influence of water about 500 m to the west, where the same fissure erupted three spatter cones, several hornitos, and – most notably of all – the copious basaltic flows that threatened Yaiza. These could have been the eruptions on 28 December 1731 that caused most of the people of Yaiza to abandon their homes. However, a few people must have stayed behind, because the whole township registered 93 births and 71 deaths between 1732 and 1736. Numbers reached their lowest in 1733, when only 9 births and 6 deaths were recorded, but there was a return to normal by 1737 (Romero 1991). It may be inferred from these statistics that the volcanic activity waned after 1733, which could, therefore, have encouraged the refugees to return home. In fact, Yaiza was spared and remained on the edge of the new malpaís.

When the eruptions ended, 11 villages had been overwhelmed and 400 houses destroyed. The fields made fertile by careful husbandry had been blanketed with rough lava, livestock had been killed, and over 30 cones formed a threatening assembly 18 km long. But although many people had to abandon their goods and their homes, not a single human life had been lost.

Although lavas undoubtedly entered the sea, and some eruptions occurred within it, most of the vents were on land. Thus, probably only a small proportion of the 200 km² that was resurfaced by the eruptions represented additions to the area of the island.

The eruptions of 1824

The calm that returned to Lanzarote in April 1736 has been broken only by three eruptions of olivine basalt from the same fissure in 1824. These eruptions were brief, mild and covered only 3 km². They were described by Father Baltasar Perdomo, the parish priest of San Bartolomé (Romero 1991). Although earthquakes had been felt in Lanzarote as early as 1822, stronger shocks occurred in the centre of the island during July 1824. Earthquakes were so strong in the early hours of both 29 and 30 July that people fled from their homes. Near Tao, the ground vibrated and rumbled "as if it was boiling" and fumes escaped from newly formed cracks. The eruption started from a fissure near Tao at 07.00 on 31 July 1824, on land belonging to the priest, Luis Duarte. Hence the cone is often known as the Volcán del Clerigo Duarte, although its more official name is the Volcán de Tao. It spent most of its life giving off nothing but fumes. The actual basaltic eruption proved to be one of the shortest ever recorded in the Canary Islands, for it had only one day of glory. Ash, spatter and several lava flows issued from as many as 25 vents that formed three small cones, one reaching 38 m high. Lines of craters developed on their summits, which followed the northeast–southwest trend of the fissure. But, from 04.00 on 1 August, activity was reduced to emissions of fumes, with rare explosions of ash. At length, on 21 August, new cracks opened in the ground after an earthquake and precipitated the most notable event of the eruption at 07.00 on

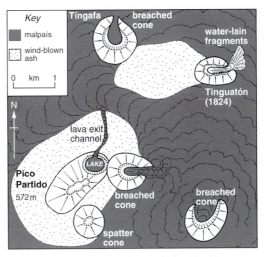

Volcanic landforms on and around Pico Partido.

The main crater of Pico Partido in the foreground, with the Montañas del Señalo beyond.

Pico Partido, with its craters forming its distinctive cloven summit. An incompletely covered lava tunnel exits from the now solidified lava lake just below the summit.

Plate 1a Vesuvius and the Forum at Pompeii. The main cone of Vesuvius rises on the horizon on the left and the ridge of Monte Somma stands out on the right.

Plate 1b Victims on the waterfront of the nuée ardente that destroyed Herculaneum (courtesy of Mario Pagano).

Plate 2 16 April 2000: the central cone of Etna (left), with the Southeast Crater (right) hidden by the fumes it is erupting. The small spatter cone below it is emitting lava flows.

Plate 3　A nocturnal eruption from the Southeast Crater of Etna (November 1998).

Plate 4a   Large lava flows from a lateral eruption of Etna (1971).

Plate 4b   Some of the active craters of Stromboli seen in mild eruption from the Pizzo Sopra la Fossa.

Plate 5a The Fossa Cone of Vulcano from Vulcanello.

Plate 5b Nísyros caldera, wit the domes on the left, hydrovolcanic accumulations in the centre, and Stéfanos crater on the right.

Plate 6 Phira and the southern walls of the caldera of Santorini.

Plate 7a   Pico de Teide, crowned by the small cone of El Pitón, with the 32 m-high Roque Cinchado (or Arbol de Piedra), which is composed of phonolitic fragments and lavas, in the foreground on the right.

Plate 7b   The crest of the Cumbre Vieja volcano, La Palma, with the El Duraznero summit and Hoyo Negro explosion crater in the centre (courtesy Juan Carlos Carracedo).

Plate 8 The Laki cone row, showing cones of various sizes formed along the fissure in 1783–4.

ROGER JONES

Above: El Golfo, on the southwest coast of Lanzarote. Weathering has revealed small variations in the resistance of the volcanic tuffs exposed by marine erosion.

Congealed lava lake, with a collapsed tunnel on the left, formed within a tuff ring on the northern flanks of Pico Partido in 1730–31.

Pico Partido

The cloven profile of Pico Partido makes it the most distinctive volcano on the island and its jagged black outline, rising 230 m above the malpaís, dominates the skyline of central Lanzarote. It is made all the more distinctive by a lava flow, outlined by pale-green lichen, that has spilt like paint down its black lapilli-strewn northern flanks. Pico Partido was, in fact, formed from a cluster of vents.

The easternmost cone formed when lava fountaining piled up spatter to a height of about 140 m. Synchronous aa lava flows prevented the eastern sector from forming and they joined similar flows to create the malpaís moulding the northern base of Pico Partido. It is marked by jagged pinnacles and several hornitos, with cowls of spatter up to 4 m high, and edged by a rugged levee of **block lava**.

More prolonged and varied eruptions gave the main mass of Pico Partido a more complex relief. The cone and its crater are elongated along the parent fissure; and its continuation is marked by a spatter rampart extending to the eastern cone. Lava fountaining scattered cowpats of spatter that welded together to form the steep upper cone, while smaller vesicular black fragments rained down over the lower slopes. The northern and western summit of Pico Partido juts more than 2 m out over the crater in a jagged lip of spatter, from which hang dribblets and stalactites up to 20 cm long. Similar features decorate the almost vertical crater walls, where reddish spatter oozed downwards under gravity to form long stalactites adhering to the walls. The wineglass shape of the crater is, of course, completely different from the funnel-shape craters of cinder cones formed from unconsolidated materials.

The elliptical crater of Pico Partido is 300 m long and about 80 m wide, and was formed by three vents as eruptions progressed eastwards along the fissure. The western crater is the shallowest and has the most scree at the foot of its confining walls, and it was partly destroyed by the larger central crater which formed after the initial eruptions. It was from these two craters that the bulk of Pico Partido was constructed. The third, easternmost crater destroyed the flank of the central cone and initiated the cleft on the north-northeastern side of Pico Partido. The third crater is much deeper than its predecessors.

When most of Pico Partido had been built up, a hydrovolcanic explosion opened a vent about 100 m up its northern flank and formed a tuff ring about 300 m across, and covered more than 1 km² to the north with small black fragments, small volcanic bombs, and xenoliths composed almost wholly of olivine crystals. Lavas then welled up and formed a lava lake that congealed within the tuff ring. Some of the fluid dark-grey basalts escaped northeastwards onto the malpaís. This lava lake also had the unusual distinction of being supplied by overspill from another lava lake high on the flanks of the eastern spatter cone. But the lavas could not keep their cohesion on the steep slopes and the flow broke into slabs of congealing lava that slid more than flowed downhill.

As the eruption waned, the crust of the lava lake congealed, only to be broken again as more lava welled up and formed arched slabs of grey lava, like a frost-heaved pavement on its surface The new molten lava escaped in an open gutter draining down the north flank of Pico Partido. At the time of maximum discharge, these fluid lavas spilt out from the gutter and formed a smooth flow splaying out from it like paint. These lavas have already been colonized by the pale green lichen that are usually among the first occupants of lavas in the Canary Islands.

the following morning. Brackish water suddenly spurted from the ground and great columns of steam and fumes soared into the air, accompanied, from time to time, by fine ash. Such hydrothermal activity had never been seen before in the Canary Islands. The Alcalde Mayor had some of the water collected and sent to Santa Cruz de Tenerife, where analysis revealed that it contained various soda salts and sulphuric acid. This hydrothermal activity lasted until 25 August, when it stopped as suddenly as it had started. Thereafter, Clerigo Duarte's volcano returned to its fumarolic somnolence until the eruption ceased altogether on 29 September.

By then the next eruption was already under way and, in contrast, was almost entirely magmatic. It began on the western end of the fissure, 13 km from its predecessor, and formed the Volcán Nuevo del Fuego. Lava in both fountains and flows immediately emerged from the fissure, but ash explosions soon followed. The sky glowed like the Aurora Borealis as the eruption entered its climax between 2 and 4 October. The ash and the sulphurous fumes made breathing difficult, even 20 km away in Arrecife. The cone attained a height of 60 m, with two craters breached on their northern sides by lava flows. The basalts arrived on the north coast, 6 km away, at 09.00 on 3 October, after having travelled at a speed of about 65 m an hour. But they did little damage because they flowed mainly across the malpaís that had formed in 1730–36. Activity came to an end with a bang. In the early hours of 5 October, there was a loud explosion and Volcán Nuevo del Fuego gave off nothing but fumes thereafter.

The baton was soon taken up by the final eruption of 1824. It formed the Volcán de Tinguatón, about 4 km along the fissure from its predecessor. It started at about 06.15 on the morning of 16 October with explosions, which created a terrifying din and seemed to light up the whole island. An hour later, three lava flows emerged.

This mixed explosive and effusive activity lasted until noon on 17 October, when the flows were hardly more than 1 km long, and the cone of cinders and spatter, elongated along the fissure, was scarcely 30 m high. No further magmatic materials erupted. However, then began perhaps the oddest episode of any eruption in the Canaries since the Spanish settlement. That afternoon, vigorous explosions sent black columns of gas, steam and old volcanic fragments soaring skywards. Then, at 16.30, hot water suddenly gushed from the crater in powerful jets reaching 16 m high. With them came billowing columns of steam, fumes and old ash – sometimes white, sometimes dark brown, and sometimes like cypress fronds. The eruption seemed to combine the hydrothermal and Surtseyan styles, although the source of the water is not clear in such an arid area, which is also 200 m above and 8 km from the sea. The water developed enough erosive power to break through the walls of the crater and then deposit a delta of fragments on the flows to the north. This extraordinary eruption ceased on Sunday 24 October 1824. Now the Volcán de Tinguatón stands like a squat ruined castle in the midst of the flows of central Lanzarote. Six deep and narrow pits on the floor of the crater mark the vents that disgorged the water: the Cuevas del Diablo or the Simas del Diablo.

The respite for Lanzarote can only be temporary. Molten magma, with a temperature probably between 900°C and 1100°C, still lies only about 4 km below the surface. A maximum temperature of 600°C has been measured at a depth of only 13 m, and over 100°C has been recorded beneath several craters and fissures (Ortiz et al. 1986). Steam emissions and temperatures high enough to kindle straw bundles, and even to cook food, continue at the Islote de Hilario in the national park.

FUERTEVENTURA

Fuerteventura is the second largest of the Canary Islands, 100 km long, less than 30 km wide, and 1725 km^2 in area. It is separated from Lanzarote only by the La Bocaina channel, which is less than 40 m deep. Lying less than 115 km from Africa, Fuerteventura is arid, and its climate bears more of the stamp of the Sahara than of the northeast trade winds. Thus, many areas have developed a caliche that has given an unusual buff colour to the lavas, although many older volcanic formations have been laid bare in the barrancos cut by ephemeral streams. The relief of Fuerteventura falls naturally into three longitudinal zones. In the west, the rolling hills of the Betancuria massif form the area where the old basement complexes are best exposed. They are transected by one of the most remarkable dyke swarms in the world, which trends from north-northeast to south-southwest. Fissures running in the same direction erupted thin, but widespreading flows, often exceeding 700 m in total thickness, which eventually buried most of their associated cinder cones. They now form an eroded tableland of horizontal basalts in the east, which rises to 807 m at the Pico de la Zorza, the highest point of the island. The possibly downfaulted central depression between them is more than 25 km long and between 5 km and 10 km broad.

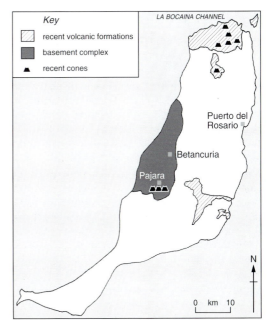

Selected volcanic features of Fuerteventura.

Fuerteventura rises in along hump from a depth of about 3000 m on the Atlantic Ocean floor, where the oceanic crust is some 180 million years old. But the rocks of Fuerteventura are altogether much younger. The eroded remnants of three ancient volcanic edifices – the northern, central and southern volcanic complexes – erupted between 22.5 and 13.2 million years ago. They now form much of the plateaux, mesas and cuchillos of the east and southeast, and in the Jandía Peninsula in the south, and they have developed a thick and widespread caliche (Fúster et al. 1968c, Abdel-Monem et al. 1971, Grunau et al. 1975, Araña & Carracedo 1979, Ancochea et al. 1996).

After a long period of calm, eruptions resumed about 2 million years ago and gave rise to basalts that were concentrated chiefly in the northern and central parts of Fuerteventura. As time went on, the emissions became smaller and less frequent and less voluminous. Eruptions along fissures produced a wealth of cinder cones and lava flows, which have been dated chiefly according to the thickness of the caliche, the depth of gullying upon the cones, and in relation to raised beaches around the island. The lavas show some variations from alkali olivine basalts towards those approaching the nature of hawaiites, although the newer basalts show tholeiitic rather than alkaline affinities.

The olivine basalts were the most extensive of these emissions. At first, they formed shield volcanoes, such as La Ventosilla, but more explosive fissure eruptions later formed large caliche-covered cinder cones and widespread aa flows, such as Piedra Sal, Temejereque, Montaña San Andreas and La Caldereta near Tetir. The more recent eruptions of olivine basalts have given rise to many cinder cones, spatter cones and rugged malpaís that are so well preserved that they cannot have long preceded the Spanish settlement of Fuerteventura in 1402. Devoid of caliche, vegetation or gullying, they have retained their original form and colours so that they stand out readily in the landscape. They form, for instance, the Pajara cones, the Malpaís de la Arena, the Malpaís Chico, in the centre, and the Malpaís Grande, in the south of the island. But their greatest contribution to Fuerteventura occurred when they formed the northern promontory of the island. Here, fissures trending from east-northeast to west-southwest gave rise to a dozen cinder cones and thin fluid flows, which pushed seawards and reduced the width of the

La Bocaina straits between Fuerteventura and Lanzarote. At this time, alternating eruptions of cinders and tuffs formed the Isla de Lobos in the straits themselves. The island erupted along the same fissure that gave rise to the cinder cones of Rebenoda, Encantada, Las Calderas and Bayuyo, which dominate the landscape of the northern promontory. Contemporaneous emissions of voluminous lava flows breached each cone down to its base. In all, the most recent episode of volcanic activity on Fuerteventura probably added some 50 km^2 to the island. And, although no eruptions have been witnessed in Fuerteventura in historical times, activity may be expected to resume, most probably in the north, perhaps within the next few millennia.

GRAN CANARIA

Gran Canaria forms the emerged centre of an eroded shield, 45 km across, that covers an area of 1532 km². It rises from a depth of 3000 m on the ocean floor to a height of 1949 m above sea level at the Pozo de las Nieves. This compact and symmetrical mass is deeply scarred by radiating barrancos and trimmed by marine cliffs where thick piles of lava are often exposed in sections approaching 100 m high. Generally speaking, the older volcanic rocks are exposed in the south-western part of the island, whereas the younger formations occur in the northeast, most notably in the peninsula of recent cones composing La Isleta north of the capital, Las Palmas.

The basal complex of Gran Canaria was masked when intense emissions occurred during a very short interval about 14.5–13.9 million years ago, at an average rate of about 5 km³ per thousand years (McDougall & Schmincke 1976, Schmincke 1981, Ancochea et al. 1990). They formed a basal shield of long thin flows, ranging from alkaline basalts to mugearites, that now constitutes more than 90 per cent of Gran Canaria.

There was a short burst of violent eruptions of trachytes, trachyphonolites and peralkaline rhyolites about 13.5 million years ago. They form lava flows, tuffs and ignimbrites and ash that were expelled when the hub of the old shield collapsed to form a caldera 15 km across and 1000 m deep. Similar eruptions continued much more feebly until about 10.5 million years ago. Gran Canaria then experienced a long period of volcanic calm and much erosion.

The third major eruptive phase took place between 5 million and 3 million years ago with the formation of the Roque Nublo stratovolcano, which covers much of the centre and northeast of the island. At first, basanites, tephrites and phonolites erupted in thin lava flows, but, as the volcano grew to about 3000 m, widespread nuées ardentes and a few phonolitic domes erupted from vents concentrated around Tejeda in the centre of Gran Canaria (Anguita et al. 1991, García Cacho et al. 1994). The stratovolcano then collapsed to form the Roque Nublo caldera, and an avalanche over 3 km³ in volume swept southwards for 28 km down its flanks. Roque Nublo itself forms a prominent pinnacle, 60 m high.

The latest phase of activity on Gran Canaria was different. Eruptions occurred from vents, often aligned on fissures, that were overwhelmingly concentrated in the northeastern half of the

Recent volcanic formations on Gran Canaria.

island. The eruptions took place in brief spurts of activity from 2.85 million to 1.5 million years ago, and they generally became less voluminous and more scattered as time went on. They mark much the smallest major volcanic episode on the island and produced fresh-looking cinder cones and lava flows (Fúster et al. 1968d). The older eruptions covered much the widest area and came from scattered vents. For example, cinder cones such as the Osorio volcano have lost their original sharp outlines, and erosion has exposed the columnar cores of their thick flows. Later eruptions, such as those forming the Pico Gáldar, were concentrated in the northeast of Gran Canaria, and the most recent have of course preserved most of their original freshness. Their basanitic cinder cones and rugged aa flows, for example, constitute much of the La Isleta peninsula near Las Palmas, the Montaña de Arucas and Las Montañetas. Hydrovolcanic eruptions also formed the well preserved maars and tuff rings of the "Calderas" of Bandama, Las Piños de Gáldar and Los Marteles. No eruptions have been recorded on Gran Canaria during historical times, but the fine state of preservation of many of its most recent features indicates that they cannot have erupted long before the Spanish Conquest of the island in 1483.

EL HIERRO

El Hierro is the smallest of the Canary Islands, covering an area of 278 km², but it is quite mountainous, reaching 1051 m at Malpasso. Rising from the ocean floor at a depth of 3000 m, El Hierro was the last island to emerge in the archipelago, and two consecutive, mainly basaltic, edifices quickly formed. The oldest dated rocks are about 1.12 million years old and they come from the El Tiñor volcanic complex, which was active for at least about 250 000 years. The second volcanic edifice, El Golfo, grew up on the eroded western flanks of El Tiñor and was active about 545 000–176 000 years ago. It began about 545 000 years ago with dyke-intruded layers of basaltic fragments and ended, about 176 000 years ago, with a predominance of flows of benmoreite and trachyte. Basaltic eruptions from rift fissures then began about 158 000 years ago, virtually as the activity of El Golfo ceased.

El Hierro is a trilobate island, and each arm is dominated by a rift zone. These three arms dominate the scenery of the island and it is here that magma has been inserted, blade like, into long parallel feeder dykes, which have thus built up the ridges by frequent and largely basaltic eruptions (Carracedo 1994). The island most probably now lies over an active branch of the Canary Island hotspot. El Hierro, indeed, has the greatest concentrations of recent and well preserved emission vents in the whole archipelago. Many small cones and lava flows, such as Julán and Orchilla volcanoes, are so little weathered that they can scarcely be more than a few thousand years old at the most. For example, near San Andrés on the central plateau the eruptions of the Montaña Chamuscada and the Montaña Entremontañas took place about 2500 years ago, and many vents near the end of the northwesternmost rift of El Hierro seem to have been active even more recently (Guillou et al. 1996). But, in spite of its youth, El Hierro has had only one possible eruption during historical times. The Lomo Negro, at the western end of the island, suffered many earthquakes between 27 March and the end of June 1793, and a small eruption could have taken place at that time (Hernández Pacheco 1982).

Coastal embayments are prominent features of El Hierro. The northern coast of the island is scalloped by the impressive embayment of El Golfo, which is 5 km across and bounded by cliffs 1100 m high. It was apparently caused by an enormous landslide less than 158 000 years ago. Lava eruptions had built up the rift ridge to such a height that its unsupported seaward flank became unstable and slipped into the sea. On the southeastern coast, a similar landslide, 3 km across, formed the Las Playas embayment, the bounding cliffs of which rise to 900 m. A third landslide forms the Julán embayment on the southwestern coast. It is still uncertain whether these landslides occurred rapidly or over many centuries, as marine erosion progressively undermined the seaward buttress of the accumulated lavas. However, although El Hierro already has a triple rift system like Tenerife, it is apparently still too young to have developed the massive central vent complex of its larger neighbour.

Recent pahoehoe lavas erupted on El Hierro.

JUAN-CARLOS CARRACEDO

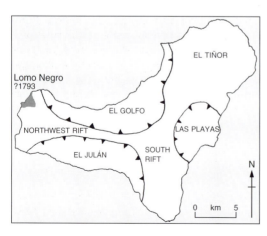

The seaward-facing embayments on El Hierro and the intervening three-arm system of rift-zone ridges.

LA PALMA

LA PALMA

La Palma is mountainous, rising to 2426 m at the Roque de las Muchados and covering 728 km². It rose above sea level during the past two million years from a base on the ocean floor about 3000 m deep. La Palma is pear shape in outline, 47 kg long. It is broadest, at 28 km across, in the north, where a high and complex shield has accumulated. Occupying the heart of this shield is the most famous landform on the island, the Caldera de Taburiente, which is 7 km wide, 15 km long and nearly 2000 m deep, and surrounded by the highest peaks in the island. In the south, it is joined to a shallower companion, the Cumbre Nueva caldera. La Palma has a volcanic sting in its tail: the modern eruptions have been concentrated on the series of parallel fissures stretching southwards along the Cumbre Vieja ridge.

In the north, the Taburiente stratovolcano grew up above a basal complex of altered basic and ultrabasic rocks about 2.0 million years ago (Ancochea et al. 1994). It is composed mainly of thick accumulations of basaltic flows interspersed with cinders and transected by innumerable basaltic dykes, whose upper layers have been dated to between 853 000 and 566 000 years old. A rift developed southwards from the Taburiente volcano and quickly grew into the steep unstable Cumbre Nueva ridge, which might have reached a height of 2500 m. About 180–200 km³ of this area then collapsed about 560 000 years ago, and another volcano, Bejenado, erupted within the collapsed area during the following 60 000 years (Carracedo et al. 1999b). There seems then to have been a quiescent period of some 370 000 years, during which the Caldera de Taburiente developed. A straight fault-guided stream, the Barranco de las Angustias, formed on the southwestern flanks of the Taburiente volcano in a particularly favoured position between the faulted edge of the Cumbre Nueva collapse structure and the new Bejanedo volcano. Vigorous headward erosion thus cut the vast hollow of the Caldera de Taburiente deep into the heart of the volcano. Consequently, fluvial erosion continued what gravitational collapse had started.

From about 125 000 years ago, the eruptive emphasis on La Palma pushed farther southwards and began to develop the present rift-fissure activity that is still forming the Cumbre Vieja volcano. However, the eruptions were much reduced between 80 000 and 20 000 years ago, during which time marine erosion pared the

Taburiente volcano, calderas and the lava flows of the historical eruptions on Cumbre Vieja volcano, La Palma.

flanks of the ridge and formed imposing cliffs around the island. Then, some 20 000 years ago, activity increased both on the north–south rift and also on a northeastward-trending rift. For the next 13 000 years, this renewed activity lengthened and widened La Palma as lava flows cascaded down the old cliffs and formed lava deltas at their bases. But 7000 years ago, activity became almost exclusively concentrated on the north–south rift. These eruptions added the latest touches to the Cumbre Vieja. It was on this volcano too that all the historic eruptions of La Palma have occurred (Plate 7b).

Historical eruptions on La Palma

All the historical eruptions on La Palma were explosions of ash and cinders and emissions of fluid, basaltic lavas. There was an eruption on the northern axis of the Cumbre Vieja ridge just before the island was settled in 1493, but its exact

125

The eruption of Teneguía in 1971: the case of the migrating breach

The latest eruption to be recorded in La Palma – and in all the Canary Islands – built the Teneguía cone between 26 October and 18 November 1971. It occurred close to where the eruptions had taken place near the San Antonio cone almost 300 years before. Earthquakes warning of the impending eruption began on 15 October 1971. They made up in frequency what they lacked in power, for over a thousand were recorded between 21 and 24 October. Then two days of relative calm ensued. But at 15.00 on 26 October, a fissure opened and lava fountains spurted from many vents along it. Basaltic flows soon emerged and reached the sea near Fuencaliente lighthouse the following dawn. On 27 and 28 October, explosions formed two successive cinder cones. The second cone to form was, in fact, known as Teneguía I because it soon became the main focus of activity. During the next two days, a lava flow breached the cone on the sector facing *up slope* to the north. The flow could not easily defy gravity for long, and, in the early hours of 31 October, the position of the breach began to migrate clockwise. The slope of the land tended to shift the flow sideways towards the east, and thus undermined the eastern arm of the cone. At the same time, ash and cinders could then pile up over the original exit and thus extend the western

arm round over the original northern breach. This migration went on until 3 November, when the flow had shifted the breach to the south-southeast, where the land sloped most steeply away from the cone and offered the easiest exit for the lavas. Thereafter, of course, the breach stayed in this optimum position. Meanwhile, at 12.39 on 31 October, an explosion 200 m north of the main cone formed Teneguía III. It remained active for only two days and formed a little pit circled by a tuff ring. It was to revive briefly once more, on 5 November. However, the effusions of the main cone continued at an increasing pace when sustained lava fountaining continued for eight days. On 8 November an explosion blasted out another pit, called Teneguía IV, on the northern flanks of the main cone. Two days later, a similar explosion alongside it created yet another pit, Teneguía V, whereupon the activity of the main cone began to wane. On 18 November, lava started to issue from the pit of Teneguía V. Like the effusions from El Duraznero in 1949, these flows heralded the end. At mid-day on 18 November, the whole eruption came to a halt, and the cone has produced nothing but fumaroles ever since. The horseshoe cone was 89 m high and the lava flows had reached the sea in four broad and rugged tongues that covered 4 km².

date, location, and even its name, have been subject to some discussion. It seems to have occurred either between 1430 and 1440 or about 1470–90. It probably formed the Montaña Quemada, a cinder cone rising 118 m high, with a breach on its northern side from which issued a lava flow about 6 km long (Romero 1991).

An eye witness, Father Alonso de Espinosa, and an Italian engineer, Leonardo Torriani, recounted the events of the next eruption in 1585. It was one of the most complex eruptions in the Canary Islands since the Spanish settlement. From late on Sunday 19 May, earthquakes of increasing vigour and frequency were felt, especially at Los Lanos, on the northern side of the Cumbre Vieja ridge. The ground began to swell up and became riddled with fissures. The magma duly reached the surface on the night of 26/27 May. Within the next two days, explosions produced masses of ash, cinders and high columns of fumes, quickly formed the Tahuya (or Teguso) cinder cone, and emitted a lava flow that eventually reached the west coast. There, said Father Espinosa, "the lava extended half a league into the sea, warmed the waters, boiled the fish therein, and melted the tar of the boats." The activity weakened considerably during the ensuing month. Then came the most extraordinary event of the eruption. The fissures began to exude

huge masses of old phonolite from the substratum, which now form the Companions of Jedey, from the name of the hamlet nearby. At the same time, about 1 km north of the main cone, more than half a dozen vents started exploding, which might have continued until 10 August 1585.

In the autumn of 1646, a much smaller eruption formed the Volcán Martín at Tigalate, on the southeastern flanks of the Cumbre Vieja ridge. On 30 September, earthquakes rumbled all over La Palma and the islands nearby. The eruption started on 1 October, and ash and cinders built a prominent cone during the next three days. They could hear the explosions in Tenerife, and, at night, the volcano "glowed like a candle". Renewed earthquakes on 4 October preceded the opening of new vents along a fissure that spurted out ash and lava flows for the next six weeks. On 15 November, yet another fissure opened close to the coast, and the flows reached the shore in a band 4 km wide. One day, the people processed to the eruption with a miraculous image of the Virgin. The following dawn, the volcano was covered in snow. On 21 December 1646, a few days later, activity ceased.

Activity resumed in 1677 in the far south of the Cumbre Vieja, between the coastal spa of Fuente Santa and the town of Fuencaliente, 5 km inland. It was long believed to have formed the large, often hydrovolcanic, San Antonio cone, which in fact is probably over 10 000 years old. The scope of the events in 1677 was much more limited: a small vent opened on the northern rim of San Antonio, and four rows of little vents erupted on its steep southwestern flanks (Carracedo et al. 1996). On 13 November, there was "a pestilential odour of sulphur in the air" and earthquakes split open fissures "that were difficult to jump across". The magma burst from the fissures at sunset on 17 November. The vent on San Antonio volcano exploded ash that formed a small cone, 30 m high, perched on its crest, but accompanying earthquakes shook most of it into the larger crater – and, no doubt, also destroyed the church tower in Fuencaliente. In December, the small cone gave off carbon dioxide, which sank into the valley near Fuencaliente and killed many head of cattle, rabbits and birds, and a peasant. And in January 1678, 27 goats succumbed to the invisible gas. While the upper vent was exploding fragments, the lower vents were emitting spatter and fluid basaltic lavas. These flows buried the spa of Fuente Santa on 26 November 1677. Indeed, basalts had covered much of the southernmost toe of La Palma when the eruption stopped on 21 January 1678. At this date, the historic eruptions on La Palma had migrated to the southern end of the island in a little over two centuries.

The following eruption paid no heed to this trend and broke out at El Charco in 1712. It was only 2.5 km from the Volcán Martín, but it occurred this time on the northwestern flanks of the Cumbre Vieja ridge. Earthquakes began to offer their usual warnings on 4 October and ash explosions duly started on Sunday 9 October from a fissure 2.5 km long. They soon built up a cinder cone that perched rather uneasily on the steep flanks of the ridge: its downslope side now rises 361 m, but the upslope side reaches no more than 25 m high. It is crowned by three craters, one of which is 135 m deep. After the cone formed, activity concentrated on the lower parts of the fissure, where fluid lava gushed out and reached the coast in a broad apron about 5 km wide, before the eruption stopped on 3 December 1712.

After this minor flurry of activity, which had seen four eruptions in 128 years, La Palma rested. Nearly 237 years later, the fissures sprang back to life on the Feast of St John the Baptist, 24 June 1949. This is why this episode of widespread activity is named after San Juan. The main focus of explosive operations lay near the crest of the Cumbre Vieja ridge, between the older volcanoes of Nambroque and El Duraznero. Notable underground rumblings and earthquakes started on 21 June. At 08.30 on 24 June, a hydrovolcanic explosion burst out north of El Duraznero. Soon ash was raining down all over southwestern La Palma, and similar pulsating emissions went on until 7 July. At 04.30 on 8 July, a more vigorous hydrovolcanic eruption heralded the opening of another fissure, about 3 km away, across the Llano del Banco. At once it became the new focus of activity and brought with it a new eruptive style: emissions of fluid basalts. They cascaded down the coastal cliffs, reached the sea, 7 km away, in a wide delta by 10 July, and gushed out until 26 July. Meanwhile, the El Duraznero fissure had two last flings. On 12 July, hydrovolcanic explosions began blasting out the Hoyo Negro, which, by the time they had finished on 30 July, had created a "Black Hole", 192 m deep, that was surrounded by a small tuff ring. At noon on 30 July, the El Duraznero fissure started spewing out fluid lavas. This phase lasted for eleven hours and eventually formed a flow 8 km long. The San Juan eruption had an unusual epilogue for a Canarian eruption. The first autumn rains on 28 November 1949 produced several volcanic mudflows, or lahars, that destroyed several bridges and a stretch of the coastal road. Luckily, they hit only a thinly populated area, and no lives were lost. The San Juan eruption also produced unusually strong seismic activity and the fissures developed during this eruption may mark the inception of another great landslide like those of the Caldera del Taburiente and the Cumbre Nueva caldera just to the north (Carracedo 1994). The latest eruption in La Palma was in 1971 at Teneguía at the southern end of the Cumbre Viejo volcano.

In all, La Palma is presently the most active of the Canary Islands and its historical eruptions have covered a total area of 37 km^2.

LA GOMERA

La Gomera, 380 km^2 in area, is the smallest of the Canary Islands apart from El Hierro and rises to a height of 1487 m at Garonjay. La Gomera is a round and broadly symmetrical island rising from a depth of some 3000 m on the Atlantic Ocean floor. It forms the summit of a vast shield volcano that has been constructed in eruptive bursts separated by long periods of erosion over several million years. La Gomera has witnessed no historical eruptions, and perhaps none for 4 million years (Cantagrel et al. 1984, Ancochea et al. 1990).

In proportion to its size, La Gomera has the largest exposed surface area of the basal complex, because the island has subsequently undergone marked uplift. As elsewhere, it is composed of peridotites, gabbros and dolerites, and it is riddled with basaltic, trachytic and phonolitic dykes that make up two thirds of its total volume. The basal complex underwent a long period of erosion lasting over a million years before activity resumed with eruptions of trachytic and phonolitic lavas some 12.5 million years ago. They were followed by the Older Basalts flows that now cover much of the island. They erupted in two main bursts: about 11 million years ago and between 8 million and 6 million years ago. After another long dormant episode, the fluid Younger Basalts covered 200 km^2 between 4.6 million and 4.0 million years ago. During the subsequent dormant period, La Gomera has undergone considerable erosion, so that clearly displayed volcanic features are rare, with the notable exception of cliffs revealing fine basaltic colonnades.

BIBLIOGRAPHY

Abdel-Monem, A., N. D. Watkins, P. W. Gast 1971. Potassium–argon ages, volcanic stratigraphy, and geomagnetic polarity history of the Canary Islands: Lanzarote, Fuerteventura, Gran Canaria and La Gomera. *American Journal of Science* **271**, 490–521.

—— 1972. Potassium–argon ages, volcanic stratigraphy and geomagnetic polarity history of the Canary Islands: Tenerife, La Palma and Hierro. *American Journal of Science* **272**, 805–825.

Ablay, G. J., G. J. Ernst, J. Martí 1995. The ~2 ka sub-Plinian eruption of Montaña Blanca, Tenerife. *Bulletin of Volcanology* **57**, 337–55.

Albert-Beltran, J. F., V. Araña, J. L. Diez, A. Valentin 1990. Physical–chemical conditions of the Teide volcanic system (Tenerife, Canary Islands). *Journal of Volcanology and Geothermal Research* **43**, 321–32.

Ancochea, E., J. M. Fúster, E. Ibarrola, A. Cendrero, J. Coello, F. Hernán, J. M. Cantagrel, C. Jamond 1990. Volcanic evolution of the island of Tenerife (Canary Islands) in the light of new K–Ar data. *Journal of Volcanology and Geothermal Research* **44**, 231–49.

Ancochea, E., F. Hernán, A. Cendrero, J. M. Cantagrel, J. M. Fúster, E. Ibarrola, J. Coello 1994. Constructive and destructive episodes in the building of a young oceanic island, La Palma, Canary Islands, and genesis of the Caldera de Taburiente. *Journal of Volcanology and Geothermal Research* **60**, 243–62.

Ancochea, E., J. L. Brändle, C. R. Cubas, F. Hernán, M. J. Hertas 1996. Volcanic complexes in the eastern ridge of the Canary Islands: the Miocene activity of the island of Fuerteventura. *Journal of Volcanology and Geothermal Research* **70**, 183–204.

Anguita, F. & F. Hernán 1975. A propagating fracture model versus a hotspot origin for the Canary Islands. *Earth and Planetary Science Letters* **27**, 11–19.

Anguita, F., L. García Cacho, F. Colombo, A. G. Camacho, R. Vieira 1991. Roque Nublo Caldera: a new strato-cone caldera in Gran Canaria, Canary Islands. *Journal of Volcanology and Geothermal Research* **47**, 45–63.

Araña, V. & J. C. Carracedo 1978. *Los volcanes de las Islas Canarias, I: Tenerife*. Madrid: Rueda.

—— 1979. *Los volcanes de las Islas Canarias, II: Lanzarote y Fuerteventura*. Madrid: Rueda.

Araña, V. & J. Coello (eds) 1989. *Los volcanes y la caldera del Parque Nacional del Teide (Tenerife, Islas Canarias)*. Serie Tecnica 7, Instituto para la Conservación de la Naturaleza, Madrid.

Araña, V. & J. M. Fúster 1974. La erupción del volcán Teneguía, La Palma, Islas Canarias. *Estudios Geológicos* **30**, 15–18.

Booth, B. 1973. The Granadilla pumice deposit of southern Tenerife, Canary Islands. *Proceedings of the Geologists' Association* **84**, 353–69.

Borley, G. A. 1974. Aspects of the volcanic history and petrology of the island of Tenerife, Canary Islands.

Proceedings of the Geologists' Association **85**, 259–79.

Buch, L. von 1825. *Physikalische Beschreibung der Kanarischen Inseln*. Berlin: Academie der Wissenschaften. [1836: published in French as *Description physique des Iles Canaries*, translated by C. Boulanger. Paris: Levrault].

Camacho, A. G., R. Vieira, C. de Toro 1991. Microgravimetric model of the Las Cañadas caldera (Tenerife). *Journal of Volcanology and Geothermal Research* **47**, 75–88.

Cantagrel, J. M., A. Cendrero, J. M. Fúster, E. Ibarrola, C. Jamond 1984. K–Ar chronology of the volcanic eruptions in the Canarian archipelago: island of Gomera. *Bulletin Volcanologique* **47**, 597–609.

Carracedo, J. C. 1994. The Canary Islands: an example of structural control on the growth of large oceanic island volcanoes. *Journal of Volcanology and Geothermal Research* **60**, 225–41.

—— 1996. Morphological and structural evolution of the western Canary Islands: hotspot-induced three-armed rifts or regional tectonic trends? *Journal of Volcanology and Geothermal Research* **72**, 151–62.

—— 1999. Growth, structure, instability and collapse of Canarian volcanoes and comparisons with Hawaiian volcanoes. *Journal of Volcanology and Geothermal Research* **94**, 1–19.

Carracedo, J. C. & E. Rodríguez Badiola 1991. *Lanzarote, la erupción volcánica de 1730*. Las Palmas de Gran Canaria: Servicio de Publicaciones, Cabildo Insular de Lanzarote.

Carracedo, J. C., E. Rodríguez Badiola, V. Soler 1992. The 1730–1736 eruption of Lanzarote, Canary Islands: a long, high-magnitude basaltic fissure eruption. *Journal of Volcanology and Geothermal Research* **53**, 239–50.

Carracedo, J. C., S. J. Day, H Guillou, E. Rodríguez Badiola 1996. The 1677 eruption of La Palma, Canary Islands. *Estudios Geológicos* **52**, 103–114.

Carracedo, J. C., S. J. Day, H. Guillou, P. J. Gravestock 1999a. Later stages of volcanic evolution of La Palma, Canary Islands: rift evolution, giant landslides, and the genesis of the Caldera de Taburiente. *Geological Society of America, Bulletin* **111**, 2–16.

Carracedo, J. C., S. J. Day, H. Guillou, F. J. Pérez Torrado 1999b. Giant Quaternary landslides in the evolution of La Palma and El Hierro, Canary Islands. *Journal of Volcanology and Geothermal Research* **94**, 169–90.

Chaigneau, M. & J. M. Fúster 1972. L'éruption du Teneguía (La Palma) et la composition des laves et gaz fuméroliens. *Académie des Sciences à Paris, Comptes Rendus* **D274**, 2948–51.

Day, S. J., J. C. Carracedo, H. Gillou, P. J. Gravestock 1999. Recent structural evolution of the Cumbre Vieja volcano, La Palma, Canary Islands: volcanic rift zone reconfiguration as a precursor to volcano flank instability? *Journal of Volcanology and Geothermal Research* **94**, 135–67.

Driscoll, E. M., G. L. Hendry, K. J. Tinkler 1965. The geology and geomorphology of Los Ajaches, Lanzarote. *Geological Journal* **4**, 321–34.

Elsworth, D. & S. J. Day 1999. Flank collapse triggered by intrusion: the Canarian and Cape Verde archipelagoes. *Journal of Volcanology and Geothermal Research* **94**, 323–40.

Féraud, G., G. Giannerini, R. Campredon, C. J. Stillman 1985. Geochronology of some Canarian dike swarms: contributions to the volcano-tectonic evolution of the archipelago. *Journal of Volcanology and Geothermal Research* **25**, 29–52.

Fernández-Navarro, L. 1919. Las erupciones de fecha histórica en Canarias. *Real Sociedad Española de Historia Natural, Memorias* **11**, 37–75.

Fritsch, K. von & W. Reiss 1868. *Geologische Beschreibung der Insel Tenerife*. Winterthur [Switzerland]: Wurster.

Fúster, J. M., S. S. Fernández, J. Sagredo 1968a. *Geología y volcanología de las Islas Canarias. Lanzarote*. Madrid: Consejo Superior de Investigaciones Cientificas.

Fúster, J. M., V. Araña, J. L. Brandle, J. M. Navarro, U. Alonso, A. Apararicio 1968b. *Geología y volcanología de las Islas Canarias: Tenerife*. Madrid: Consejo Superior de Investigaciones Cientificas.

Fúster, J. M., A. Cendrero, P. Gastesi, E. Ibarrola, J. Lopez Ruiz 1968c. *Geología y volcanología de las Islas Canarias: Fuerteventura*. Madrid: Consejo Superior de Investigaciones Cientificas.

Fúster, J. M., A. Hernández-Pacheco, M. Muñoz-García, E. Rodríguez Badiola, L. García Cacho 1968d. *Geología y volcanología de las Islas Canarias: Gran Canaria*. Madrid: Consejo Superior de Investigaciones Cientificas.

Gagel, C. 1910. *Die Mittelatlantischen Vulkaninseln*, vol. 7. Berlin: Handbuch der regionales Geologie.

García Cacho, L. J. L. Díez-Gil, V. Araña 1994. A large volcanic debris avalanche in the Pliocene Roque Nublo stratovolcano, Gran Canaria, Canary Islands. *Journal of Volcanology and Geothermal Research* **63**, 217–29.

Gastesi, P. 1973. Is the Betancuria Massif, Fuerteventura, an uplifted piece of oceanic crust? *Nature* **246**, 102–104.

Guillou, H., J. C. Carracedo, F. Pérez Torrado, E. Rodríguez Badiola 1996. K–Ar ages and magnetic stratigraphy of a hotspot-induced fast grown oceanic island: El Hierro, Canary Islands. *Journal of Volcanology and Geothermal Research* **73**, 141–55.

Hausen, H. 1956. Contributions to the geology of Tenerife (Canary Islands). *Societa Scientifica Fennica Communications in Physics and Mathematics* **18**, 1–254.

—— 1959. On the geology of Lanzarote, Graciosa and

the Isletas (Canary Islands). *Societa Scientifica Fennica Communications in Physics and Mathematics* **23**, 1–116.

Hernández-Pacheco, A. 1982. Sobre una posible erupción en la isla del Hierro (Canarias). *Estudios Geológicos* **38**, 15–25.

Hoernle, L. & H. U. Schmincke 1993. The role of partial melting in the 15 Ma geochemical evolution of Gran Canaria: a blob model for the Canary hotspot. *Journal of Petrology* **34**, 599–626.

Humboldt, A. von 1814. *Voyages aux régions équinoxiales du nouveau continent 1799–1809*, vol. I. Paris: Dufour.

Ibarrola, E. & J. Lopez Ruiz 1967. Estudio petrografica y quimíco de las erupciones recientes (series IV) de Lanzarote. *Estudios Geológicos* **23**, 203–213.

Klügel, A., H. U. Schminke, J. D. L. White, K. A. Hoernle 1999. Chronology and volcanology of the 1949 multi-vent rift-zone eruption on La Palma (Canary Islands). *Journal of Volcanology and Geothermal Research* **94**, 267–82.

Krafft, M. & F. D. de Larouzière 1999. *Guide des volcans d'Europe et des Canaries*, 3rd edn. Lausanne: Delachaux et Niestlé.

Krejci-Graf, K. 1961. Vertikal-Bewegungen der Makaronesen. *Geologische Rundschau* **51**, 73–122.

Laughton, A. S., D. G. Roberts, R. Graves 1975. Bathymetry of the northeastern Atlantic: Mid-Atlantic Ridge to southwest Europe. *Deep-Sea Research* **22**, 791–810.

Lyell, Sir C. 1855. *A manual of elementary geology*. London: John Murray.

Martí, J., J. Mitjavila, V. Araña 1994. Stratigraphy, structure, age and origin of the Cañadas Caldera (Tenerife, Canary Islands). *Geological Magazine* **131**, 715–27.

Martí, J., G. J. Ablay, S. Bryan 1996. Comment on "The Canary Islands: an example of structural control on the growth of large oceanic island volcanoes" by J. C. Carracedo. *Journal of Volcanology and Geothermal Research* **72**, 143–9.

Masson, D. G. 1996. Catastrophic collapse of the volcanic island of El Hierro 15 ka ago and the history of landslides in the Canary Islands. *Geology* **24**, 231–4.

McDougall, I. & H. U. Schmincke 1976. Geochronology of Gran Canaria, Canary Islands: age of shield building, volcanism and other magmatic phases. *Bulletin Volcanologique* **40**, 57–77.

Mehl, K. W. & H. U. Schmincke 1999. Structure and emplacement of the Pliocene Roque Nublo debris avalanche deposit, Gran Canaria, Spain. *Journal of Volcanology and Geothermal Research* **94**, 105–134.

Mitchell-Thomé, R. C. 1981. Volcanicity of historic times in the middle Atlantic islands. *Bulletin Volcanologique* **44**, 57–69.

Morgan, W. J. 1972. Deep mantle convection plumes and plate motions. *American Association of Petroleum Geologists, Bulletin* **56**, 203–9.

Moss, J. L., W. J. McGuire, D. Page 1999. Ground deformation of a potential landslide at La Palma, Canary Islands. *Journal of Volcanology and Geothermal Research* **94**, 251–65.

Ortiz, R., V. Araña, M. Astiz, A. García 1986. Magnetotelluric study of the Teide (Tenerife) and Timanfaya (Lanzarote) volcanic areas. *Journal of Volcanology and Geothermal Research* **30**, 351–77.

Perret, F. A. 1914. The volcanic eruption of Tenerife in autumn 1909. *Zeitschrift für Vulkanologie* **1**, 20–31.

—— 1925. *Volcanological observations*. Washington DC: Carnegie Institution.

Pitman, W. C. & M. Talwani 1972. Sea-floor spreading in the North Atlantic. *Geological Society of America, Bulletin* **83**, 619–46.

Real Audiencia de Canarias 1731. *Copia de las Ordones y providencias dadas para el alivio de los Vezinos de la Isla de Lanzarote en su dilatado padezer a causa del prodigioso volcán, que en ella rebentó el primer dia de Septiembre del año immediato pasado de 1730, y continúa hasta el dia de la fecha . . . Canaria y Abril 4 de 1731*. Legajo 89, Gracia y Justicia. Simancas [Spain]: Archivo General de Simancas.

Ridley, W. I. 1970a. The petrology of the las Cañadas volcanics, Tenerife, Canary Islands. *Contributions to Mineralogy and Petrology* **26**, 124–60.

—— 1970b. Abundance of rock types on Tenerife. *Bulletin of Volcanology* **34**, 196–204.

—— 1971a. The origin of some collapse structures in the Canary Islands. *Geological Magazine* **108**, 477–84.

—— 1971b. The field relations of the Cañadas volcanoes, Tenerife, Canary Islands. *Bulletin Volcanologique* **35**(2), 318–34.

Romero, C. 1991. *Las manifestaciones volcánicas del archipiélago Canario* [2 vols]. La Laguna, Tenerife: Gobierno de Canarias.

San Miguel de la Camara, M., J. M. Fúster, M. Martel 1952. Las erupciones y materiales arrojados par ellas en la Isla de La Palma, Junio–Julio 1949. *Bulletin Volcanologique* **12**, 145–63.

San Miguel de la Camara, M. & T. Bravo 1967. Active volcanoes of the Canary Islands. In *The Atlantic Ocean*, part I: *catalogue of the active volcanoes of the world*, M. Newmann Van Padang, A. F. Richards, F. Mochedo, T. Bravo, P. E. Baker, R. W. Le Maitre (eds), 55–105. Rome: International Association of Volcanology.

Scarth, A. 1994. *Volcanoes*. London: UCL Press.

—— 1999. *Vulcan's fury: man against the volcano*. London: Yale University Press.

Schmincke, H. U. 1976. The geology of the Canary Islands. In *Biogeography and ecology of the Canary Islands*, G. Kunkel (ed.), 67–184. The Hague: Junk.

—— 1982. Volcanic and chemical evolution of the Canary Islands. In *Geology of the northwest African continental margin*, V. von Rad, K. Hinz, M. Sarnthein, E. Seibold (eds), 273–306. New York: Springer.

Siebert, L. 1984. Large volcanic debris avalanches: characteristics of sources, areas, deposits and associated eruptions. *Journal of Volcanology and Geothermal Research* **22**, 163–97.

Soler, V. & J. C. Carracedo 1984. Geomagnetic secular variation in historical lavas in the Canary Islands. *Royal Astronomical Society, Geophysical Journal* **78**, 313–18.

—— 1986. Aplicación de las técnicas paleomagneticas de corto período a la datación del volcanismo subhistorico de la isla de Tenerife. *Geogaceta* **1**, 33–5.

Stillman, C. J. 1999. Giant Miocene landslides and the evolution of Fuerteventura, Canary Islands. *Journal of Volcanology and Geothermal Research* **94**, 89–104.

Tinkler, K. J. & B. Booth 1973. The Granadilla pumice deposit of southern Tenerife, Canary Islands. *Proceedings of the Geologists' Association* **84**, 353–70.

Torriani, L. 1592. *Descripción de las Islas Canarias* [translated by A. Cioranescu, 1978]. Santa Cruz de Tenerife: Goya.

Valentin, A., J. F. Albert-Beltran, J. L. Piez 1990. Geochemical and geothermal constraints on magma bodies associated with historic activity, Tenerife (Canary Islands). *Journal of Volcanology and Geothermal Research* **44**, 251–64.

Viera y Clavijo, J. 1982. *Noticias de la historia general de las Islas Canarias* [2 vols]. Santa Cruz de Tenerife: Goya.

Watts, A. B. & D. G. Masson 1995. A giant landslide on the north flank of Tenerife, Canary Islands. *Journal of Geophysical Research* **100**, 24487–98.

Webb, B. & S. Berthelot 1839. *Histoire naturelle des isles Canaries* [2 vols]. Paris: Béthune.

White, J. D. L. & H. U. Schminke 1999. Phreatomagmatic eruptive and depositional processes during the 1949 eruption on La Palma (Canary Islands). *Journal of Volcanology and Geothermal Research* **94**, 283–304.

Williams, H. 1941. *Calderas and their origin*. Publications in Geological Science 25 (pp. 239–346), University of California, Berkeley.

Wolff, J. A. 1985. Zonation mixing and eruptions of silica-undersaturated alkaline magma: a case study from Tenerife, Canary Islands. *Geological Magazine* **122**, 623–40.

5 Portugal: the Azores Islands

The Azores are an autonomous region of Portugal situated in the Atlantic Ocean 1500 km west of Lisbon. The archipelago of nine islands is scattered over 600 km and falls into three groups: Flores and Corvo in the west, Santa Maria and São Miguel in the east, and a central cluster of five islands – Terceira, Graciosa, São Jorge, Pico and Faial. Seven of the islands rise from the Azores Platform on the eastern flanks of the Mid-Atlantic Ridge. Only Flores and Corvo rise on its western flanks. All the islands are active except Santa Maria, which is the farthest from the Ridge. All except São Jorge have stratovolcanoes and all except one have been decapitated by calderas. The Pico do Pico, soaring to 2351 m above sea level, is the only stratovolcano that still retains the pristine glory that makes it the incomparable landmark of the Azores. Although the Azores are almost entirely volcanic, they are far from stark and bare: laurels, hydrangeas and azaleas dominate a floral extravaganza that is rarely seen in temperate climates.

The Azores were uninhabited when the Portuguese explorers were led to them by the goshawks, the Açores, that were flying about the islands. Settlement began first on Santa Maria and São Miguel in 1439 and then on Terceira in 1450, on Pico and Faial in 1466, on Graciosa and São Jorge in 1480, and on Flores and Corvo at the start of the next century. Historical records of eruptions therefore extend back only between 500 and 600 years, but over 30 eruptions have taken place during this period, either on the islands or off shore. However, the Azores are not growing rapidly; Iceland has a much higher output of lava per visible eruption (Self 1976); eruptions occur four times more often in Iceland

(Ridley et al. 1974); and eruptions of similar composition have built the Canary Islands five to ten times faster (Moore 1991). But, of course, Iceland is 40 times larger than the Azores islands, and many submarine eruptions must have passed unnoticed on the Azores Platform.

The Azores and their submarine plinth grew up on the Mid-Atlantic Ridge near the triple junction of the North American, Eurasian and African plates. Their activity is probably related to a hotspot and perhaps also to a secondary band of seafloor spreading (McKenzie 1972, Laughton & Whitmarsh 1974, Weijermars 1987). The Mid-Atlantic Ridge, forming the boundary between the North American and Eurasian plates, passes through the Azores. The western islands of Flores and Corvo belong to the North American plate, but the location of the boundary between the Eurasian and African plates is not at all clear. The East Azores fracture zone runs from the Mid-Atlantic Ridge along the southern edge of the Azores Platform. Thus, if the fracture zone

The Azores and the Mid-Atlantic Ridge.

forms the main plate boundary, then the central and eastern Azores must belong to the Eurasian plate. However, an axis of secondary seafloor spreading runs through the central Azores along the Terceira Rift, which probably passes through Graciosa, Terceira and São Miguel (Krause & Watkins 1970, Self 1976, Searle 1980). In either case, parts of the central and eastern Azores could therefore belong to the African plate or to an Azores microplate. The volcanic activity in the Azores was most probably also intensified by one large, or several small, hotspot plumes (Féraud et al. 1980). Whatever the reasons behind the growth of the central and eastern Azores, all display a common and impressive predominance of stratovolcanoes, rifts, faults, fissures and volcanic alignments, running parallel to the spreading axis, that have been the leitmotifs in the development of their scenery.

The stratovolcanoes in the Azores are mostly gently sloping cones, usually more than 10 km in diameter and rising about 1000 m above sea level. They are crowned by beautiful deep calderas, which so impressed the early settlers that they also rather confusingly gave the name Caldeira (cauldron) to the whole mountain. These calderas are the hallmarks of the Azores.

The eruptions that have taken place along fissures running from northwest to southeast are the second most striking characteristic of the Azores and they completely dominate the landscape of São Jorge. What these eruptions lack in volume they make up for in number, for there are over a thousand cinder cones in the Azores. They occur on the flanks of stratovolcanoes and dominate the scenery on the plains and plateaux, and they are commonly associated with basaltic, hawaiitic or sometimes mugearitic lava flows. The younger flows often have rugged black aa surfaces, which the early settlers called mistérios, because they had frightening and mysterious associations that the eruptions during historical times did nothing to dispel. Although they have now often been planted with woodlands, they still stand out among the meadowlands of the Azores, and they reach their finest expression on Pico.

The fissure eruptions in the Azores extend well below sea level. Deeper eruptions, stifled by the water pressure, reached the surface as bubbling gas emissions and hot discoloured seas. Such eruptions have marked the activity of the Don João de Castro Bank. But, in the shallower coastal waters, Surtseyan eruptions

built bulky tuff cones such as those protecting both Angra do Heroísmo in Terceira, and Horta in Faial. Older Surtseyan tuff cones are dotted about the coasts of the Azores, but, because they formed in such vulnerable positions, marine erosion soon reduced them to picturesque islets such as the Ilhéus das Cabras, off Terceira, and the Ilhéus dos Mosteiros off São Miguel.

Several islands have been marked by faulting and rifting, which was probably associated with the zone of secondary spreading branching from the Mid-Atlantic Ridge. The Terceira Rift transects that island and São Miguel, and another rift in eastern Faial forms a distinct fault trough. On the other hand, uplifted blocks seem to delimit most of both São Jorge and eastern Pico. The archipelago still suffers from typical mid-ocean-ridge earthquakes that are less than magnitude 5.0 on the Richter scale and have shallow epicentres, less than 30 km deep.

Although six of the islands are still active, fumaroles are their only persistent manifestation. They make their best displays in the Caldeira das Furnas in São Miguel, where tourists can bathe or have meals specially cooked – in different vents, of course. Weak fumes usually issue from the summit of Pico, and the earthquake in May 1958 briefly revived those in the caldera of Faial. Magmatic emissions are much less frequent; the latest formed Capelinhos in Faial in 1957–8, although other eruptions have since occurred below sea level. At all events, volcanic features are never out of sight in the Azores, and volcanic activity plays a role in the local place names exceeded only by religion. Mistério and caldeira have already been mentioned, but other common names include biscoitos (rugged aa lava flow), bagacina (cinders), queimado (burnt), fogo (fire), furna (oven), timão (arched yoke) and cabeço (head). Thus, volcanic activity is never far from the consciousness of the Azoreans.

The Azores are young. The eastern group contains the oldest rocks, which reach about 5 million years old in Santa Maria and 4 million years old in eastern São Miguel (Abdel-Monem et al. 1975). No lavas on any of the remaining islands are apparently more than a million years old. The oldest rocks commonly outcrop in the southeast of each island, and the more recent eruptions have tended to occur in the northwest, nearest the Mid-Atlantic Ridge. Nevertheless, there was no regular progression of eruptions from island to island towards this ridge. Thus, for instance, Faial, in the west, and São Miguel, in the east:

both have some of the oldest and some of the youngest lavas in the archipelago (Booth & Croasdale 1978).

The eruptions in the Azores produced a predominance of alkali basalts, hawaiites and mugearites, but some evolution to trachytes and pantellerites was associated with the violent explosions that formed the calderas. The basaltic rocks, commonly erupted from fissures, formed many cinder cones and lava flows, and Surtseyan eruptions took place where these fissures extended into shallow water.

SÃO MIGUEL

São Miguel is the largest and most varied island in the Azores. It covers an area of 747 km² and is 65 km long, with a maximum width of 15 km, culminating at a peak of 1103 m at the Pico da Vara. It is notable for its rich vegetation, fertile soils, four stratovolcanoes with majestic calderas, dozens of cinder cones, aligned for the most part on fissures, thermal springs, and a beautiful coastline dominated in many places by high cliffs. However, large tracts of bare lava are rare, for most of the flows are either weathered or have been blanketed by trachytic pumice exploded from the stratovolcanoes.

The eruptions that gave rise to São Miguel began about 4 million years ago at a depth of 2000 m on the floor of the Atlantic Ocean, and the island has some of the oldest exposed lavas in the archipelago. However, five eruptions on land, and a further seven offshore, have been recorded since the island was first settled in 1439 (Canto 1880b, Weston 1964, Mitchell-Thomé 1981). In general, activity on São Miguel spread westwards and has now probably ceased in the east.

Four stratovolcanoes form the backbone of the island. Povoação and Furnas are contiguous in the east; the Água de Pau, or Fogo volcano, occupies the centre; and the Sete Cidades volcano forms much of the northwestern part of the island. Sete Cidades was a separate island until it was joined to the rest of São Miguel by the eruptions of the Região dos Picos (Booth et al. 1978).

The calderas and the recent eruptions on and around São Miguel.

Povoação and the Nordeste

The northeastern region contains the oldest rocks exposed on São Miguel, 4.00–0.95 million years old (Abdel-Monem et al. 1975, Forjaz 1997). It is a broad upland, composed of thick piles of basaltic flows, which are probably the remains of a shield that culminated at the Pico da Vara. However, this upland has been slashed by narrow gorges and pared by marine erosion into cliffs commonly over 100 m high. Povoação, the oldest stratovolcano on São Miguel, grew up on the southern fringe of these uplands. The eruptions started with basalts, continued with hawaiites and mugearites, and ended with trachytic emissions. Plinian explosions of trachytic pumice accompanied the formation of the Povoação caldera on the southern flanks of the stratovolcano. It seems to have occurred in two stages, about 820 000 and 700 000 years ago. The southern wall of the caldera is missing. It could have foundered along the fault that seems to delimit the adjacent coast, but, more likely, it could have been swept seawards by a massive landslide. Neither the northeastern uplands nor Povoação volcano have experienced any eruptions during the past few thousand years.

Furnas stratovolcano

Furnas is an indistinct stratovolcano with an imposing caldera. It emerged from the Atlantic Ocean more than 100 000 years ago and its products cover some 75 km^2 and have a volume of about 60 km^3 (Moore 1990, 1991, Cole et al. 1995). Most of the rocks are trachytes that occur in lava flows, pumice, ash and cinders, as well as in nuées ardentes and mudflows.

The best view of Furnas caldera can be obtained from the trachytic dome of the Pico do Ferro, perched on its northern rim. It is the youngest caldera on São Miguel and, when it collapsed about 12 000 years ago, massive eruptions scattered trachytic pumice over much of the island. The caldera covers an area of 35 km^2 and the southern rim is low, but the remaining walls rise 100 m–250 m almost vertically from its floor. Its central part later foundered even further, perhaps about 11 000 years ago, to form an inner caldera, which has a diameter of 5 km, and its western parts are occupied by the Lagoa das Furnas. Plentiful hot springs are still concentrated on the northern lakeshore, at the foot of the Pico do Ferro, and especially in the spas at Terra Nostra and Caldeiras, on the outskirts of Furnas.

The Lagoa das Furnas, looking southwards from Pico do Ferro.

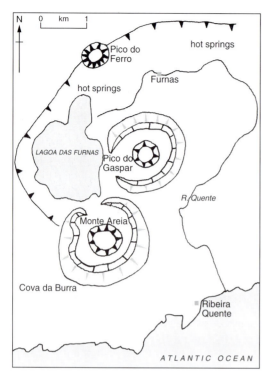

The main volcanic features in Furnas caldera.

Trachytic eruptions have already started to fill the caldera. The Pico do Ferro itself is about 11 200 years old and it could have extruded before the caldera collapsed completely. But the main volcanic events in the caldera were the ten

explosive eruptions, designated Furnas A to Furnas J, that took place during the past 5000 years (Booth et al. 1978). Most of these eruptions were hydrovolcanic and related, no doubt, to the three lakes lying within the caldera at that time. About 2900 years ago, the third and largest eruption (Furnas C) expelled a widespread blanket of 1.5 km^3 of pumice from a vent located on the flat plain where the village of Furnas now stands. This plain might therefore be the filled remains of a maar. In fact, the subsequent eruptions occurred on a ring fracture and they formed a crescent of tuff rings into which trachytic domes have extruded – and where many of the chief fumaroles are found. The twin domes and tuff rings of the Pico das Marcondas, for instance, first formed during the fifth (Furnas E) and eighth (Furnas H) episodes. About 1100 years ago, the dome and tuff rings of the Pico do Gaspar erupted and also produced the extensive pumice deposits of Furnas F, G and I. Pico do Gaspar also had the important morphological role of impounding the Lagoa das Furnas, the sole successor of the three former lakes. Thus, the Lagoa das Furnas has the unusual distinction of not occupying the lowest parts of the Furnas caldera, for the plain of Furnas lies almost 100 m below it. The lake is only 14 m deep and could itself be an old maar, now largely filled by pumice.

About 1445, there was an eruption within the caldera, but the only magmatic eruption recorded here during historical times occurred

The eruption in Furnas caldera in 1630

At about 20.00 on 2 September 1630, earthquakes began to shake the whole of São Miguel, virtually destroying the villages of Ponta Garça and Povoação, near Furnas, causing landslides on the south coast at the mouth of the Ribeira Quente and making the church bells ring 30 km away in Ponta Delgada. The magma duly erupted inside the caldera between 02.00 and 03.00 on 3 September; a small column of ash and fumes rose skywards, and the wind blew the ash southwestwards. Soon, more violent hydrovolcanic eruptions and nuées ardentes blasted out from the caldera lakes, and an eye-witness said that two of the lakes gave off "clouds of fire". One of these nuées ardentes swept down the gorge of the Ribeira Quente and surged seawards to form a small delta. Another formed "a burning stream" that spread through the woods to the village of Ponta Garça. On 4 September, a sub-Plinian climax started three days of full fury, sending a column of ash and steam 14 km into the air, spreading pumice, ash and lapilli over a wide area, burning the forests, and plunging the whole island into

darkness. Ash even fell 550 km away to the west on the island of Corvo. The wind changed direction for the last, most violent and purely magmatic phase of the eruption. It lasted for 15 hours and expelled nearly one third of all the magma ejected. Ash and lapilli showered down mainly to the northeast of the vent, and deposited a blanket of fragments that reached at least 2.5 m thick in Ponta Garça and even 10 cm in Ponta Delgada. When the climax drew to a close on 6 September 1630, the Lagao das Furnas was the only lake left in the caldera. The eruption had built a distinctly asymmetrical tuff ring, the Cova da Burra, inside which the trachytic dome now forming Monte Areia extruded during the next two months. The eruption expelled about 2.5 km^3 of fragments and killed 195 people. Between 80 and 115 people – half the population of the village – died in Ponta Garça; some 30 people were killed who had lived in huts beside the lakes; and the remainder met their deaths in and around the villages of Furnas and Povoação. (Canto 1880b)

The Lagoa do Fogo caldera, Água de Pau volcano. Its scalloped rim may indicate that the caldera was formed by intermittent collapse

near its southern edge in 1630 (Cole et al. 1995). Furnas caldera is clearly not extinct. The average interval between eruptions during the past 2900 years has been 362 years. The caldera has only to continue that pattern for another eruption to happen there within the next few decades.

Achada das Furnas

The plateau of Achada das Furnas abuts onto both the Água de Pau and Furnas volcanoes. It forms a moorland saddle of basaltic lava flows, between about 10000 and 26000 years old, which have been blanketed with trachytic pumice expelled mainly from Água de Pau stratovolcano. There are also cinder cones on the plateau, about 100 m high, such as the Pico do Meirim, the Pico das Tres Lagoas and the Pico de el Rei. They were formed along fissures at the same time as the later eruptions of Água de Pau and Furnas volcanoes. About 3800 years ago, a hydrovolcanic explosion of pumice and lithic debris also formed a tuff ring enclosing a maar, inside which a trachytic dome later arose (Moore & Rubin 1991). The Lagoa do Congro now occupies the maar.

Água de Pau stratovolcano

The low pyramidal outline of the Água de Pau stratovolcano dominates central São Miguel. It is 15 km in diameter and reaches 947 m at the Pico da Barrosa. Its elliptical crest rises in a wild isolated moorland that is quite unlike any other on the island. It surrounds a caldera, with a maximum width of 3.5 km from northwest to southeast, which is the smallest in São Miguel. Some 250 m below its rim lies the green Lagoa do Fogo ("lake of fire"), which has been the source of five major eruptions in the past 5000 years (Walker & Croasdale 1971, Moore 1991).

Água de Pau began erupting on the ocean floor about 290000 years ago and the volcano now covers an area above sea level of about 166 km^2 and has a volume of some 80 km^3. Água de Pau was more explosive than its neighbours, giving off a large proportion of trachytic pumice and nuées ardentes in addition to its basaltic and trachytic flows. One section on the south coast near Ribeira Chà, for instance, shows that 65 explosive eruptions occurred over a period of 34000 years. Trachytic pumice also now mantles the flanks of Água de Pau and hides much of the detail of its geological history. Most of the volcano seems to have grown between about 100000 and 40000 years ago, and the summit may have collapsed several times during this period.

The two most recent calderas have been identified. The larger, outer, caldera formed during a great trachytic eruption about 33 000 years ago and was 7 km long and 4 km across, and probably had a volume of 5 km³. The eruption inaugurated about 15 000 years of calm, which was disturbed only when three trachytic pumice rings erupted 18 600 years ago. Then, about 15 200 years ago, came another great trachytic outburst that created the caldera enclosing the Lagoa do Fogo, whose white trachytic beaches form an unusual scenic feature in the Azores. Like its predecessor, this eruption also heralded a dormant period, which lasted for 10 000 years. Fogo caldera probably collapsed intermittently because it has steep sides that are often coated with pumice fragments, its edges are scalloped into distinct lobes, and the lake occupies three intersecting hollows that mark the shifting focus of the eruptions.

Activity resumed within the caldera between about 4675 and 4435 years ago with the short Plinian outbursts, perhaps spread over decades or centuries, that expelled the Fogo-A pumice. Nuées ardentes and ashfalls distributed a thick blanket of buff-grey pumice and made a valuable indicator bed that is still 5 m thick on the south coast and covers about 200 km². Mudflows then also swept down to both coasts and covered about 24 km². Four similar but less powerful eruptions followed. The Fogo-B pumice came from a vent just north of the caldera some 3242 years ago. The Fogo-C ochre pumice, which is notably rich in lithic fragments, was expelled from a vent in the south of the caldera. Fogo-D erupted masses of pumice from the centre of the caldera. The latest Plinian eruption of Fogo took place in 1563. Severe earthquakes began on 24 June and lasted for five days. On 29 June, a huge Plinian column burst from the Fogo caldera. The wind spread most of the pale trachytic ash and pumice that covered more than 200 km² over the east of the island before the eruption ended on 3 July. This eruption may have induced the latest episode of collapse in the caldera.

Água de Pau also has more satellite cones than its companions, with the trachytic Cerrado Novo and the hawaiitic Monte Escuro among the most prominent (Zbyszewski et al. 1958). About 15 eruptions have occurred within the past 10 000 years on its western flanks; and some of the older cones and flows, for example, have been dated to 8700 and 6500 years ago. The hawaiitic flow that forms a lava delta on the north coast at Ribeira Seca erupted 1790 years ago, the Pico das Mos

about 1250 years ago, and the Mata das Feiticeiras only about 1000 years ago (Moore 1991). Other un-named cinder cones erupted near the south coast, including the cone rising to 153 m, which emitted the fine delta of alkali olivine basalt on which the village of Caloura now stands. This cone has been dated to the period between the Fogo-A eruption and the D eruption from Sete Cidades caldera, that is, between about 5000 and 3500 years ago. The vitality of Água de Pau volcano is also shown on its northern flanks where there are hot springs of sufficient power to be harnessed by the geothermal station of Caldeiras, 4 km southeast of Ribeira Grande.

The saddle of the Região dos Picos

The saddle of the Região dos Picos, stretching across São Miguel between Sete Cidades and Àgua de Pau, marks the greatest concentration of recent basaltic activity on the island. Although it began to form about 50 000 years ago, its surface features are much younger. It is dominated by about 250 cinder cones and their cultivated ash-covered lava flows, which originated along en echelon fissures trending slightly obliquely to the main northwest–southeast grain of the western half of the island. Many vents were so closely spaced that the cinder cones are joined together and also tend to be smaller than their more widely separated companions, which are often nearly 200 m high. Although several lava flows reached either the north or south coasts, most of them are thin and less than 5 km long.

The older cones, erupted before the Fogo-A explosion 5000 years ago, occupy a broad band parallel to the south coast and lying 5–6 km inland, which extends up to Sete Cidades volcano. Several cones erupted after the Fogo-A eruption and before the Sete D eruption of about 3500 years ago. These include the Serra Gorda, which has, no doubt, dominated the saddle ever since it erupted about 4400 years ago. As its name suggests, this large cinder cone stands over 200 m high and 1 km across, and is easily recognized by its double-hump summit slashed by a fault-guided breach. The vent also expelled several basaltic lava flows that spread in an irregular apron around its southern base. At about the same time, Surtseyan eruptions formed the Capelas cone on the north coast and the Ilhéu Rosto de Cão off the south coast.

At least 18 eruptions have occurred from

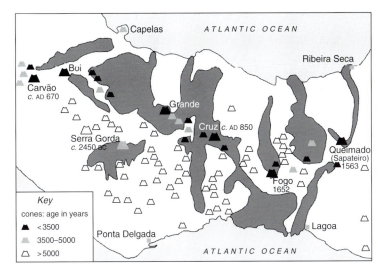

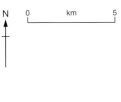

Recent eruptions in the Região dos Picos, São Miguel.

fissures in this area within the past 3000 years. The fissure some 2 km long between the Pico do Cedro and the Pico do Enforçado gave rise to eight cinder cones and a broad basaltic flow spreading to the northeast. However, the backbone of the saddle is the fissure stretching 5 km long that formed the Pico Grande, Mato do Leal and Pico Cruz, as well as seven other smaller cinder cones and two explosion pits. Pico Cruz erupted trachytic fragments some 1100 years ago. But perhaps the main contributions of this fissure were the large basaltic flows that spread to both the north and south coasts, where they form a delta, 4 km wide, west of Lagoa.

Activity here since the settlement of the island has been on a smaller scale. In June 1563, a basaltic eruption took place at the same time as the great trachytic explosion in the Fogo caldera. It occurred when a short fissure cracked open the trachytic dome of Queimado (or Pico Sapateiro), formed several little craters on its crest and gave off a thin basanitic flow that swamped part of the village of Ribeira Seca on the north coast. On 19 October 1652, an eruption from a vent at the base of Fogo 2 cone formed a conelet, Fogo 1, and a mugearite flow that reached the sea 4 km away.

Sete Cidades stratovolcano

Sete Cidades stratovolcano covers all the northwest of São Miguel. It grew up from a base 2000 m deep on the ocean floor and was a separate island until eruptions in the central saddle joined it to the rest of São Miguel (Zbyszewski 1980). It is now 14 km across at sea level, with a volume above sea level of some 70 km³. Its smooth slopes rarely exceed 10° and are covered with farmland and plantations of the conifer *Cryptomeria*. The Sete Cidades caldera crowns the stratovolcano and owes its name to the legend that seven cities were engulfed when it collapsed. Almost 5 km in diameter and 6 km³ in volume, the caldera is home to rich vegetation, ranging from laurel and cryptomeria to hydrangea and the now rampant ginger lily (*Hedychium gardnerianum*). Several squat wooded cones and the village of Sete Cidades occupy its floor, but its most striking features are two lakes, Lagoa Azul and Lagoa Verde, which often reflect the blue and green of their names. This majestic caldera, whose extraordinary serenity belies its violent origins, is the deepest and perhaps the most beautiful in the Azores. The scene from the Visto do Rei ("king's view") is the most photographed in the islands.

Sete Cidades stratovolcano began to grow up about 290 000 years ago, at about the same time as Água de Pau. The early eruptions of basalts were gradually succeeded by trachytes and it was only in later life that the volcano turned to the violent trachytic explosions that have blanketed its slopes with layers of pumice. The Sete Cidades caldera collapsed about 22 000 years ago during a Plinian outburst of trachytic pumice. Its appearance bears out its youth: the rim is undissected and the inner scarps have not weathered enough to form large basal screes. Many nuées ardentes surged out during the eruption, especially to the southeast, and massive deposits of pumice are still 60 m thick on the coast, 13 km from the caldera. After the caldera formed, a mugearite flow erupted near Feteiras on the

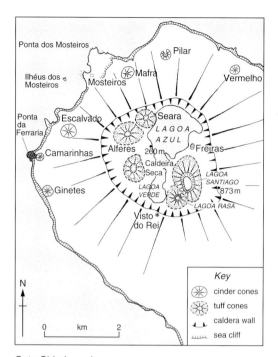

Sete Cidades volcano.

Just over 5000 years ago, a series of a dozen violent trachytic eruptions began, which sprang from vents along a ring fracture situated about 1 km inside the caldera walls. They formed a ring of squat cones – Seara, the Caldeira do Alferes, the Caldeira Seca – and the un-named steeper tuff cones containing the crater lakes of Lagoa Rasa and Lagoa de Santiago. In addition, the Lagoa Azul and Lagoa Verde are probably maars resulting from hydrovolcanic explosions. These seven features may have been at the origin of the Seven Cities legend.

These eruptions have been designated Sete A–L (Booth et al. 1978). About 5000 years ago, a Surtseyan eruption from a lake within the caldera built the Seara tuff cone and distributed the first (Sete A) layer of fine grey ash over the stratovolcano. An eruption on dry land formed the Alferes trachytic cinder cone perhaps about 4000 years ago. Most of the remaining eruptions came either from the craters of Lagoa Rasa or Lagoa de Santiago, or from both of them. They not only ejected thick layers of ash and pumice well beyond the caldera but also built two prominent tuff cones within it. For all its eruptive vigour, however, the Lagoa Rasa crater now occupies a shallow basin, choked with thick pumice and ash. But the Lagoa de Santiago is a fine explosion crater, made sinister by the steep wooded slopes of pumice rising 200 m around it. Some of the

southeastern flanks of the stratovolcano about 20 890 years ago, and domes and nuées ardentes erupted about 17 160 years ago near the Ponta do Escavaldo on the west coast. But the caldera itself stayed quiet for 17 000 years.

Sete Cidades caldera, looking northwards from the Visto do Rei.

later eruptions from these craters have been dated. The Lagoa de Santiago expelled the Sete J layer 1860 years ago, and the Lagoa Rasa expelled the Sete K layer 1570 years ago. The last ash layer (Sete L) erupted in a lake from the vent that built the Caldeira Seca at the same time. Its slopes are scarred by many closely spaced gullies, which make it look much older than it is, because carbon-14 measurements indicate that it erupted either 660 or 500 years ago (Moore 1991). These dates could correspond with an eruption that is said to have occurred in 1444. The small cinder cone of Freiras, beside the Lagoa Azul, has also been proposed as a source of this eruption (Machado 1967). However, the description is a much better match to a more violent hydrovolcanic eruption. Indeed, an early chronicler, Gaspar Fructuoso (1591), recounted how the first Portuguese explorers had made a note from their boat of a distinctive peak at Sete Cidades to aid future navigation. When they returned with a group of colonists a year later, in 1439, they saw that the sea was cluttered with pumice and tree trunks, and that the distinctive peak had disappeared, because "fire had arisen and burned it in the interim. And, for most of their first year of settlement, the colonists could hear, from their straw huts, the roars and bellows given out by the Earth, with great tremors from the subsidence and burning of the peak that had disappeared." But the attribution of this eruption to the Caldeira Seca also poses problems. The first settlers could not have seen the Caldeira Seca from their ship because it would have been hidden within the Sete Cidades caldera: it could not have been "the distinctive peak" that they noted. It has thus been suggested that the eruption might have come from the Lagoa do Canário, on the upper eastern flanks of the stratovolcano (Forjaz 1997a).

In view of all the commotion in the caldera during the past 5000 years, it is surprising that no volcanic activity has disturbed the caldera since those early days of Portuguese settlement. However, the earthquakes of low magnitude during the summer of 1998 could have been the first indications of a revival.

Activity on the Sete Cidades volcano during the past few millennia was not confined solely to the caldera. Even more basaltic emissions erupted from satellite vents all around the lower flanks of the volcano and in Maciço das Lagoas, which stretches southeastwards from the caldera. Eruptions have also been recorded offshore.

Maciço das Lagoas

Maciço das Lagoas (the Lakes Massif) was named thus because many craters are filled with lakes, and yet more lakes have formed where the drainage has been impounded between the cones. The eruptions occurred where three major fissures extend southeastwards from the caldera of Sete Cidades. In fact, it forms the highest upland area on the stratovolcano, culminating in the Pico de Eguas at 873 m, and the vents were so closely spaced that the basaltic cinder cones often merge together. The three fissures were active at different times. The central fissure erupted about 3500 years ago, just before the Sete D layer exploded from the caldera; the Pico de Eguas erupted 2700 years ago on the southwestern fissure; and the 200 m-high cinder cone of the Pico do Carvão erupted on the northeastern fissure 1280 years ago (Moore 1990). The cinder cones are up to 200 m high and their craters are unusually deep, which suggests some hydrovolcanic influence in their construction.

Eruptions and erosion on the coast around Sete Cidades volcano

The eruptions of the coastal fringe have formed cones and flows of basalt, mugearite and trachyte. Surtseyan activity on the northeastern coast formed the Capelas volcano, which is now reduced to a half-eroded promontory. However, the remaining eruptions usually took place, on radial fissures, on top of a 100 m cliff. Many of these cones, such as Várzea, Lagoa do Pilar, and Mafra, are accompanied by lava flows that cascaded down the cliffs and formed lava deltas along the shore. These eruptions have taken place throughout the past 5000 years, but their blunt outlines suggest that they are not very young. The most recent eruption on the stratovolcano on land probably took place 840 years ago. It built up the basanite cone of the Pico das Camarinhas and emitted the flow that now extends seawards as the Ponta da Ferraria, the westernmost point of São Miguel.

One of the most active fissures in the Azores during historical times extends northwestwards from the Ponta da Ferraria. It has produced three recorded submarine eruptions. On 3 July 1638, after a week of earthquakes, "a great fire hurled masses of black sand as high as three church steeples placed one upon the other, and the

Don João de Castro Bank

Don João de Castro Bank marks the crest of a growing stratovolcano and could be the site of one of the next major islands to form in the Azores. In 1941, a shoal called the Don João de Castro Bank was discovered 64 km northwest of São Miguel and 51 km southeast of Terceira. A spectacular submarine Surtseyan eruption had begun there on 9 December 1720 and could be seen from both islands. In four days, it gave birth to an islet, which apparently reached nearly 1 km across and 180 m high during the ensuing two weeks. Towards the end of December, the activity calmed down enough to enable visitors to approach by boat. They observed that the island was composed of ash and fragments and that one vent was still operating. Activity waned during the following months and the winter waves had destroyed the island by March 1721, leaving only the submerged shoal behind. This shoal was the location of several earthquake spasms in 1988 and 1989. A magnitude 5.5 earthquake occurred on 27 June 1997 and was followed by 45 earthquakes exceeding magnitude 4.0 and some 2000 weaker tremors during the following two months. Although no fragments reached the open air, such seismic activity suggests that another eruption could have occurred at that time.

billowing vapours joined the clouds. The sand that fell back down formed an island that only disappeared with the advent of winter, and left behind a vast and dangerous shoal." (Weston 1964). Calm returned on 28 July. In December 1682, violent earthquakes rocked all the island. "The earthquake, while the morning preachers were preaching, on 13 December was so powerful that people thought that they were going to be cast from the Earth. But it pleased God to raise up fire in the sea during the following week, almost four leagues off Ferraria . . . roasting a quantity of fish that were thrown up onto the coast. So much pumice was floating on the sea that a big sailing ship en route from Angra [in Terceira] could not force a passage through it." (Montalverne, in Weston 1964). This brief, but powerful, eruption lasted about two or three days and seems to have taken place at a depth of about 300 m.

After seven months of earthquakes, there was another eruption on 31 January 1811, on the same fissure, but in shallower water, less than 2 km offshore. It produced a Surtseyan display of explosions, and jets of ash and cinders that lasted for eight days and spread ash and sulphurous fumes as far as Ponta Delgada, 20 km away. But it never formed an islet. This lapse was rectified when another Surtseyan eruption broke out just to the northwest, on the morning of 14 June 1811, after four days of vigorous earthquakes. Great clouds of ash, fumes and cinders thundered from the sea, and an islet started to form on 16 June.

Cinder cone eroded by the sea, north of Mosteiros. Weaker fragments have been eroded much more than the consolidated bands of cinder and spatter that now form the curved reefs. The original vent was near the left edge of the view.

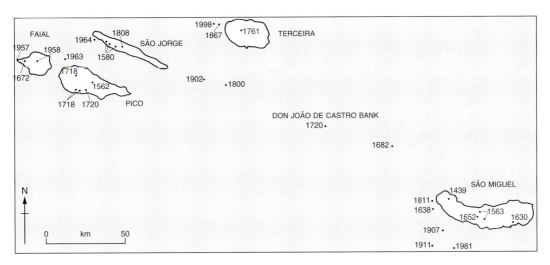

Locations and dates of eruptions during historical time in and around the Azores.

It was 1600 m in circumference and about 70 m high when the eruption waned on 22 June and began to emit nothing but fumes. The British frigate *Sabrina* was in São Miguel at the time and, on 4 July, Captain Tillard landed on the island, planted the British flag upon it, claimed it for King George III, and named it after his ship. However, his imperial enterprise was not rewarded, for Sabrina's fragile tuffs succumbed to the Atlantic waves before October was out.

The relationships between coastal volcanic activity and marine erosion are beautifully displayed at Mosteiros in the far northwest of São Miguel. Off shore, the four islets of Ilhéus dos Mosteiros are all that remains of the eastern half of a Surtseyan tuff cone. Off the northern end of Mosteiros village lies an older and much more eroded cinder cone, which must have erupted on dry land when sea level was lower than at present. When sea level rose, the waves planed off the cone to such an extent that only its concentric black layers of spatter jutted out from the sea. Then, Mafra and a smaller vent nearby erupted and their lavas formed a rugged delta that covered the southern part of the eroded cinder cone on the shore. In turn, the Atlantic waves planed off the surface of the lavas, when the sea stood some 15 m higher than today. The village of Mosteiros stands on this surface. Now, the present waves are starting their work all over again and attacking the exposed fringes of the lava delta.

SANTA MARIA

Santa Maria is a small island, 16 km long and 8 km wide and rising only to 590 m. It has shown no sign of activity since it became the first island in the Azores to be settled in 1439. Basaltic eruptions built up the island from a depth of about 2000 m on the floor of the Atlantic Ocean. The oldest basalts yet dated, near Vila do Porto, are 5.14 million years old, and similar hawaiitic lavas from Pico Alto in the centre of the island are 5.11 million years old (Féraud et al. 1980). Some flows contain layers of marine sandstones and limestones, which mark a period of subsidence in the history of Santa Maria, which has apparently no parallel in other islands. About 4.2 million years ago, renewed activity formed the Pico do Facho in the west and much of the undulating zone in the centre and east. All these volcanic masses have been deeply weathered and eroded by both fluvial and marine erosion, and São Lourenço volcano, for example, has been almost removed by the sea. Santa Maria has witnessed neither earthquakes nor thermal-spring activity during historical times.

TERCEIRA

Terceira is so named because it was the third island in the Azores to be discovered and settled about 1450. It forms an ellipse, 30 km long from east to west, and 15 km wide from north to south, with an area of 406 km². On the fertile coastal plain, the volcanic rocks have weathered enough to provide viable farmland and support an almost unbroken necklace of settlements around the island. It is often separated from the sea by fretted lava cliffs, about 20 m high, but the beautiful capital of Terceira, Angra do Heroísmo, grew up where the Surtseyan cone of Monte Brasil protects the most sheltered bay on the south coast. The eastern third of the island is low and gently undulating, but its higher and hillier central and western parts in the interior form grassy moorlands that rise to 1021 m in the Serra de Santa Bárbara. It is here that the lavas are younger, where the latest eruption took place on the island in 1761, and where fumaroles, such as the Furnas do Enxofre, are still active (Zbyszewski et al. 1971). Terceira has also suffered several earthquakes since the Portuguese settlement; the one on 1 January 1980 caused much damage, especially in Angra do Heroísmo.

Terceira rises from a depth of over 1500 m on the floor of the Atlantic Ocean, and only just over a tenth of the volcanic pile rises above sea level. The submarine part forms a basal shield of olivine basalts and hawaiites that were expelled during the past million years (Self 1976). However, most of the exposed rocks of Terceira are much younger and the oldest dated lavas are 300 000 years old (Féraud et al. 1980). The rocks also evolved from olivine basalts through hawaiites and mugearites and benmoreites to trachytes and even to pantellerites, which outcrop nowhere

Recent cones and lava flows, and caldera walls.

else in the Azores (Self & Gunn 1976). In general, the peralkaline eruptions give rise to lava flows and domes rather than to fragments. For most of the history of the island, basalts and more alkaline lavas erupted in almost equal proportions; however, during the past 23 000 years, eruptions of alkaline lavas were notably more voluminous than basalts, although they were less frequent.

The Terceira Rift curves across the island from northwest to southeast. At present, it forms an active fissure zone, about 2 km wide, with over two dozen recent cinder cones, basaltic flows, spatter cones, intruded dykes, and elongated clefts. But it is a rather enigmatic feature, because it hardly ever forms a distinct morphological unit, especially where it has been recently less active in the east. However, the four stratovolcanoes that dominate the scenery of Terceira all lie on, or close to, the rift zone. They formed in succession from east to west, caldera-forming eruptions have decapitated their summits, and the western pair is still active.

Cinquo Picos stratovolcano

Cinquo Picos stratovolcano occupies much of southeastern Terceira, and its caldera, 7 km across, is one of the largest in the Azores. It makes little visual impact because it is the oldest, lowest, smoothest, most gently sloping and most eroded on the island. Only two sectors survive in the landscape: the smooth Serra do Cume, in the northeast, and the even smaller Serra da Ribeirinha is all that remains of the opposite rim. The stratovolcano is composed chiefly of mugearite lavas, with some flows of olivine basalt, hawaiite and trachyte on its flanks. Hawaiites near the summit of the Serra do Cume have been dated to 300 000 years ago. Some time afterwards, much pumice erupted and the caldera formed. The floor of the caldera is covered by much more recent ash and flows, and makes a vast chequerboard of meadows, which is decorated by the five much younger basaltic cinder cones – the "cinquo picos" – that gave the caldera its name.

Guilherme Moniz stratovolcano

Guilherme Moniz stratovolcano and the rim of its caldera are represented by the crescent of the Serra do Morião, which reaches 632 m and forms the northern backcloth to Angra do Heroísmo.

The outlines of Guilherme Moniz caldera are altogether sharper than those of its neighbour. The inward-facing scarps of the caldera are often vertical and more than 150 m high. They are composed of thick flows and domes of trachyte, with few layers of fragments. Thus, effusive activity must have dominated the life of the stratovolcano until the violent eruptions that created the caldera. This event has not been accurately dated, but on morphological grounds it could well have taken place less than 100 000 years ago. The fringes of Pico Alto volcano have covered the northern sector, and recent basalts from satellite vents on its lower flanks invaded the Guilherme Moniz caldera, which therefore became a rare example of a caldera floored by lavas that erupted from another volcano.

The wooded lava flow forming the Ponta do Mistério and the flanks of the Pico Alto volcano.

Pico Alto stratovolcano

Pico Alto has perhaps the most complicated history and structure of any volcano in Terceira, or, indeed, in the Azores. It occupies much of north-central Terceira, culminates at 808 m, and is composed almost entirely of trachytes and pantellerites. It is probably less than 60 000 years old. It developed a caldera, but subsequent eruptions from over a dozen vents filled it so quickly and completely that the initial hollow has been replaced by an isolated upland wilderness of confused relief around Pico Alto itself. Several explosions of pumice occurred and one of the largest formed the narrow train of the Angra ignimbrite about 23 000 years ago. The more widespread Lajes Ignimbrite has been dated to 18 600 years ago at São Mateus on the south coast and to 19 680 years ago at Lajes, which could indicate that there were two separate eruptions. Subsequent eruptions emitted trachytic flows, reaching 50 m thick, and extruded domes, some of which sagged down to form additional thick flows. The southernmost flanks of Pico Alto impinge upon the Terceira Rift, where some vents are still active; the Furnas do Enxofre, for instance, now constitutes the major area of solfataras and fumaroles on the island.

Santa Bárbara stratovolcano

As Pico Alto grew to maturity, Santa Bárbara stratovolcano began to develop 10 km to the west. Santa Bárbara occupies the western part of

Terceira and its diameter at sea level is about 12 km. It is the most clearly defined, highest, steepest and youngest stratovolcano in Terceira. Flows of mugearite and hawaiite form a shield up to about 700 m that is surmounted by a steeper central cone that reaches 1021 m, where the benmoreites found near the summit are less than 29 000 years old. A violent eruption destroyed the crest of Santa Bárbara and formed a small caldera about 25 000 years ago. Much of it was then filled with trachytic domes and pitons, but about 15 000 years ago a small caldera formed a narrow hollow within its predecessor. Several trachytic domes have already extruded within it, and it now also contains a small lake.

During the past few thousand years, several cinder cones and domes have erupted on the flanks of Santa Bárbara. In the south, for example, the Cerrado das Sete and the Pico dos Enes probably lie on one of its radial fissures, and the Pico da Catarina Vieira and the Pico dos Padres arose on another. The prominent trachytic dome and lava flow of the Pico Rachado, which stands 200 m high on the upper northern flanks of Santa Bárbara, could also belong to a radial fissure. However, the radial fissures on both the eastern and western flanks of the volcano merge with those developed on the Terceira Rift zone. These eruptions are probably among the youngest on the stratovolcano and form both trachytic domes, such as Serreta and Negrão in the west, and cinder cones such as the Pico da Candela in the east.

Pico do Gaspar spatter cone and the Lagoa do Negro.

The Terceira Rift

The Terceira Rift has its finest expression – and its latest eruptions – in the upland saddle between the Pico Alto and Santa Bárbara volcanoes. They gave rise to basaltic cinder cones that are typified by Bagacina, whose very name is the local word for volcanic cinders. It is a fresh cone, 120 m high, that was apparently constructed within the past few thousand years at most, and it is breached by a hawaiite flow that reaches the south coast 5 km away near Angra do Heroísmo. Pico Gordo, its counterpart some 2 km to the northwest, expelled an even younger tongue of olivine basalt to the north coast, where it now juts out to sea in the spectacularly jagged reefs of the Ponta dos Biscoitos. Between Pico Gordo and Bagacina is a fine and unusually large spatter cone, the Pico do Gaspar, where a wall of spatter encircles a crater about 100 m deep. Its horned profile, forming the background to the Lagoa do Negro, is one of the most striking in Terceira. Farther east, the Algar do Carvão erupted 2115 years ago on the lower flanks of the Pico Alto volcano. It still retains both the gaping abyss of the main vent and the magnificently rugged pinnacles on its lava flow. The basalts invaded the Guilherme Moniz caldera, where they were ponded back in a lake of lava. At length, part of the flow spilled from the caldera, swept eastwards around the slightly older cones of the Pico da Cruz, Areeiro and Gualpanar, and escaped in two fine rugged tongues, one extending to the north coast and the other to the south coast. Even the calmer easternmost part of the rift zone has also experienced

some recent basaltic eruptions, and the latest flows jut out to sea in a lobe at Porto Martins.

All this activity probably occurred within a few millennia of the Portuguese settlement, but the only historically recorded volcanic activity in Terceira took place in April 1761 (Canto 1882a, Zbyszewski et al. 1971). It occurred in two distinct areas lying just to the north of the rift zone. After three days of earthquakes, the mistério Negro extruded in the west on 17 April and oozed down the lowest fringes of Santa Bárbara volcano. It now forms a squat rounded hump of black pantelleritic trachyte made up of domes and short lava flows about 50 m thick. On 21 April, an entirely different eruption began some 2 km away to the east. It produced three small basaltic cinder cones, of which the appropriately named Pico do Fogo is the largest. It emitted a short lava flow to the west and a flow some 2 km long to the north-northeast. But the longest flow of all, a tongue of hawaiite, travelled 5 km as far as the church at Biscoitos and partly covered the slightly older alkali olivine basalts that had previously been expelled from the Pico Gordo. These eruptions ended on 28 April 1761.

Submarine eruptions

Four submarine eruptions have been recorded off Terceira. A small eruption began on 24 June 1800, when the sea glowed, gave off fumes, and rumbled like thunder for three days. In 1867, five months of earthquakes culminated in a strong shock on 1 June that damaged 200 houses in and

Pico do Fogo

The Terceira Rift in the central part of the island.

around the coastal village of Serreta. That night, an eruption boomed out some 6 km from the Ponta da Serreta and, during the next week, great fountains of water, steam and occasional masses of cinders gushed skywards from six or seven vents. The third eruption, which took place nearby, was even more discrete. It was identified only when the telegraph cable between Pico and Terceira was broken, burned and buried in tuffs at depths varying between 450 m and 1400 m. It caused little comment, because it occurred during the night of 7–8 May 1902 between the notorious eruptions of the Soufrière of St Vincent and Montagne Pelée in Martinique. The latest activity has been called the Serreta eruption because it occurred out to sea, 9–14 km northwest of that village. It began with small earthquakes in November 1998. During the following four months, white fumes rose intermittently from vents at a depth of about 500 m, and orange lights were observed on 23 December. On 8 January 1999, incandescent lava blocks, up to 3 m across, kept on rising to the surface in half a dozen places, and each floated for several minutes. They were composed of alkali basalts that were riddled with gas holes, and it seems that they were emitted as hot lava balloons, rather like detached pillow lavas, that lost their enclosed

The lavas of Biscoitos and the half-eroded cone of Matias Simão, on the coast of northern Terceira.

Monte Brasil sheltering the harbour of Angra do Heroísmo.

gas and sank when the congealed skin cracked on contact with the air (Forjaz et al. 2000).

These eruptions formed no islets, presumably because they occur in waters that are still deep enough to stifle any Surtseyan activity. However, the southern coast of Terceira has experienced several Surtseyan eruptions in the recent past. The remains of two tuff cones lie off Angra do Heroísmo; and the Ilhéus das Cabras, 6 km away to the east, mark the remnants of an older and more exposed Surtseyan cone. It has already been cut into two by the waves, although its original conical structures are still evident.

In terms of settlement, the formation of Monte Brasil was by far the most important of all the recent Surtseyan eruptions in Terceira, because it created the sheltered bay (or angra) that led to the foundation of the capital, Angra do Heroísmo. A basaltic eruption in shallow water formed the bulky Surtseyan tuff cone of Monte Brasil, which rises 205 m above the waves and is over 1 km in diameter. Its crater is 170 m deep and powerful explosions kept open the southern sector of the cone. The whole mass is composed of thin beds of fine yellow and brown tuffs, and a high isthmus of windblown fragments joins the cone to the mainland. Although the sea has cut cliffs around much of Monte Brasil, it has also been partly protected by a skirt of black basaltic flows that emerged during its later Strombolian phase. As a result, Monte Brasil still forms a worthy bastion for Angra do Heroísmo, in spite of centuries of marine attack.

GRACIOSA

Graciosa has a greater proportion of cultivated land than any other island in the Azores, which is a reflection of its lower altitude, drier climate, deeper soils and gentler slopes. The centre, and oldest part, of the island is composed of two upland blocks that face each other across a broad axial valley. To the north lies an extensive plain scattered with fresh cinder cones, but to the south the scenery is dominated by the strato-volcano, Caldeira. Basaltic flows and fragments often predominate throughout the island, but notable amounts of hawaiite, mugearite and trachyte also erupted, especially as the strato-volcano developed. No volcanic eruptions have been recorded since Graciosa was first settled in about 1480 and, until the repercussions of the Terceira earthquake were felt in 1980, the only previous major seismological event had occurred when Praia and Luz were damaged in 1730 (Zbyszewski et al. 1972, Scarth 1994).

Eruptions began at a depth of 1500 m or more on the ocean floor, but the oldest emerged rocks occur in the piles of horizontal flows comprising the central upland blocks. Hawaiites forming the base of the Serra das Fontes have been dated to 620 000 years ago, trachytes at the base of the Serra Dormida are 350 000 years old, and those near the summit of the adjacent Serra Branca date from 270 000 years ago. These uplands are demarcated by steep straight scarps, often over 100 m high, that seem to have formed by faulting. Thus, the two older upland areas of Graciosa seem to form upthrown fault blocks that are now separated by a central fault trough.

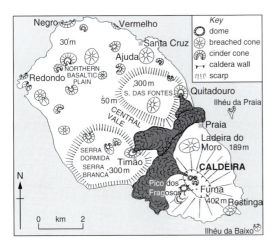

The chief volcanic features of Graciosa.

Caldeira is now the most striking natural feature of Graciosa. It probably first emerged as a separate island to the south of the upthrown lava blocks. Rising to only 402 m and stretching only 5 km across, the stratovolcano is relatively young, and erosion has been limited to valleys etched only 2 m or 3 m deep near the summit, and to rugged sea cliffs that are usually less than 20 m high, except around Restinga in the far southeast.

The stratovolcano was built up by eruptions of basaltic flows and layers of cinders, lapilli, ash and tuffs, 20 cm to 8 m thick. They enclose only two thin fossil soils, which indicates that the volcano grew up rapidly (Zbyszewski et al. 1972).The inner walls of the caldera are composed almost entirely of lava flows, which suggests that effusive eruptions predominated for much of the life of the volcano. The transition to violent eruptions was therefore sudden and belated. The caldera is elliptical, stretching 1600 m from northwest to southeast, but it is only 875 m broad, and its rim is notched in the northwest. It is certainly less than 200 000 years old, and the sharp rim, steep walls and paucity of weathered material all suggest that it would be surprising if the caldera were, in fact, more than 20 000 years old.

The original amphitheatre might have been more than 400 m deep, but later eruptions have already covered its floor. In the northwest, two basaltic domes rise 100 m high and a flank of the western dome has collapsed to reveal its typical onion-skin layering. In the centre of the caldera, three small cinder cones form grassy mounds between 50 m and 100 m high. A basaltic flow also erupted and escaped through the notch on the northwestern rim of the caldera, and then spread around the flanks of the stratovolcano. The smaller branch flowed eastwards to the sea near Praia, and the larger western branch stretched out into Folga Bay. It is possible, indeed, that this flow first joined Caldeira to central Graciosa. However, the most original, and probably the latest, feature in the caldera lies in its lowest southeastern part in the Furna do Enxofre ("sulphur oven"). This is a dark vent, 47 m deep, with two main fumaroles and an underground lake with a floor below sea level.

Meanwhile, other eruptions had taken place on the flanks of the stratovolcano. The Pico da Ladeira do Moro formed an impressive double cinder cone, some 150 m high and nearly 1 km across, that now dominates Praia. Close to the southwestern rim of the caldera, twin domes

CALDEIRA

Pico dos Fragosos

LADEIRA DO MORO

LAVA FLOWS
FROM TIMÃO

The town of Praia, Ladeira do Moro and the flanks of Caldeira.

form the Pico dos Fragosos, which have bristling rounded summits of hawaiite and mugearite. The eastern dome collapsed and formed a viscous lava flow that congealed in a thick rugged steep-sided tongue stretching 1 km down slope. Other satellite vents of the stratovolcano, erupted the cones, the much eroded remains of which now form the islets of Ilhéu de Praia and the Ilhéu de Baixo. Although they now lie in the sea, they are composed of ash and cinders that erupted on land, probably when sea level was lower several thousand years ago.

The northwestern third of Graciosa is a low-lying plain, covering about 25 km². It is composed of many basaltic flows that are often weathered deeply enough to support lava-walled vineyards, meadows and even arable land. The flows reach the coast in intricate and very rugged minor cliffs. Dotted along fissures generally trending from northwest to southeast are 20 or so breached cones 50–100 m high. Most of these are cinder cones, with sharply defined outlines that are probably only a few thousand, or possibly even only a few hundred, years old. Although the majority are Strombolian cones, one of the most prominent, Ajuda, is a typical Surtseyan cone, which dominates the skyline of the capital, Santa Cruz. Within several other coastal cones, such as Redondo, Negro and Vermelho in the northwest, layers of yellow tuffs between the cinders reveal Surtseyan episodes in their formation. These cones have already lost two thirds of their volume to the sea, but yet have scarcely been

eroded more than Capelinhos in Faial, which erupted as recently in 1957–8.

About 15 breached cinder cones also erupted, at much the same time, on the plateaux of the Serra das Fontes, the Serra Branca and the Serra Dormida. Here, relief has played an important role, for several cones formed on the crests of the bounding scarps. Quitadouro, for example, erupted at the summit of the southern edge of the Serra das Fontes. The southern sector could not form because lavas cascaded down the scarp towards Praia. The lava flow and the slumped cinders now form a chaotic tongue issuing from the reddish-brown armchair of Quitadouro. Its counterpart, the Pico Timão, is a fresh cone, about 100 m high and with a clearly marked crater rim, which occupies a commanding position at the top of the edge of the Serra Dormida. Timão forms a crescent of bare reddish cinders that is breached where a lava flow escaped down the scarp and prevented the northeastern sector of the cone from forming. The basalts spread southeastwards and entered the sea, 4 km from the vent, in a 1 km-wide lobe north of Praia, and they have maintained their rugged aa surface that plantations of eucalyptus, cryptomeria and laurel cannot disguise. In their progress, they also covered parts of the flow that had issued from the caldera, as well as the snout of the flow from Quitadouro. The morphology suggests that Timão cannot have erupted long before the first colonization of Graciosa in about 1480.

SÃO JORGE

Seen from Graciosa, the island of São Jorge looks like a defensive wall protecting the conical fortress of Pico beyond. Its crest rises about 800 m above sea level and culminates at 1053 m in the Pico da Esperança. São Jorge forms a spine, rarely more than 6 km wide, running in a straight line for 54 km from northwest to southeast, and it is the only island in the Azores where fissures have played such an overwhelming role in the scenery. The straight and steep cliffs demarcating the island further emphasize the rectilinear theme in the landscape. São Jorge bears no trace of a stratovolcano or a caldera.

Activity began at about 1000 m down on the floor of the Atlantic Ocean. The eruptions above sea level produced fluid flows of alkali basalt, with some hawaiites and a few mugearites, whereas rather more explosive eruptions formed the many cinder cones, about 100 m high, that dominate the backbone of the island (Forjaz & Fernandes 1975). The older parts of São Jorge lie in the east and the most recent parts in the west. There have been two terrestrial and perhaps two submarine eruptions in São Jorge during historical times.

São Jorge is divided into two unequal parts by a band of faults trending from north-northwest to south-southeast, which cut slightly obliquely across the predominant tectonic trends on the island. The Ribeira Seca stream has cut deeply into this band of fractured rock. The eastern third of São Jorge forms the Serra do Topo, which is composed of piles of thin flows of alkali basalt and hawaiite. They are sometimes interspersed with layers of weathered cinders, exhibiting various phases of repose, and yellowish tuffs, that

suggest hydrovolcanic eruptions. The dissected relief of the plateau lavas and the smooth outlines of the cinder cones, which have almost invariably lost trace of their craters, indicate that the Serra do Topo is the oldest part of São Jorge. The morphological inference is corroborated by dates ranging from 550 000 to 110 000 years in age (Féraud et al. 1980). Both the northern and southern coasts are apparently faulted down towards the sea, and the Serra do Topo formations are also brought to an abrupt end where the Ribeira Seca faults cross the island.

The remaining two thirds of São Jorge lying to the west of the Ribeira Seca faults are clearly younger than the area to the east. But the same eruptive style continued. The Rosais complex, about 30 000 years old, and the younger Manadas complex, about 24 000 years old, make up most of the broad ridge forming the centre and west of São Jorge. Thin alkali basalt and hawaiite flows once again emerged repeatedly from fissures and piled up in a total thickness of over 500 m. Older cinder cones were often buried beneath newer accumulations, but the fresh outlines and sharp craters of the youngest cones dominate the scenery along the spinal crest of the island. The Manadas cones, for example, near the northwestern end of São Jorge, were probably formed only during the past few thousand years. The cones are larger and more closely spaced in the centre of the island, but they become smaller and more scattered in the northwest. As in the eastern zone, most of the lava piles end abruptly in majestic straight cliffs that were probably created by faulting. These cliffs are so sheer that the few coastal settlements cling to a succession of **fajãs** – lava deltas or lobes of rubble that have fallen from cliffs.

These eruptions have continued into historical

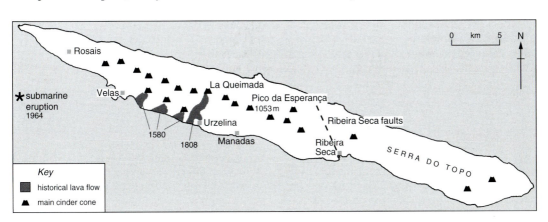

Recent features and historical lava flows of São Jorge.

times (Fouqué 1873, Canto 1880a, 1884). After over a hundred earthquakes had rocked the island during the previous two days, the first historic eruption on São Jorge began just east of Velas on 1 May 1580. The most remarkable event of this eruption was the appearance of nuées ardentes. Not only are they unusual in such basaltic fissure eruptions, but this was in fact the first time that they were named. The local eye witnesses described them as nubes ardentes, or scorching incandescent clouds, but the term was largely forgotten until the famous French geologist Lacroix described their more lethal counterparts that destroyed Saint-Pierre in Martinique in 1902. The basaltic eruption seems to have pursued an habitual course at first: two cinder cones quickly formed at Queimada, and lava flows made their way quickly to the shore. At this juncture, about a dozen men took a boat to rescue valuables from a house on the coast. Some of them were still searching the house when it was suddenly enshrouded by a nuée ardente. One of the men ran out towards the boat and he was so badly burned that his skin peeled off. Those left in the house perished. Another nuée ardente then caught the five men who had stayed behind on the boat, and their burns were so extensive that they only just escaped with their lives. The first two vents were active only for a few days, but the third went on erupting until the end of

August. It is said that 4000 head of cattle were lost in the eruption. Other nuées ardentes killed more people as they tried to rescue their possessions from the far less dangerous threat of the advancing basaltic flows. When the islanders realized that the volcano was unleashing still more of these unpredictable, inescapable and lethal clouds, they fled in terror from São Jorge. The lavas formed the mistério da Queimada.

Some believe that 18 volcanic islets formed off São Jorge during an eruption in 1757. However, it seems more likely that the large earthquake on 9 July 1757 provoked major landslides that were mistaken for volcanic eruptions when they crashed into the sea (Forjaz & Fernandes 1975). The major eruption on São Jorge was in 1808.

The latest volcanic event on São Jorge began on 15 February 1964, when more than 500 tremors shook the island during the next nine days. At the same time, a strong smell of sulphur fumes permeated the air on the west coast and the sea became discoloured. Although no cinders were thrown into the air from the ocean, it is most likely that a submarine eruption occurred about 100 m deep off Rosais (Forjaz & Fernandes 1975).

The eruption at Urzelina on São Jorge in 1808

The chief historical eruption on São Jorge began, exactly 228 years after the first, on 1 May 1808, when three spatter cones and lava flows gushed out northwest of Urzelina (Canto 1880a, 1884). But the main activity took place a little afterwards from the vent of La Queimada, to the northeast of the village. There was a thunderous rumbling of the ground, and no less than eight large, and five smaller, cinder cones exploded forth for several days. The mistério of alkali basalt poured down the scarp bordering the plateau and eventually reached the sea. It is said that some men took the poles that were used to hold the cover the Holy Sacrament during processions and dipped their ends into the molten lava in a propitiatory gesture. Others took the lava-coated sticks to use as divining rods.

However, the greatest danger was again to come from nuées ardentes. By the middle of May, the craters had gone silent, and it seemed that the lavas had done their worst. But, on 17 May, the first nuée ardente suddenly burst out from La Queimada, and surprised and killed some goat-herds and their flock up on the ridge. Some of those who tried to run away were burned when their breadbaskets caught fire. With prodigious force and a

terrifying noise, the nuée ardente surged over the ground, igniting everything in its wake, and descended in a huge black cloud onto Urzelina. In the church, the village priest thought that his last hour had come, but the building protected him from the worst of the hot blast. When he emerged from the church, he wandered, in shock, through the devastation until he met two priests and a group of victims; all had been badly burned. A few were fit enough to return home, but many others sought shelter in the church and in the houses nearby. Many had lost the skin from their exposed limbs; some people were so swollen and blackened that nobody could recognize them; and some had even had their legs broken by the blast and flying masonry. But all those who had breathed the fiery cloud had died at once. In all, more than 30 people were killed, including an elderly couple who were blasted away along with their home, and one victim that the nuée ardente threw into the sea. A little later, a lava flow swept down from the new volcano and half buried the church at Urzelina. The dire combination of flows and nuées ardentes plagued the area until 10 June. The survivors claimed that they had seen a vision of hell.

PICO

Pico is the second-largest island in the Azores, and its name is proof enough that its outstanding landmark is a majestic volcano, rising to 2351 m – the highest summit in Portugal, and twice as tall as any other peak in the Azores. Pico is the only stratovolcano in the archipelago that has not been decapitated by a caldera-forming eruption. The island of Pico is almost 50 km long, and the stratovolcano forms its broadest part in the west, where it reaches a width of 16 km, but its eastern tail is mostly less than 8 km across. These distinctive parts reflect different eruptive styles: a cluster of vents has built up the stratovolcano, but fissure eruptions have predominated in the east, where aligned cones decorate a high basaltic plateau, from which lava flows have cascaded down the steep cliffs to the sea. The eastern parts of Pico contain the oldest lavas, but they are apparently less than 37 000 years old (Woodhall 1974, Féraud et al. 1980). The whole island is young: Pico is the newest addition to the Azores, and it is higher, rockier and more obviously volcanic than any of its neighbours. It is also by far the stoniest place in the archipelago, largely because it has witnessed some of its most recent eruptions. They have made Pico the main home of the mistérios, the black lava flows that were so awe inspiring and inexplicable to the early settlers and which still seem to emerge menacingly from the recesses of the hills.

The Pico do Pico

The stratovolcano of Pico was built up from the ocean floor at a depth of 2000 m and it has grown into the bulkiest volcanic pile in the Azores. Except for a few short barrancos on its southern flanks, vast tracts of Pico have retained the slopes inherited from the latest eruptions. The Pico do Pico has two distinct parts: a gently sloping basal shield, decked with many cinder cones, forms its plinth, and a sharp break of slope at about 1100 m demarcates the steeper summit cone (Zbyszewski 1980). The whole volcano is composed mainly of cindery flows of alkali basalt, which are blanketed at times by basaltic ash, lapilli, cinders and scree. They give the bare upper parts of Pico their typical blue-grey colour and offer a precarious home for grass, myrtle and heather. Farther down, the flanks are wooded and scattered with laurel, tree heaths and vineyards with high walls. There is no hint of a caldera.

The basal shield or plinth of Pico

The basal shield of Pico reaches 16 km across at sea level and its slopes range from 10° to 20° and rise to about 1100 m. It contains most of the 125 or so cinder cones that have sprouted on Pico in recent millennia. Most of them rise on the lower reaches of the shield, near the present coast, but it is possible that later eruptions buried any older cinder cones in the upper areas of the shield. The cones commonly rise about 100 m and they are often breached by lava flows on their downslope sides. They are often aligned on either radial or regional fissures. In the southeast, for instance,

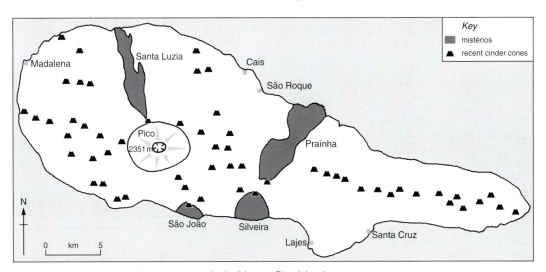

Pico stratovolcano, recent cinder cones and mistérios on Pico island.

The mistérios

The mistérios, the grim, black, rough and rugged aa lava flows, reach great sizes on Pico. Even from a distance, their presence is betrayed by planted eucalyptus, cryptomeria or laurel, or by stone-walled Verdelho vineyards, by the biscoitos of clinkery lava, or by rugged low promontories jutting out to sea. These lava flows inspired the awe of the early settlers, who called them mistérios long before the first eruptions were recorded on Pico in 1562. This suggests that several such flows retained their original and awesome appearance, and therefore had probably erupted shortly before the island was first settled in 1466. One of these eruptions took place between the Pico da Urze and the Cabeço da Fajã, and caused a mistério that displaced a stream to the west. Similarly, the three fresh un-named cinder cones, rising to 137 m, 182 m and 139 m above sea level between São Mateus and São Caetano on the south coast, clearly cannot have erupted long before the settlement of the island, and their wide mistério bulges 1 km out into the sea. At about the same time, the Cabeço Grande and the Cabeço Pequeno erupted the lavas around the well named hamlet of Biscoitos, near Madalena.

The four eruptions recorded during historical times on Pico emitted fluid flows of alkali basalt that are mostly less than 3 m thick (Canto 1879, 1882b, 1883). Each covered more than 10 km², and each created a large mistério. The first eruption to be reliably recorded in the Azores formed the broad mistério da Praínha (Weston 1964, Mitchell-Thomé 1981). After three weeks of earthquakes that shook the whole island, explosions of ash and fumes started the eruption early on 21 September 1562 and built the Cabeços do Mistério cinder cones in about a week. On 24 September, fluid basaltic lavas, in a mass 3 km wide, set fire to the dry scrub and woodland, cascaded down the 500 m-high scarp bordering the plateau, and entered the sea in the wide lobe that now forms the Ponta do Mistério. At times, 40 glowing streams of lava were lighting up the whole island at night. This eruption is reputed to have lasted for two years.

On 1 February 1718, a fissure trending almost from north to south split across the basal shield of Pico, but passed underneath the summit cone apparently without disturbing it. The eruptions in the northern sector built up seven small cinder cones, of which Laurenço Nuñes is the largest. The vents emitted the mistério da Santa Luzia in about two weeks. Its black alkali basalts spread 5 km as far as Santa Luzia and entered the sea at Cachorro in a magnificent array of black pinnacles. At the same time, the southern sector of the fissure opened, and four vents erupted the cinder cones of Cabeço de Cima and Cabeço de Baixo and emitted the mistério de São João, which destroyed the original village of that name. It now forms a black basaltic headland at the Ponta de São João. Two people are said to have died in this eruption, which apparently lasted for ten months. On 11 February 1718, an undersea eruption also began along the same fissure, 100 m off São João. It formed an islet that was joined to the coast before activity stopped ten months later.

On 10 July 1720, another vent opened at Soldão, on the edge of the eastern plateaux, only 3 km northeast of São João. It formed the cinder cone of the Cabeço do Fogo and gave off noxious gases that poisoned much livestock in the vicinity. It also emitted the mistério da Silveira – or Soldão – which spread sideways so easily that it formed a lobe, 3 km across, which scarcely makes an impression on the coastline as it enters the sea. This was the most recent eruption to occur on Pico.

The Pico do Pico, crowned by Piquinho, seen from Madalena.

154

one major radial fissure clearly gave rise to the cinder cones of Forçado, Bois, Cabeço de Cima and Cabeço de Baixa near São João. In the west, one of the longest fissures gave rise to Tamusgo, Manuel João and das Casas. Although there have been no Surtseyan eruptions around the fringes of Pico in recent times, the two islets off Madalena (the Ilhéus Deitado and Em–Pé), are the eroded remains of an older Surtseyan cone.

The basal shield is broader in the north than in the south, because its southern flanks have been faulted down along a straight scarp, 500 m high and 5 km long, that runs westwards from São João towards São Caetano. The fault scarp is so well preserved that it could have formed within the past few millennia, although the most recent lavas have covered both its western and eastern ends. The eastern edges of the basal shield lie on top of, and are thus younger than, the lavas of the eastern plateaux of Pico.

The summit cone

The growth of the upper cone of the Pico do Pico marked the concentration of eruptions into a central cluster of vents, and its slopes approach 30°. However, similar alkali basalts built both the basal shield and the upper cone, and lava flows mantle the surface, along with coarse lava screes and cinders. Thus, if the surface exposures are typical, the concentration of eruptions into a central focus was not accompanied by an increase in their explosive power.

The summit of Pico is instantly recognizable. The smooth rim of the crater encloses a hollow that is 500 m wide, but it is now only about 30 m deep, because pahoehoe lavas have filled it almost to the brim. The cinder cone of Piquinho rises 60m above these lavas to decorate the crest of the volcano, and a basaltic cornice wraps around its northeastern rim. Piquinho, in turn, has a small crater that gives off occasional fumes. Although the crest of Pico may not have erupted in historical times, the fresh lavas at the summit are probably less than a thousand years old.

Few satellite cones interrupt the regularity of the stratovolcano. However, Cabras, Capitão, Lomba de São Mateus, Queiro, João Duarte and Torrinhas, for example, erupted at the base of the summit cone, some 2 km from the main vent. Between them rise lines of three or four deeply cratered basaltic spatter cones, and their accompanying lava flows, that were all emitted from small fissures radiating from the main cone. They all have the sharp outlines of recent formations.

Eastern Pico

Eastern Pico stretches out eastwards like a comet's tail for 25 km from the stratovolcano and it makes one of the emptiest and most isolated areas in the Azores, where settlement is restricted to the narrow coastal belt. Many grassy basaltic cinder cones rise from the plateau and they are surrounded by weathered lava flows that blend imperceptibly into its surface and are thus clearly older than their counterparts on the flanks of Pico. The eruptions took place along two sets of fissures trending from northwest to southeast, slightly oblique to the overall trend of the island. Where the island narrows in the east, the fissures are short and close together, and they produced breached cinder cones less than 100 m high. Explosions also formed small maars that are so well preserved that they are seem to be among the youngest volcanic features in the area. In the broader western and central parts of the plateau, the fissures are fewer and wider apart, and have given birth to larger cinder cones, such as Grotoes and Caveiro, which are over 100 m high. They are surrounded by lava flows that have impounded many lakes.

The backbone of eastern Pico ends abruptly on both sides along a series of fault scarps, with remarkably straight sections, 500 m or more in height. They rise direct from the coast at Terra Alta in the north and between Santa Cruz and Calheta in the South. Valley exposures in the scarps reveal the piles of horizontal hawaiite and mugearite lava flows that make up the whole plateau and have been dated to less than 25 000 years ago at Terra Alta and to less than 37 000 years ago at Arrife in the south. Farther west, from Santo Amaro to São Roque on the north coast, and from Santa Cruz to Arrife in the south, the scarps rise between 1 km and 2 km inland, but often approach 700 m in height. Here, lava flows have tumbled down the scarps and then spread out in extensive aprons at their base. On the south coast, for instance, a few thousand years ago at most, a narrow flow filled an old valley cut into the scarp and spread its rugged black basalts into the sea to form the Ponta dos Biscoitos at Santa Cruz. On the north coast, the towns of Praínha, São Roque and Cais do Pico are built on similar recently erupted lobes. Sometimes, copious lava emissions, like the mistério da Praínha, the mistério da Silveira and the steep accumulations near Cais do Pico have buried the original scarps under fans of lava that splayed

The lava delta of Ponta dos Biscoitos at Santa Cruz das Ribeiras, Pico.

seawards in wide promontories. All these lava lobes can be more than 20 m thick and they form extremely rugged shores, but they have undergone little marine erosion and there is no evidence of raised beaches upon them. It is most likely, then, that they came from the most recent eruptions during the past few thousand years.

FAIAL

Faial is a compact island, 21 km long and 12 km wide, whose scenery is dominated by the stratovolcano called Caldeira, rising to 1043 m. A string of villages forms an almost complete ring on the gentle slopes that form a balcony above the rugged lava cliffs, and only Horta, in the shelter of Monte Guia, really touches the sea. Faial epitomizes the predominant traits of the Azores. It has lush meadowland, thick hedges of hydrangea; coppices of cryptomeria, laurel and eucalyptus; a large stratovolcano with a summit caldera; a well defined rift zone; long fissures, with lines of recent cinder cones; and – best of all – the birth of Capelinhos.

Faial grew up from a depth of 1500 m on the flanks of the Mid-Atlantic Ridge. It sprang from the same set of fissures, trending northwest to southeast, that also gave rise to Pico. In fact, the Canal do Faial, between Faial and Pico, is only 100 m deep and 7 km across, and the two islands could be the emerged parts of a single volcanic complex. Of all the central group of the Azores, Faial lies closest to the Mid-Atlantic Ridge, which passes 100 km to the west. Probably as a result, Faial has registered many earthquakes that have usually been the mild shocks typical of the mid-ocean ridges. Nevertheless, the tremors in May 1958, during the eruption of Capelinhos, destroyed the nearby village of Praia do Norte, and on 9 July 1998 an earthquake of magnitude 5.8 on the Richter scale killed 8 people and wounded 154 others.

Faulting has apparently played a major role in delimiting the coasts of Faial. The interplay of fault blocks defines the promontories and bays of the east coast alongside the Canal do Faial. On the west coast, faults have thrown the fringes of the stratovolcano down below sea level and they now form majestic straight cliffs often over 200 m high. On the other hand, although Faial lies close to the Mid-Atlantic Ridge, it is surprising that it has witnessed only two eruptions since the island was first settled in 1466. Both occurred on the western peninsula. The Cabeço do Fogo and Picarito erupted in 1672–3, and Capelinhos produced its display in 1957–8 at the western end of the self-same fissure system (Canto 1881, Machado 1959, Zbyszewski et al. 1959). A small submarine eruption was also noticed in 1963 in the Canal do Faial, but no lavas reached the surface.

Olivine-rich alkali basalts predominated in the areas of most recent activity in the west and southeast. Hawaiites and mugearites occur especially in the central rift zone and on the stratovolcano. Trachytes and benmoreites outcrop on the lower flanks of the stratovolcano and in its caldera (Brousse et al. 1981, Métrich et al. 1981).

The growth of Faial

The basalts that floor the Mid-Atlantic Ridge around Faial are about 5.5 million years old. Faial began to grow up from this basement much later, and the oldest lavas, exposed along parts of the Canal do Faial, are about 730 000 years old (Féraud et al. 1980). These are the oldest rocks exposed in the central Azores and they prove that the individual islands were not formed in succession towards the Mid-Atlantic Ridge. The basal lavas accumulated over a long period, for one of their youngest flows is only about 30 000 years old. The fault trough, forming the central rift zone, foundered after these lavas erupted; and its clarity in eastern Faial also suggests a relatively recent origin. It is about 7 km wide and has three, and sometimes four, fault scarps, each rising between 60 m and 130 m high. The rift zone is made up of three segments, each offset by about 1 km, and it can be traced right across the island from the Canal do Faial to Capelinhos.

Caldeira, the stratovolcano, grew up within the rift zone after it had foundered. It is composed of flows and fragments of hawaiite, mugearite, benmoreite and trachyte, which spill beyond the rift zone and reach the sea. Its flanks are blanketed with trachytic pumice from the Plinian eruption that formed the central caldera. It is 1850 m in diameter and 470 m deep, and the undissected inner walls are almost vertical, with few screes of weathered debris – suggesting that the caldera probably collapsed much less than

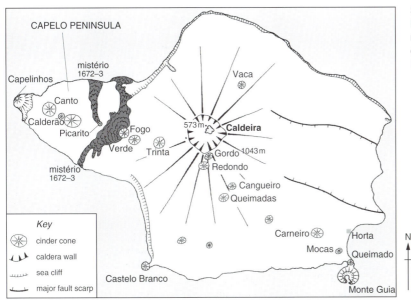

Some volcanic features of Faial, with the zones of most recent activity concentrated in the Capelo Peninsula in the west.

30 000 years ago. But the caldera is old enough for small eruptions to have already resumed within it, for its floor contains a cinder cone and a dome of trachyte that are both less than 30 m high. Meanwhile, several basaltic cinder cones, each about 100 m high, erupted on the upper flanks of the stratovolcano. They include Cabeço Gordo (1043 m), which now forms the highest point of the caldera rim, as well as Cabeço Redondo to the south, Rinquim to the northwest and Cangueiro and Queimadas in the southeast. Castelo Branco, the Surtseyan cone of white trachyte fragments resembling an offshore fortress, probably also belongs to this period, for it has been much eroded by the sea and only a small blunted crater survives on its summit.

About 11 000 years ago, the focus of activity shifted to the southeastern corner of Faial, when about a dozen basalt and hawaiite cinder cones and lava flows erupted. Most of the flows are weathered and covered by Horta and its suburbs, but their original ruggedness is revealed where they form the intensely fretted black cliffs on the southern coast. The eruptions took place along three fissures that trend from northwest to southeast – and the longest continues, in fact, from the Canal do Faial right to the western peninsula. It gave rise to a line of cones including Monte das Mocas and Carneiro, which dominates Horta.

Where one of these fissures reaches the Canal do Faial, basaltic eruptions formed Monte Queimado and Monte Guia, which stand side by side, making the promontory that protects Horta harbour. They are a dissimilar couple. Monte Guia is a typical Surtseyan cone, 1 km in diameter, that erupted in shallow water and is composed of innumerable thin layers of fine, yellowish and often welded tuffs. The southern sector is missing – and probably never formed – and its wide double crater is deep enough to be invaded by the sea. Monte Guia probably erupted during the past few thousand – perhaps even few hundred – years. It is, in fact, better preserved than Capelinhos, which was formed in 1957–8, albeit in a more exposed position. Its companion, Monte Queimado, is quite different. Its burnt appearance comes from the black and reddish cinders of a cone that erupted on land. But it could not have formed under present conditions, because its base now lies below sea level. Most probably, therefore, Monte Queimado erupted before Monte Guia, when the sea was lower, and could not reach the vent. When the sea level rose, Monte Guia erupted, and at the same time, the waves trimmed the edges of Monte Queimado, and redistributed the fragments to form the beach that now joins both cones to Faial.

The Capelo Peninsula

Both the northwestern and southwestern fringes of the stratovolcano foundered below sea level along two faults that intersect where they meet the line of fissures on the axis of the western, or Capelo, peninsula. It juts out 7 km westwards into the Atlantic Ocean and it ended, before 1957, in a smattering of islands called the Ilhéus

Horta and Monte Queimado, looking north from Monte Guia.

Flooded by the sea, the double crater – foreground and middle ground – of Monte Guia.

dos Capelinhos. Its eruptions have now replaced some of the land lost when the flanks of the stratovolcano foundered below sea level. The eruptions formed a broad prong of 16 cinder cones, usually called cabeços ("heads"), which arose on five en echelon fissures, each about 2 km long and 500 m apart, that trend from northwest to southeast, obliquely across the spine of the peninsula.

Most of the peninsula grew out westwards as a result of eruptions on land, where water could not usually invade the erupting vents. Its base consists of a thick cake of black basalt and hawaiite flows that forms a plinth tapering from a height of 600 m near the stratovolcano to 94 m at the old Capelinhos lighthouse. Higher sea levels formed cliffs on some of the older flows, like those due west of Capelo, but others, such as those near the Ponta da Varadouro in the south, still form rugged, scarcely eroded lava deltas. All of these flows have formed an effective buffer against marine attack and they illustrate a net gain of volcanic materials at the expense of the ocean. Only Capelinhos and the Costa da Nau volcano, at the very tip of the peninsula, were created primarily by Surtseyan activity, although, before Capelinhos erupted, the Ilhéus dos Capelinhos also marked the wreck of another

Surtseyan cone that had lost its battle with the waves.

The cones crowning the Capelo peninsula were all built by lapilli and cinder eruptions, and their bases stand at least 100 m above present sea level. The oldest eruptions, whose products are still visible, formed the Cabeço do Pacheco, Caldeirão, Furno Ruim and, most notably, the Cabeço do Capelo, which is over 200 m high, and is young enough for its crater still to be apparent. At the same time, a Surtseyan eruption formed the Costa da Nau tuff cone, off what was then the westernmost point of the peninsula. The Cabeço da Trinta erupted later on the flanks of the stratovolcano, and the Cabeço do Canto and Caldeirina probably followed soon afterwards. Their hawaiite lava flows partly covered the Costa da Nau and formed most of the western snout of the peninsula before 1957. In turn, the flows from the Cabeço do Canto were covered by tuffs from the Surtseyan eruption of the cone, the remains of which still survive as the Ilhéus dos Capelinhos. The next eruptions occurred inland, where they formed the Cabeço Verde and Pingarotes, but the true sequence of the remaining eruptions on the peninsula has not yet been established.

The ruins of Capelinhos village and the lighthouse, with the new volcano on the left horizon.

Historical eruptions

The eruptions recorded during historical times illustrate the same theme. In the east, the Cabeço do Fogo and Picarito erupted between 24 April 1672 and 28 February 1673 (Machado 1959). Seven months of intermittent earthquakes preceded the arrival of the magma at the surface "with such a din that it seemed as if the end of the world had come" (Canto 1881). Ash spread all over Faial, and flows escaped from the base of the cone. Cabeço do Fogo stopped erupting at the end of April 1672, but the baton was at once taken up by Picarito. Its flows buried 307 houses in the villages of Praia do Norte and Ribeira Brava (which was rebuilt as Ribeira do Cabo), and made 1200 people both homeless and landless. Three men were surrounded and incinerated by the flows when they went to inspect the flows at close quarters. Two new mistérios were expelled: one to the north coast; and the other to the south coast. They spread, some 4 km across and 5 km long, over an area already well armoured with lava flows. The place name Biscoitos, on the

The Costa da Nau, with the Capelo Peninsula just visible in the background, seen from the summit of Capelhinos.

The eruption of Capelhinos in 1957–8

The second eruption in Faial during historical times started in waters 80 m deep near the Ilhéus dos Capelinhos. Thus, the new eruption was only regaining ground previously lost to the Atlantic Ocean. Capelinhos arose between 27 September 1957 and 24 October 1958, during which Surtseyan activity gradually gave way to Strombolian activity (Tazieff 1958, Camus et al. 1981, Forjaz 1997b).

There were over 200 tremors between 16 and 27 September 1957. At 08.00 on 27 September, the sea began to boil about 400 m west of the Ilhéus dos Capelinhos. On 29 September, black basaltic ash was fired 1 km into the air in typical Surtseyan pointed jets, shaped like cypress fronds or cocks' tails, and billowing white clouds of steam often rose 4 km high around them. Each explosive interaction of magma and water happened at two-hourly intervals and lasted about half an hour. When this activity stopped on 29 October, a tuff cone, 100 m high, had been built. But by 1 November, all that remained of it was a shoal around the Ilhéus dos Capelinhos.

After a week of total calm, eruptions resumed on 7 November 1957, from a new vent, about 500 m east of the first. A new tuff cone quickly grew up and windblown ash built an isthmus that joined the new volcano to Faial for the first time on 12 November 1957. Then, as the eruptions began to wane in early December, this second cone began to slump into the sea, and its days, too, seemed numbered.

However, at 22.30 on 16 December 1957, molten lavas emerged and indicated that sea water had failed to penetrate the vent for the first time. But the vent was blocked only intermittently and Surtseyan explosions continued to dominate during the spring of 1958. However, the effusive phases gave off enough lava to form an apron around the tuff cone that both retarded erosion and kept the sea out of the vent. By the end of March, the tuff cone had grown to a height of 150 m and a diameter of 1 km, and had covered the Ilhéus dos Capelinhos in the process. The Surtseyan jets were still firing 1800 m into the air, and lava bombs were whistling down over the promontory. The villages had to be abandoned. A thick and noxious blanket of ash, and the sodium chloride that had evaporated from the ocean, destroyed crops and buried houses. Soon, only the now useless and badly damaged Capelinhos lighthouse remained defiant.

The last episode of the eruption began when over 450 earthquakes were recorded between 12 to 14 May 1958. They were centred on the caldera, not on Capelinhos, but the effect on the new volcano was radical. The Surtseyan activity virtually ceased. The most spectacular phase of the eruption began when huge lava fountains spurted from within the crater of the tuff cone and constructed a basaltic spatter cone that had attained 75 m before the end of May 1958. This phase continued until the eruption suddenly stopped on 24 October 1958. By then, about 2.4 km^2 had been added to the area of Faial. The core of Capelinhos was a steep spatter cone of jagged wine-red basalt, 160 m high. It was encircled by a smooth and more gently sloping tuff cone, 150 m high and 1 km across, that was composed of fine tuffs like buff sand. Its surface was strewn with remarkable basaltic bombs the size and shape of tortoises, where the "legs" had burst from the molten interior as they hit the ground. The tuff cone was also almost completely surrounded by lava flows.

But, Capelinhos did not produce enough lava to protect itself properly. Fifty years later, only 1 km^2 of the new land remained. Almost half the spatter cone had been removed and its jagged crater stood on the brink of towering cliffs. The sea had swept away part of the tuff cone, revealing the main islet of the old Ilhéus dos Capelinhos once again. Capelinhos needs another eruption to save it from destruction.

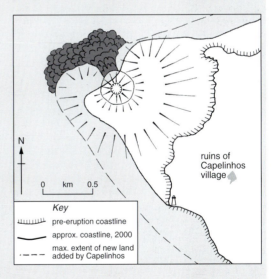

Key
- ⏝⏝⏝ pre-eruption coastline
- ── approx. coastline, 2000
- ·-·-· max. extent of new land added by Capelinhos

Capelinhos.

south coast, reveals the rugged nature of their surfaces and, although they are now wooded, they still make the most impressive mistérios outside Pico. The lavas completely cut off some 70 families who had unwisely escaped westwards along the peninsula and they had to be supplied with food by boat until a new track could be made long after the eruption ended on 28 February 1673. The eruption added some 360 million m^3 of volcanic material to the peninsula, and little has been eroded away. Capelinhos has lost much more.

Capelinhos, Faial, with the Strombolian cone in the centre, surrounded by the Surtseyan tuff cone, much eroded by the sea on the left.

FLORES AND CORVO

Flores and Corvo are west of the Mid-Atlantic Ridge and belong to the North American plate. The main trends of both islands, emphasized by dykes running from north-northeast to south-southwest, are thus quite different from those of their neighbours to the east. Flores is a compact elliptical island, 17 km from north to south and 12 km from east to west. It is unique in the Azores because it is dominated neither by long fissures nor by a large caldera. The Funda caldera is less than 1 km across; the fissures are subsumed within a complex volcanic pile that issued from an array of eruptive centres of similar size; and waterfalls cascading over innumerable colonnades of ancient lava flows attract the eye as much as the volcanic hills. Flores forms a cluster of small stratovolcanoes that crown a large accumulation rising more than 1000 m from its base on the ocean floor. There was a broad succession of basalt, mugearite and trachyte eruptions, with a return to basaltic emissions in recent times (Zbyszewski 1980). The oldest rocks of Flores tend to be exposed around the coast and are dated to over 600 000 years old (Féraud et al. 1980). The central zone is younger, especially where a blanket of fine exploded fragments surrounds the Morro Grande, the highest summit on the island at 942 m. No eruptions have occurred on Flores in historical times, and the advanced age of the dated lavas, coupled with their deeply weathered and eroded surfaces, suggests that the volcanoes of Flores may now be extinct.

Corvo is the smallest island in the Azores, only 7 km from north to south and 4 km from east to west. It belongs to the North American plate, but its structure is apparently much simpler than its companion, Flores. The island is the emerged summit of a basaltic stratovolcano, rising to 718 m, which culminates in a caldera 300 m deep and 2 km across. Subsequent eruptions have begun filling the caldera with small cinder cones that are now interspersed with small lakes. The smooth, gently sloping, inner slopes of the caldera indicate that it could be among the oldest in the Azores, and no eruptions have been recorded on the island during historical times. Thus, Corvo may not be active.

BIBLIOGRAPHY

Abdel-Monem, A. A., L. A. Fernandez, G. M. Boone 1975. K–Ar ages from the eastern Azores (Santa Maria, São Miguel and the Formigas Islands). *Lithos* **8**, 247–54.

Agostinho, J. 1931. The volcanoes of the Azores Islands. *Bulletin Volcanologique* **14**, 123–38.

—— 1937. Volcanic activity in the Azores. *Bulletin Volcanologique* **2**, 183–92.

—— 1950. O Monte Brasil. Esboco monografico. *Açoreana* **4**(4), 343–55.

—— 1959. Actividade vulcânica nos Açores. *Boletim Sociedade A. Chaves* **5**, 362–478.

Booth, B., R. Croasdale, G. P. L. Walker 1978. A quantitative study of and five thousand years of volcanism on São Miguel, Azores. *Royal Society, Philosophical Transactions* **A288**, 271–319.

Brousse, R., H. Bizouard, N. Métrich 1981. Fayal dans l'Atlantique et Rapa dans le Pacifique: deux series faiblement alcalines évoluant sous conditions anhydres. *Bulletin Volcanologique* **44**(3), 393–410.

Camus, G., P. Boivin, A. de Goër de Hervé, A. Gourgaud, G. Kieffer, J. Mergoil, P. M. Vincent 1981. Le Capelinhos (Faial, Açores) vingt ans après son éruption: le modèle eruptif "surtseyen" et les anneaux de tufs hyaloclastiques. *Bulletin Volcanologique* **44**(1), 31–42.

Canto, E. P. 1879. Vulcanismo nos Açores. Anno de 1562: erupção na Ilha do Pico. *Archivos dos Açores* **1**, 360–7.

—— 1880a. Vulcanismo nos Açores. Anno de 1580: erupção na Ilha de São Jorge. *Archivos dos Açores* **2**, 188–93.

—— 1880b. Vulcanismo nos Açores. Anno de 1630: erupção na valle das Furnas, Ilha de São Miguel. *Archivos dos Açores* **2**, 527–47.

—— 1881. Vulcanismo nos Açores. Anno de 1672: erupção na Ilha do Faial. *Archivos dos Açores* **3**, 344–51, 426–34.

—— 1882a. Vulcanismo nos Açores. Anno de 1760–1761: terremote e erupçoes na Ilha Terceira. *Archivos dos Açores* **4**, 362–5.

—— 1882b. Vulcanismo nos Açores. Anno de 1718: erupção na Ilha do Pico. *Archivos dos Açores* **4**, 497–506.

—— 1883a. Vulcanismo nos Açores. Anno de 1720: erupção na Ilha do Pico. *Archivos dos Açores* **5**, 343–45.

—— 1883b. Vulcanismo nos Açores. Anno de 1811: erupção submarina em São Miguel. *Archivos dos Açores* **5**, 448–54.

—— 1884. Vulcanismo nos Açores. Anno de 1808: erupção na Ilha de São Jorge. *Archivos dos Açores* **6**, 437–47.

—— 1887. Vulcanismo nos Açores. Anno de 1672: erupção no Capelo, Ilha do Faial. *Archivos dos*

Açores **9**, 425–32.

Capaccioni, B. V. H. Forjaz, M. Martini 1994. Pyroclastic flow hazard at Água de Pau volcano (São Miguel island, Azores archipelago) inferred from the Fogo-A eruptive unit. *Acta Vulcanologica* **5**, 41–8.

Castello Branco, A., G. Zbyszewski, F. M. de Almeida, O. de V. Ferreira 1959. Le volcanisme de l'Ile de Faial et l'éruption du volcan de Capelinhos: rapport de la première mission géologique. *Serviços Geológicos de Portugal* **4**, 9–29.

Cole, P. D., G. Quieroz, N. Wallenstein, J. L. Gaspar, A. M. Duncan, J. L. Guest 1995. An historic subplinian/phreatomagmatic eruption: the 1630 AD eruption of Furnas volcano, São Miguel, Azores. *Journal of Volcanology and Geothermal Research* **69**, 117–35.

Davies, G. R., M. J. Norry, D. C. Gerlach, R. A. Cliff 1989. A combined chemical and Pb–Sr–Nd isotope study of the Azores and Cape Verde hotspots: the geodynamic implications. In *Magmatism in the ocean basins*, A. D. Saunders & M. J. Norry (eds), 231–55. London: Geological Society [Special Publication 42].

Féraud, G., I. Kaneoka, C. J. Allègre 1980. K–Ar ages and stress and pattern in the Azores: geodynamic implications. *Earth and Planetary Science Letters* **46**, 275–86.

Féraud, G., H. U. Schmincke, J. Lietz, J. Gastaud, G. Pritchard, U. Bleil 1981. New K–Ar ages, chemical analyses and magnetic data of rocks from the islands of Santa Maria (Azores), Porto Santo and Madeira (Madeira Archipelago) and Gran Canaria (Canary Islands). *Bulletin Volcanologique* **44**(3), 359–75.

Forjaz, V. H. 1997a. *Alguns vulcões da Ilha de São Miguel*. Ponta Delgada [São Miguel]: Observatório Vulcanológico Geotérmico dos Açores.

—— (ed.) 1997b. *Vulcão dos Capelinhos. Retrospectivas: vol. I*. Ponta Delgada [São Miguel]: Observatório Vulcanológico Geotérmico dos Açores.

Forjaz, V. H. & N. S. M. Fernandes 1975. *Noticia explicativa das folhas "A" e "B": Ilha de São Jorge*. Lisbon: Serviços Geológicos, Carta Geológica de Portugal.

Forjaz, V. H., F. M. Rocha, J. M. Medeiros, L. F. Meneses, C. Sousa 2000. *Noticias sobre o vulcão oceânico da Serreta, Ilha Terceira dos Açores*. Ponta Delgada [São Miguel]: Observatório Vulcanológico e Geotérmico dos Açores.

Fouqué, F. 1867. Sur les phénomènes volcaniques observés aux Açores. *Académie des Sciences à Paris, Comptes Rendus* **65**, 1050–1053.

—— 1873. Voyages géologiques aux Açores. *Revue des deux mondes* (2ème année, 2ème période) **103**, 40–65.

Fructuoso, G. 1591. *Saudades da terra (livro IV)* [reprinted 1966]. Ponta Delgada [São Miguel]: Instituto Cultural de Ponta Delgada.

Krause, D. C. & N. D. Watkins 1970. North Atlantic

crustal genesis in the vicinity of the Azores. *Royal Astronomical Society, Geophysical Journal* **19**, 261–83.

Krejci-Graf, K. 1961. Zur Geologie der Makoronesen 8 – Die Caldeira von Graciosa, Azoren. *Zeitschrift Deutschen Geologische Gesellschaft* **113**, 85–95.

Laughton, A. S. & R. B. Whitmarsh 1974. The Azores/Gibraltar plate boundary. In *Geodynamics of Iceland and the North Atlantic area*, L. Kristjansson (ed.), 63–81. Dordrecht: Reidel.

McKenzie, D. P. 1972. Active tectonics in the Mediterranean region. *Royal Astronomical Society, Geophysical Journal* **30**, 109–185.

Machado, F. 1955. The fracture pattern of the Azorean volcanoes. *Bulletin Volcanologique* **17**, 119–25.

—— 1957. Caldeiras vulcânicas nos Açores. *Atlântida* I(5), 275–8.

—— 1959. A erupção do Faial em 1672. *Serviços Geológicos de Portugal* **4**, 89–99.

—— 1962. Sobre o mecanismo da erupção dos Capelinhos. *Serviços Geológicos de Portugal* **9**, 9–19.

—— 1967. Active volcanoes of the Azores. *Catalogue of the active volcanoes of the world part* 21: *the Atlantic Ocean*, 9–52. Rome: International Volcanological Association.

—— 1972. Acid volcanoes of San Miguel, Azores. *Bulletin Volcanologique* **36**(2), 319–27.

—— [undated] *c.* 1980. The Capelinhos volcanic eruption [translations by F. S. Weston of four articles entitled "Activadade vulcânica do ilha do Faial"]. *Atlântida* **2**, 225–34 and 305–315, and **3**, 40–55 and 153–9. Horta [Faial, Azores]: Tourist Delegation of Faial.

Machado, F. & T. Freire 1976. Erosão marina no cone vulcânico do Capelinhos. *Atlântida* **20**, 206–209.

Machado, F., W. H. Parsons, A. F. Richards, J. W. Mulford 1962. Capelinhos eruption of Faial volcano, Azores, 1957–58. *Journal of Geophysical Research* **67**, 3519–29.

Métrich, N. & H. Bizouard, J. Varet 1981. Pétrologie de la série volcanique de l'Ile de Fayal (Açores). *Bulletin Volcanologique* **44**(1), 71–93.

Mitchell-Thomé, R. C. 1981. Vulcanicity of historic times in the Middle Atlantic islands. *Bulletin Volcanologique* **44**(1), 57–69.

Moore, R. B. 1990. Volcanic geology and eruption frequency, São Miguel, Azores. *Bulletin of Volcanology* **52**, 602–614.

—— 1991. Geology of three late Quaternary stratovolcanoes on São Miguel, Azores. *United States Geological Survey, Bulletin* **1900**, 1–46.

Moore, R. B. & M. Rubin 1991. Radiocarbon dates for lava flows and pyroclastic deposits on São Miguel, Azores. *Radiocarbon* **33**, 151–64.

Ridley, W. I., N. D. Watkins, D. J. McFarlane 1974. The oceanic islands: Azores. In *The ocean basins and margins* (vol. 2), A. E. M. Nairn & F. G. Stehli (eds), 445–83. New York: Plenum.

Saemundsson, K. 1986. Subaerial volcanism in the western North Atlantic. In *The geology of North America*, vol. M: *the western North Atlantic region*, P. R. Vogt & B. E. Tucholke (eds), 69–86. Boulder, Colorado: Geological Society of America.

Scarth, A. 1994. *Volcanoes*. London: UCL Press.

Schmincke, H. U. 1973. Magmatic evolution and tectonic regime in the Canary, Madeira and Azores island groups. *Geological Society of America, Bulletin* **84**, 633–48.

Searle, R. C. 1980. Tectonic pattern of the Azores spreading centre and triple junction. *Earth and Planetary Science Letters* **51**, 415–34.

Self, S. 1976. The recent volcanology of Terceira, Azores. *Geological Society of London, Quarterly Journal* **132**, 645–66.

Self, S. & B. M. Gunn 1976. Petrology, volume and age relations of alkaline and saturated peralkaline volcanics from Terceira, Azores. *Contributions to Mineralogy and Petrology* **54**, 293–313.

Serralheiro, A., V. H. Forjaz, C. A. de M. Alves, B. Rodrigues 1989. *Carta vulcanológica dos Açores, Ilha do Faial*. Ponta Delgada [São Miguel]: Centro de Vulcanológica.

Storey, M., J. A. Wolff, M. J. Norry, G. F. Marriner 1989. Origin of hybrid lavas from Agua de Pau volcano, São Miguel, Azores. In *Magmatism in the ocean basins*, A. D. Saunders & M. J. Norry (eds), 161–80. London: Geological Society [Special Publication 42].

Tazieff, H. 1958 and 1959. L'éruption de 1957–58 et la tectonique de Faial (Açores). *Bulletin de la Société Belge de Géologie* **67**, 13–49 and *Serviços Geológicos de Portugal* **4** (1959), 71–8.

Udias, A., A. Lopez-Arroyo, J. Mezcua 1976. Seismotectonics of the Azores–Alboran region. *Tectonophysics* **31**, 259–89.

Walker, G. P. L. & R. Croasdale 1971. Two Plinian-type eruptions in the Azores. *Geological Society of London, Quarterly Journal* **127**, 17–55.

Waters, R. V. & A. C. Fischer 1971. Base surges and their deposits: Capelinhos and Taal volcanoes. *Journal of Geophysical Research* **76**, 5596–614.

Weijermars, R. 1987. A revision of the Eurasian/African plate boundary in the western Mediterranean. *Geologische Rundschau* **76**(3), 667–76.

Weston, F. S. 1964. List of recorded volcanic eruptions in the Azores with brief reports. *Boletim do Museu e Laboratório Mineralógico e Geológico da Faculdade de Ciências, Universidade de Lisboa* **10**, 3–18.

Woodhall, D. 1974. Geology and volcanic history of Pico Island Volcano, Azores. *Nature* **248**, 663–5.

Zbyszewski, G. 1980. Géologie des Iles Atlantiques. In

Géologie des pays Européens, 157–71. Paris: Dunod.

Zbyszewski, G., F. M. de Almeida, O. da V. Ferreira, C. T. de Assuncao 1958. *Noticia explicativa da folha B, São Miguel (Açores)*. Lisbon: Serviços Geológicos, Carta Geológica de Portugal.

Zybszewski, G. & O. da V. Ferreira 1959a. Le volcanisme de l'Isle Faial et l'éruption du volcan Capelinhos: rapport de la deuxième mission géologique. *Serviços Geológicos de Portugal* **4**, 339–45. Lisbon.

Zbyszewski, G. F. M. de Alameida, O. da V. Ferreira, C. T. de Assunção 1959b. *Noticia explicative da folha Faial (Açores)*. Lisbon: Serviços Geológicos, Carta Geológica de Portugal.

Zbyszewski, G., O. da V. Ferreira, C. T. de Assunçao 1961. *Noticia explicativa da folha de Ilha de Santa Maria (Açores)*. Lisbon: Serviços Geológicos, Carta Geológica de Portugal.

Zbyszewski, G., A. D. de Medeiros, O. da V. Ferreira, C. T. de Assunçao 1971. *Noticia explicativa da folha Ilha Terceira (Açores)*. Lisbon: Serviços Geológicos, Carta Geológica de Portugal.

6 Iceland

Iceland covers an area of 102 846 km^2 and it is almost completely volcanic in origin. It constitutes the longest emerged segment along any of the world's mid-ocean ridges and it is the largest volcanic area in Europe. The landscape of Iceland offers a glimpse of how mid-ocean ridges develop, especially in the central axial zone of contemporary activity, which curves across the island from north to southwest in a swathe 60–80 km wide. Here, the North American plate on the west is diverging from the Eurasian plate on the east. Crustal accretion occurs at their edges through the injection of great swarms of sheeted dykes, often accompanied by eruptions of basalts onto the land surface. Broad expanses of plateaux basalts stretch from both sides of the band of contemporary activity and cover three quarters of Iceland, which were themselves erupted on older axial zones during the past 16 million years, but have been carried to extinction away from the axis, as crustal divergence, rifting and accretion have continued. Thus, in northwest Iceland, the exposed basalts are about 15 million years old, whereas those in eastern Iceland date from 13 million years ago (Saemundsson 1986). They are bordered on their inner

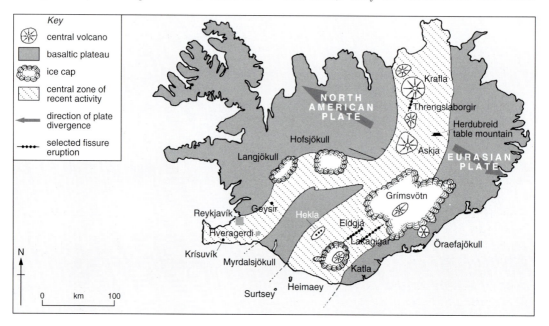

The main volcanic features of Iceland, with the central zone of recent volcanic activity between the diverging Eurasian and North American plates.

edges, alongside the band of present activity, by younger basalts 3.1–0.7 million years old. In the zone of contemporary activity, all the eruptions occurred less than 700 000 years ago. All of these basalts erupted in a broadly similar fashion and, therefore, the activity of the contemporary axial zone clearly indicates how most of Iceland was formed. Iceland has widened by some 400 km from east to west since the divergence began. The axial zone is widening at an average rate of 1.5–2.0 cm per year (Tryggvason 1984).

Iceland has been built up on a segment of the Mid-Atlantic Ridge, which lies between the Kolbeinsey Ridge to the north and the Reykjanes Ridge to the southwest. The 400 km-long Icelandic segment is offset by two transform-fracture zones where most of the major earthquakes have occurred in historical times: the Tjörnes fracture zone off the north coast and the south Iceland seismic zone crossing southwestern Iceland (Saemundsson 1979, 1986).

These tectonic aspects of Iceland are fundamentally similar to those displayed on submarine spreading ridges. Nevertheless, Iceland represents an unusual volume of volcanic materials – more than twice the thickness of the average ridge – built above sea level by the presence of a hotspot beneath the spreading axis, which has been active for at least 25 million years and perhaps as much as 44 million years (Sigurdsson & Loebner 1981). Volcanic rocks about 1500 m thick are exposed in Iceland and they lie on a volcanic basement three times thicker (Saemundsson 1979). The Icelandic hotspot seems to be a cylindrical zone about 300–400 km in diameter and more than 400 km deep (Wolfe et al. 1997). It is probably now centred beneath the zone of maximum lava accumulation that lies below the highest parts of the axial zone under Vatnajökull (Saemundsson 1986).

Of the exposed volcanic sequences, basalts represent 80–85 per cent, silicic and intermediate materials about 10 per cent, and the remainder comprises fluvial or glacial sediments that are themselves derived from volcanic rocks. The basalts form distinct petrological and morphological units. Porphyritic basalts arise most often in large rapid eruptions from fissures and they form characteristically massive lava flows. Olivine-rich tholeiitic basalts are emitted as thin individual pahoehoe flows (called helluhraun in Iceland), which repeated eruptions often pile up in lava shields. They seem to come from magmas lying more than 10 km deep. On the other hand,

olivine-poor tholeiitic basalts emerge chiefly from fissures and the central volcanoes, after having spent time in shallow reservoirs. They form as aa lava flows (called apalhraun in Iceland) Alkali olivine basalts erupt from reservoirs about 3 km deep outside the main zones of rifting, and tended to develop after, and cover the products of, the tholeiitic activity. The greatest differentiation, probably derived after the longest periods in shallow reservoirs, is represented by the silicic eruptions, which are often associated with violent explosions that have reached Plinian proportions at Hekla, Askja and Öraefajökull. Among the more silicic rocks, about two thirds form lava flows or intrusions, but one third comprises layers of welded tuffs and ash that are often of a rhyolitic or rhyodacitic nature. Hekla, Öraefajökull and Askja have been their main sources in historical times.

Many of the characteristic genetic, tectonic and petrological features of the mid-ocean ridges imparted to Iceland can be studied in the open-air landscape, without recourse to the expensive equipment required where the ridges remain submerged. However, high precipitation and proximity to the Arctic Circle have helped maintain four main ice caps. Vatnajökull, with an area of 8400 km^2, is the largest ice sheet in Europe. The smaller ice caps of Langjökull, Mýrdalsjökull and Hofsjökull together are about 2500 km^2 in area. They create additional problems when eruptions take place beneath them, for they often generate destructive **jökulhlaups** or glacier bursts. Much of the rest of the country lies in a periglacial subpolar climatic environment, where the general absence of vegetation

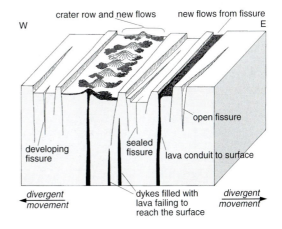

A typical Icelandic rift and fissure swarm (after Imsland 1989).

and thick soils reveals the eruptive features with a stark clarity that makes Iceland one of the finest volcanic exhibitions on Earth. Moreover, for a mid-ocean ridge environment, the volcanic landforms display a surprising variety and, although basaltic effusions have been predominant, almost every kind of volcano can be found (Thórarinsson & Saemundsson 1979). Volcanic systems composed of an elongated fissure swarm, with a central volcano in its midst that might even have developed a caldera, are some of the most striking elements of contemporary activity (Jakobsson 1979). They are accompanied by broad basaltic shields, much steeper table mountains formed beneath the old ice sheets, occasional domes of silicic lava, as well as maars and tuff rings, and a whole gamut of hydrothermal features.

The Norse settlement of Iceland began in AD 874 and thus historical records of volcanic activity, however vague or conflicting in the early centuries, are available for more than a thousand years. On average, a magmatic eruption is under way in Iceland every fifth year (Thórarinsson 1981). However, most **Icelandic eruptions** are among the least dangerous in the volcanic repertoire, although they are often of great scientific interest. The formation of Surtsey in 1963 not only created a new island in the Vestmannaeyar but also did much to clarify ideas about these particular Surtseyan eruptions. The eruption on the neighbouring island of Heimaey ten years later marked an important step in the development of procedures for civil protection and damage limitation. Finally, the studies of the rifting and eruptive episode that began at Krafla volcano in 1975 gave greater insights into the way that active mid-ocean ridges operate.

The plateau basalts

The plateau basalts cover about three quarters of Iceland, forming open and rather stark uplands of rounded hills and ridges, between 750 m and 1250 m high, each scarred by the outcrops of layered lava flows. They have been scraped bare by repeated glaciations, and eroded by streams in the non-glacial periods. The lava flows now dip gently down towards the axial zone from both west and east (Saemundsson 1986). Most of the plateaux developed from eruptions of tholeiitic basalts that emerged from deep fissures as widespreading fluid lava flows, 5–15 m thick. Central vent volcanoes may also amount to as much as

Piles of lava flows forming craggy escarpments in the Icelandic plateau basalts, Reykjarfjördur.

half of the volume of the plateaux, although they have often been buried by later lava flows (Walker 1963). There are perhaps as many as 50 or more such central volcanoes. They mainly form shields and occasional stratovolcanoes composed chiefly of basalts, but also have some andesites or even rhyolites in their make up. They have gentle slopes and modest heights. Breiddalur, in eastern Iceland, for instance, had a volume of $400 \, km^3$ and gentle slopes of 9°, but it probably never stood more than 600 m above the surrounding lava plains. As they grew, the central volcanoes were periodically swamped by thicker and more fluid flood basalts disgorged from the fissures around them, so that they now form insignificant features in the landscape.

An essentially similar pattern of eruptions continued within the more recent series, initiated 3.1 million years ago, that covers a quarter of Iceland along each side of the active zone and is more than 500 m thick. Glaciations introduced greater variety, not only by depositing glacial debris between the lavas but more importantly by provoking the formation of pillow lavas, pillow breccias and **palagonites** during **subglacial eruptions**. These palagonites are known as móberg in Iceland.

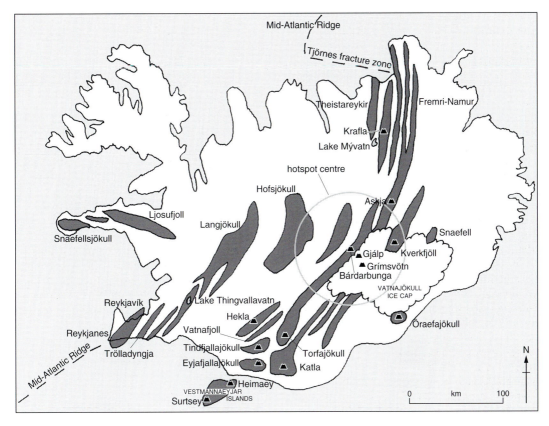

The Mid-Atlantic Ridge, the Icelandic hotspot focus, volcanic systems and recently active volcanoes (▲).

The active volcanic zone

The axial zone of contemporary activity is demarcated geographically by the inner borders of the basaltic plateaux lavas and defined geologically by eruptions in the past 700000 years. During the glacial periods, eruptions occurred under the ice sheets, which led to the formation of subglacial table mountains and thick layers of palagonitic basalts and tuffs. Apart from activity beneath present ice caps, post-glacial eruptions have taken place in the open air.

The same general pattern of eruptions has continued into modern times and, as in the past, basalts have composed the vast bulk of the output. However, the active zone is more complex than it at first appears. It curves southwestwards in a belt 60–80 km wide from the north coast and then branches into two arms: a western rift zone, running across the Reykjanes Peninsula to join the Reykjanes segment of the Mid-Atlantic Ridge; and an eastern rift zone, which is propagating towards the Vestmannaeyar Islands (Gudmundsson 1995). The eastern arm may be the site

of contemporary rift jumping at the expense of the western arm (Saemundsson 1986).

The rifting zones cover about 18000 km² (60%) of the present active area and they account for 77 per cent of its volcanic production (Imsland 1989). They are dominated by fissure swarms arranged en echelon and running parallel to the trend of the active zone. The swarms of fissures have often developed into more complex volcanic systems, where a low and broad elliptical central volcano has grown up in their midst. Some 29 volcanic systems have been active and they have produced between 400 km³ and 500 km³ of lava, covering about 12000 km². Southern Iceland has been the most productive region (Saemundsson 1979). These volcanic systems are 40–100 km long and 2–10 km wide. In the active zone as a whole, plate divergence first causes continual crustal stretching for 100–150 years. It is then brought to an end by a short episode of active rifting, commonly limited to one volcanic system at a time, which lasts for less than a decade (Saemundsson 1979, Sigvaldason 1983, Tryggvason 1984, 1986). The rifting opens

Thingvellir rifts, marking one of the most recent zones of plate divergence in Iceland.

fissures and faults that usually develop parallel to the trend of the active zone, at right angles to the directions of plate divergence. From these fissures the basalts often emerge during the eruptions that distinguish Icelandic activity.

In their initial stages, the fissure swarms produce voluminous hot effusions of fluid tholeiitic basalts. In time, slightly more evolved tholeiitic lavas are emitted, which are rather more viscous and explosive in character and which tend to be expelled from vents concentrated in the centre of the fissure swarms or volcanic systems (Imsland 1989). This concentration is associated with the development of a shallow reservoir in the crust, which facilitates the evolution of the magma and builds up a broad central volcano. Although eruptions apparently increase in frequency, the volume expelled during each active episode decreases. However, the central volcano may develop a caldera, as a result of either an explosive eruption of dacitic or rhyolitic fragments or of the lateral migration of magma from the reservoir into fissures. The Askja caldera was apparently formed by lateral magma migration, but the Öskjuvatn caldera collapsed within it in 1875 after a large rhyolitic explosion (Annertz et al. 1985). The fissure eruptions of Krafla between 1975 and 1984 perhaps represent an intermediate stage of small eruptions accompanying

marked fissure formation on the flanks of a central volcano that has already developed a caldera.

Not all of the rifting causes eruptions at the surface, although magma almost certainly invades the fissures at depth. Near Reykjavík at Thingvellir, rift fissures are particularly well marked. They first developed about 9000 years ago. Now, fissures, often bordered by recently formed cliffs, continue for 2–10 km across country. Small earthquakes sometimes form gaping clefts at the surface and it is said that a larger earthquake in 1789 lowered the whole area by 67 cm. At all events, the extent of fault-trough subsidence in this area may be judged by the fact that Lake Thingvallavatn lies below sea level.

Fissure eruptions in Iceland are beautiful, often spectacular, usually brief and rarely dangerous. With the notable exception of outbursts on the scale of the Lakagígar (Laki) eruption in 1783, they soon reach a brief crescendo, followed by several weeks of waning emissions. Their activity usually begins with lava fountaining, often in continuous curtains 1 km or more in length. The lavas are hot and fluid, reaching the surface at temperatures of about 1200°C. After

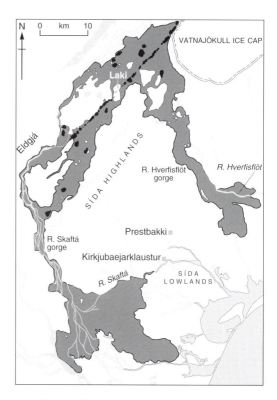

Lava flows emitted during the Laki eruption of 1783–4 (after Thordarson 1995, and personal communication).

The Skaftár Fires: the Laki eruption in 1783–4

Most famous of Icelandic fissure eruptions was the Skaftár Fires or Laki eruption of 1783–4, also notorious for its far-reaching effects (Thórarinsson 1970a, Jackson 1982, Thordarson & Self 1993, Thordarson et al. 1996, Scarth 1999). It sprang from Lakagígar, the fissure extending northeastwards across the Sída Highlands, through the 818 m-high mountain of Laki, which had itself formed along older fissures extending 100 km from the subglacial volcano of Grímsvötn. The eruption lasted eight months, during which its tholeiitic basalts never varied in composition, which suggests that they originated at depth, formed a reservoir beneath the solid crust, and then rose quickly to the surface (Gudmundsson 1987). The events were described by Rev. Jón Steingrímsson, the Minister at Prestbakki, in the Sída Lowlands.

After earthquakes had rumbled for a week, the fissure opened on the southwestern flanks of Laki on Whit Sunday morning, 8 June 1783, and hot fluid tholeiitic basalts gushed forth. The initial discharge was extremely high: two thirds of the total volume were emitted – at an average discharge of 5000 m^3 per second – during the first 50 days. The lava quickly invaded the gorge of the River Skaftá and, on 12 June, emerged onto the lowlands more than 35 km from the fissure. Two days later, the lava had filled the gorge to the brim, the ground was shaking, and an odd foul-smelling blue fog was veiling the Sun. On 25 June, the fissure lengthened, and explosions and lava fountains extended the line of cones across the mountain. More eruptions followed extensions along the fissure, until 25 July. Four days later, an extension of the fissure opened up on the opposite, northeastern, flank of Laki. This section also emitted great volumes of lava, but this time they swamped the Hverfisfljót gorge to the east. The vigour of the eruption began to decline in mid-August and more markedly from the end of October, so that the emissions had weakened considerably before they stopped altogether on 7 February 1784. By then, two vast lava flows had formed prominent deltas in the Sída lowlands and more explosive activity had also formed no less than 140 cones, ranging from 30 m to 70 m high, on a fissure 27 km long (Plate 8). The lava flows reaching a volume of 14.7 km^3 overwhelmed an area of 565 km^2, and a further 8000 km^2 was covered with a thin mantle of fine ash. A few churches, farms and homesteads were damaged, but no-one was killed directly.

The factor that gave the Skaftár Fires their notoriety was the emission of poisonous gases. About 8 million tonnes of fluorine contaminated the farm animals and their pastures, and, as a result, more than three quarters of the sheep and horses, and half the cattle, died. The dreadful famine that ensued killed 20 per cent of the population (Jackson 1982). About 122 million tonnes of sulphur dioxide was released in columns that rose 6–13 km high and generated some 250 million tonnes of sulphuric acid aerosols that remained aloft for about two years (Thordarson et al. 1996). The aerosol spread as a curious dry blue fog and caused consternation all over Europe during the summer and autumn of 1783. Even the experts did not know where it came from or what it portended, and only one, the French naturalist Mourgue de Montredon, correctly linked it with Iceland on 8 August 1783, a year before Benjamin Franklin gained undue credit for some speculations about the event. This sulphuric aerosol certainly seems to have been responsible for the three long and abnormally cold winters in the northern hemisphere from autumn 1783 to spring 1786, although it may not have caused the abnormally hot and stormy July in western and central Europe for which it has also been blamed.

several hours, emissions become concentrated in vents, usually 250–750 m apart, where moderate explosions form a row of individual cones of spatter and cinders. At the same time, many cones are breached by quiet effusions of lava that quickly form widespread and relatively thin flows. The fissure is most active, and the greatest volumes of material are expelled, during the early stages of the eruption, which commonly last for only a few hours or a few days. Effusion rates can be high during the first few hours and sometimes exceed 1000 m^3 per second. Then, emissions become weaker and are separated by lulls that increase in length until they stop almost unnoticed. The fissure then usually becomes extinct. The lava solidifies, seals the cleft and welds together the older lavas alongside it, forming a subterranean dyke that may be 4–6 km deep and more than 10 km long (Rubin 1990). The old fissure becomes a band of strength, 1–2 m wide, that is added to the increasing bulk of Iceland. The stress of rifting is taken elsewhere until another fissure develops alongside it. Thus, a multitude of lava-clogged dykes border each new fissure. Typical fissure formation with the emission of lava flows and the growth of a cone row is well illustrated by the Threngslaborgir and Ludentsborgir cone rows east of Lake Mývatn (Rittmann 1938).

Since settlement in AD 874, the southeastern branch of the axial zone has been the most prolific source of basaltic fissure eruptions in Iceland. The fissures here are mainly related to two volcanic systems, both of which are partly masked by contemporary ice caps: Grímsvötn lies beneath Vatnajökull, and Katla lies below Mýrdalsjökull. The Eldgjá eruption of about AD 935 came from a fissure that extended northeastwards from the Katla volcanic system; the Lakagígar eruption of 1783 occurred on a fissure

Eldgjá fissure, which opened in about AD 935.

THE NORTHERN ACTIVE VOLCANIC ZONE

The active volcanic zone finds perhaps its clearest expression in northern Iceland between Vatnajökull and the Atlantic Ocean. Several active volcanic systems range en echelon along the axial zone. Askja–Dingjufjöll erupted in 1875, sporadically between 1921 and 1926, and again in 1961. The Krafla system was active between 1724 and 1729, and again from 1975 to 1984. On the other hand, the Theistareykir system underwent rifting without eruption in 1618; the Fremri–Namur system has not yet been active in historical times. Thus, the results of plate divergence have not been uniform, at least within the human timescale. The rifting at Krafla, between 1975 and 1984 in particular, threw light on problems associated with the mechanisms of rifting, the formation of magma reservoirs and the lateral transfer of magma to fissure swarms (Sigvaldson 1983, Tryggvason 1984, 1986, 1989).

Askja–Dyngjufjöll

Dyngjufjöll lies in the Ódádahraun ("the desert of crimes"), a vast and stark region covering 3000 km^2 north of Vatnajökull. It is a broad massif, with an area of 250 km^2, that rises about 700 m from its base. It is the eroded remains of a stratovolcano, and Askja caldera crowns its summit. Dyngjufjöll is the centre of a volcanic system that is 100 km long and some 20 km wide, with two major fissure swarms, one trending north-northeastwards through the present western edge of Askja caldera, the other running northeastwards through the eastern parts of the caldera (Annertz et al. 1985, Sigvaldason et al. 1992). The eruptions became concentrated in the centre of the fissure swarm when the glacial ice sheets still covered the land. They built up a thick and widespread series of basaltic palagonites, including pillow lavas, pillow breccias and tuffs, formed in a subglacial water body. When the volcano rose above the level of the ice and meltwater, basaltic lava flows erupted and occasional layers of breccias, tuffs and rhyolitic fragments. When the ice retreated from this area about 7000 years ago, eruptions from the western fissure chiefly gave rise to extensive basaltic lava flows, and small explosion craters formed on its eastern counterpart. It was probably between 4500 and 6000 years ago that the Askja caldera formed, because

stretching southwestwards from the Grímsvötn volcanic system. Eldgjá, like Katla, erupted transitional alkali basalts, whereas Lakagígar emitted tholeiitic lavas, as did Grímsvötn. The Eldgjá and Lakagígar fissures run parallel to each other and are only 5 km apart and they produced two of the most voluminous outpourings of lava in the world in historical times (Thórarinsson 1970a, Miller 1989, Thordarson et al. in press).

The eruption along the Eldgjá "fire fissure" in about AD 935 was the largest emission of flood basalts on Earth during the past millennium. It is one of the largest fissures in Iceland, made up of four slightly offset sections, with an overall length of 57 km, that are 140 m deep and sometimes as much as 600 m wide. The eruption featured at least eight distinct episodes and may have lasted from three to eight years. The cleft opened, lava fountaining followed, and activity became concentrated in individual vents that formed cinder cones and spatter cones, and gave off much sulphur dioxide. At the same time and immediately afterwards, 19.6 km^3 of fluid flows were discharged that covered 780 km^2. The fissures released about 184 million tonnes of sulphur dioxide into the atmosphere, troposphere and lower stratosphere, and the lava flows also released about 35 million tonnes into the lower troposphere. Thus, Eldgjá produced the greatest volcanic pollution during historical times, and about 1.8 times as much as its only rival, Laki, in 1783. However, the climatic effects of the Eldgjá eruption were not as intense, perhaps because emissions were distributed over several years. Nevertheless, the date of about AD 935 coincides with an acidity peak in cores from the Greenland ice sheet (Hammer 1984, Zielinski et al. 1995).

it took place before the deposition of ash erupted from Hekla about 3800 years ago. On the whole, lava production was about 30 times higher 8000–4500 years ago, when it expelled 5.98 km^3 of lava per thousand years, than during the period from 2900 years ago, when it emitted lava at a rate of only 0.32 km^3 per thousand years. This period of rapid volcanic accumulation could be explained by unloading of the pressure of the ice on the crust, which therefore reduced the pressures upon the crustal magmas below.

Askja ("the box") is one of the most impressive sites in the whole of Iceland. It is a circular caldera with an area of almost 50 km^2 and a volume of 8.3 km^3, and its walls are between 100 m and 400 m high. The collapse of Askja was apparently not associated with an explosive eruption, but rather with migration of magma from the reservoir out into the fissure swarm. After a rest of several thousand years, rifting and basaltic eruptions resumed within the Askja caldera and sometimes flowed out via the downfaulted Öskjuop col onto the flanks of Dyngjufjöll. However, these eruptions were neither frequent nor voluminous, but they continued until medieval times. They were then succeeded by an interlude of almost total quiescence lasting 400 years.

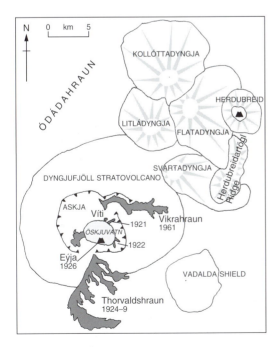

Askja, Dyngjufjöll, with the most recent lava flows, neighbouring shield volcanoes, and the table mountain of Herdubreid.

The Plinian eruption of Askja in 1875 was the first eruption recorded with certainty in the area. It was preceded by vigorous fumarole activity, rifting and earthquakes, and tholeiitic basalts entered the magma reservoir during the latter part of 1874 and early 1875. Basaltic eruptions occurred on 18 February, 40 km north of Askja, where they formed the Sveinagjá lava flow, and 25 km south of Askja, where they formed the Holuhraun lava flow.

Early on Easter Monday, 29 March 1875, a Plinian eruption expelled a large column of ash and pumice (Thórarinsson 1967a). It expelled 2 km^3 of rhyolitic pumice in less than 12 hours, which eventually spread as far as Scandinavia and even formed ephemeral pumice islets in the Atlantic Ocean. Much of the surface of Askja caldera is strewn with black ash and white pumice from this outburst. Towards the close of the eruption came the formation of the small hydrovolcanic Víti ("hell") crater, which is still filled with warm waters.

After this great expulsion of material, the northeastern part of Askja caldera collapsed at intervals until 1907. This new caldera, nested within the larger Askja caldera, covered an area of 3 km by 4 km and was by then largely filled by Lake Öskjuvatn. Small basaltic flows erupted alongside it in 1921 and 1922, and between 1924 and 1929 rifting formed a 6 km-long fissure trending north-northeastwards on the flanks of Dyngjufjöll outside the Askja caldera. It emitted a large lava flow covering 16 km^2. The fissure also extended into both calderas and erupted a small basaltic cinder cone in 1926 that now forms the Eyja islet in Lake Öskjuvatn.

The eruptions of 1961 within Askja caldera were heralded by tremors from 6 October (Thórarinsson & Sigvaldason 1962). Solfataras started up just north of the Víti crater on 10 October, and powerful geysers and mudpots developed, all of which are rare at the inception of an Icelandic eruption. On 26 October an 800 m-long fissure opened nearby that produced lava fountains rising 500 m, and then the Vikrahraun lava flow, which eventually grew to a length of 7.5 km. The lava had an aa surface at first, but its later emissions developed fine pahoehoe forms. In addition, three craters were formed on the fissure that together expelled 3 million m^3 of fragments. When the eruptions stopped in early December 1961, both the Askja caldera, and the smaller Öskjuvatn caldera within it, had foundered a little. No subsequent eruptions have been as violent as that in 1875, but they are slowly filling Askja caldera. However, they are being counter-balanced to some extent by continued foundering, punctuated by periods of uplift, which are both centred in the midst of the caldera. It has been suggested that these deformations could be related to pressure fluctuations in the upper part of the Iceland mantle plume. Moreover, prolonged activity at Krafla between 1975 and 1984 apparently also caused continuous subsidence at Askja (Tryggvason 1989).

Krafla

Krafla is the core of a major volcanic system in northern Iceland, which rises to 818 m and has a diameter of some 20 km. Its centre is marked by a caldera about 10 km across. Its fissure swarm, about 80 km long and up to 10 km wide, stretches from Axarfjordur on the north coast and extends to the area east of Lake Mývatn.

The eruptions of the Krafla volcanic system

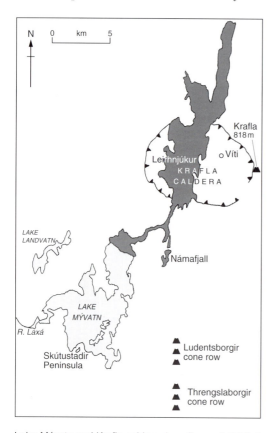

Lake Mývatn and Krafla caldera: lava flows of 1725–9 and 1975–84.

began under the ice, and its basement is thus composed of subglacial tuffs and palagonitic basalts, as well as of voluminous tholeiitic basalt flows. They are interspersed by layers of welded dacitic fragments that were expelled when the broad summit caldera collapsed. The caldera probably formed during an interglacial period and has since been almost filled by repeated emissions of basaltic flows (Sigvaldason 1983). About 18 eruptions have occurred in and around the caldera and a further 15 in the south of the volcanic system. The first main burst of eruptions occurred just after the ice sheets melted. A quiescent period then lasted at least 4000 years. The second main burst of activity has lasted until the present time. It began about 2500 years ago when a hydrovolcanic eruption formed the maar and tuff ring of Hverfjall. About 2100 years ago, eruptions along two major fissures formed the great pair of cone rows of Ludentsborgir and Threngslaborgir, both of which run from north to south and are only 250 m apart (Rittmann 1938). Threngslaborgir is 4 km long, and its extremities are offset some 250 m to the west of its main section. Ludentsborgir curves slightly north-northeastwards and reaches a length of 2.5 km. Both fissures apparently operated simultaneously and formed two rows of about 30 often-intersecting cones less than 30 m high. Most of the cones were breached as they formed when the vigorous effusions of the Younger Laxá basalts gushed from the fissures. They had, for Krafla, an unusually large volume of 2–3 km³ and they covered an area of 220 km², ponded back Lake Mývatn, flowed northwards for 63 km, and almost reached the north coast. The lavas entered Lake Mývatn and formed a whole series of **rootless cones** that are scattered across the islets in the lake and form most of the arc-like Skútustadir Peninsula on its southern shores (Rittmann 1938).

A similar pattern was repeated between 1724 and 1729 during the Mývatn Fires, which sprang from the Leirhnjúkur fissure on Krafla (Grönvold 1984). The episode was inaugurated on 17 May 1724 with a brief hydrovolcanic explosion that was possibly caused when snowmelt penetrated the fissure and encountered the rising magma. First, rhyolitic ash and then basaltic cinders and lapilli exploded from the vent and piled up in a tuff ring about 100 m high, and the Víti maar, 300 m across and 33 m deep, formed in the hollow during the summer. The rifting caused many earthquakes, which developed fault troughs and upthrown fault blocks and lowered Lake Mývatn by about 2 m. Six smaller hydrovolcanic eruptions formed aligned craters in the immediate neighbourhood at the same time.

The real Mývatn Fires started on 11 January 1725, when very minor eruptions of mud and lava began on the Leirhnjúkur fissure in the western part of Krafla caldera. This fissure was the main source of activity when the Mývatn display began in earnest in August 1727. The lava fountaining, in both curtains and individual jets, was accompanied by the emission of copious helluhraun (pahoehoe) basaltic lava flows that covered 35 km² northeast of Lake Mývatn. From time to time, other parallel fissures nearby joined the display, including the Hrossaldur cone row in 1728 and the Bjarnarflag cone row in 1729. After the Mývatn Fires burnt out in August 1729, their sole magmatic aftermath was the small eruption north of Leirhnjúkur in 1746. However, mudpots and solfataras continued to issue from many of the vents for the rest of the eighteenth century; and some have maintained extensive fumarole activity until the present day (Óskarsson 1984). Two centuries of repose then followed.

The recrudescence of rifting at Krafla was betrayed by a marked increase in earthquake activity and uplift in the latter part of 1975. The rifting episode of Krafla from 1975 to 1984 was characterized by repeated slow uplift or inflation, followed by shorter, quicker periods of subsidence or deflation. At the same time, repeated rifting and widening took place in a narrow highly fissured band within the Krafla fissure swarm. Generally speaking, the uplifts and subsidences were less marked after 1980.

The episode began on 20 December 1975 with rapid subsidence and a small basaltic eruption some 2 km along the fissure stretching northwards from Leirhnjúkur. The accompanying earthquakes migrated along the fissure as far as the north coast, 30–70 km from Krafla. It was here that the greatest crustal widening took place and formed gaping fissures and faults up to 2 m high. This general pattern was to repeat itself about 20 times during the whole rifting episode. Slow inflation (or uplift) was followed by rapid deflation (or subsidence), rifting, and sometimes by eruptions. In some cases, however, especially during 1977 and 1978, deflation occurred without lava emissions onto the surface, although earthquakes, fissures and subterranean dyke formation were always registered. The periods of inflations, amounting to about 2 m or less in the central area of Krafla, usually lasted from one to seven months, whereas the deflation periods lasted only from one to twenty days. In general, the eruptions seem to have been linked with the fastest deflation rates.

The eruptions occurred on 20 December 1975, 28 April 1977, 8 September 1977, 16 March 1980, 10 July 1980, 18 October 1980, 30 January 1981, 18 November 1981 and 4 September 1984. When emissions were most frequent from 10 July 1980 to 18 November 1981, each was also more voluminous than usual. The eruption that started on 8 September 1977 had an unusual consequence about 9 km south of Leirhnjúkur at the centre of the Krafla caldera (Larsen et al. 1979). Here, in the Námafjall geothermal field, five boreholes had been drilled to extract steam for the power station. Borehole 4 was 1138 m deep, and, at 23.45 on 8 September, it began to erupt. A roaring eruptive column of steam and ash, about 15–25 m high, lasted for about one minute. Calm ensued for the next 10–20 minutes, broken by occasional red flashes. The episode then concluded with a minute of explosions of glowing cinders. The magma had apparently been injected from a fault intersecting the borehole at a depth of 1038 m. The little eruption gave off a thin cover of ash and cinders some 50 m long and 25 m wide, which had a volume of about 26 m³. In contrast, the largest eruption of the series was the last, on 4 September 1984. Deflation started at 20.10 and the eruption ensued at 23.49. Lava fountains expelled some 24 km² of lava flows from a fissure stretching 9 km north of Leirhnjúkur, which was fed by a dyke 1 m wide and 2.5 km deep. Most of the eruption was over in 12 hours, but weaker activity persisted from one vent for two more weeks (Tryggvason 1986).

During the rifting episode, magma rose into a shallow reservoir about 2.6 km or 3 km that may have been supplied by a magma layer lying 8–30 km below (Einarsson 1978, Beblo & Björnsson 1980, Gudmundsson 1987). Alternatively, the shallow reservoir could have been supplied by three individual reservoirs relaying magma towards the surface and lying at depths of about 5 km, 20–25 km and 30 km. The rifting fissures opened below the surface and magma was quickly inserted, blade-like, into them at a rate approaching 500 m³ per second, to the accompaniment of earthquakes. The fissures were 4 km or more deep and often over 50 km long in a major swarm and were arranged more or less vertically (Marquart & Jacoby 1985). The eruptions usually reached the surface 10–30 km from the reservoir.

Rootless cones, formed about 2100 years ago when the Younger Laxá flows entered the shallow eastern shore of Lake Mývatn.

Víti maar, Krafla, formed by the hydrovolcanic explosion on 17 May 1724 that initiated the Mývatn Fires.

The reduction of volume and magma pressure entailed by the lateral migration of the magma from the shallow reservoir immediately caused a rapid deflation or subsidence at the surface. Magma nevertheless continued to rise into the shallow reservoir so that it was replenished in about three to five months. Thus, it filled more quickly than it emptied and, consequently, periods of inflation were much longer than those of deflation. Each filling, of course, slowly inflated the central zone of Krafla and the whole process was repeated. In all, these episodes of magma intrusion widened the fissure swarm by about 5 m, with a maximum of some 9 m in the centre. The Krafla fissure swarm, as a whole, has apparently widened by an average of about 10–15 m during every millennium of the past 10 000 years. The whole series of eruptions covered some 36 km^2, but their volume amounted to less than 0.3 km^3.

SHIELD VOLCANOES

Although the volcanic systems composed of central volcanoes and fissure swarms comprise the major features of the active volcanic zone, smaller independent central volcanoes also occur, as well as independent fissure systems related to rifts. Shield volcanoes are probably the most widespread landforms created by concentrated vents; similar subglacial eruptions formed the table mountains; and apparently independent rift systems are best represented by Thingvellir. The shields and independent fissures commonly expel olivine tholeiite basalts, whereas the volcanic centres erupt quartz tholeiitic basalts (Óskarsson et al. 1982).

Most of the shields formed 9000–7000 years ago, and they seem to have continued the eruptions that formed the subglacial table mountains. Shield-forming activity may possibly be waning, because the offshore eruption of Surtsey in 1963–4 was the only such episode during historical times. They were formed by accumulations of basaltic flows that were emitted at the low average rate of about 5 m^3 per second (Saemundsson 1979). The shields have gentle slopes, broad basal diameters, ranging from 5 km to 10 km across, and their heights rarely exceed 500 m. They are almost entirely composed of copious outpourings of olivine tholeiitic basalts, which were erupted in a hot and fluid condition.

Skjaldbreidur is perhaps the most conspicuous shield volcano in Iceland. It forms a characteristic accumulation of fluid lava flows with gentle slopes radiating from a central hub rising to 1060 m, marked by a crater 350 m across. Ketildyngja and Kollóttadyngja, each of which has emitted copious lava flows, provide other examples in the Ódádahraun, to the northwest of Askja. About 1900 BC, Ketildyngja erupted the enormous Older Laxá lava flow, which covered 330 km^2, reached the sea 82 km away, and also formed the first lava dam that impounded Lake Mývatn. These lavas, erupted in a hot fluid state, now have a distinctive helluhraun (pahoehoe) surface. Kollóttadyngja is a similar shield. It is 5 km across and its gentle slopes rise to a height of 460 m. It is probably about 7000 years old.

PLINIAN ERUPTIONS IN ICELAND

The Plinian eruptions in Iceland are generated when more evolved lavas develop after relatively long sojourns in magma reservoirs. They are richer in volatiles and more viscous in character, and are marked by a transition to alkali basalts, dacites and even occasionally to rhyolites. Fragments and lava flows are expelled, but overall production rates are low. Their lava flows typically develop apalhraun (or aa) surfaces. This activity usually springs from vents clustered on very short fissures, which form slightly elongated, often rather small and compact, shields or stratovolcanoes. They frequently lie on top of an older volcanic basement that had been generated predominantly by eruptions from fissures (Imsland 1989).

These violent eruptions occur in four regions on the flanks alongside the main areas of fissure activity: the southernmost area of the eastern arm of the active volcanic zone, the Snaefellsnes Peninsula, the alignment of stratovolcanoes stretching northeastwards beneath the Vatnajökull ice cap from Öraefajökull to Snaefell, and the two volcanic areas under and around the Langjökull and Hofsjökull ice caps in central Iceland.

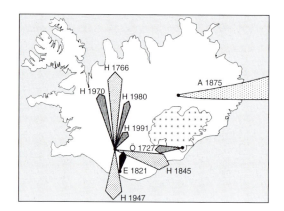

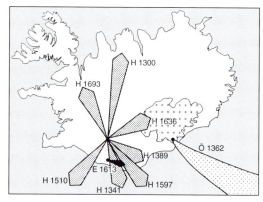

Torfajökull

Torfajökull is probably one of the most active silicic volcanoes in the world. It covers an area of about 450 km² and forms the largest expanse of rhyolitic rocks in Iceland (McGarvie 1984, McGarvie et al. 1990). A ring fracture, some 30 km across along its major axis, may in fact represent the rim of an old caldera. Since the most recent glacial period, Torfajökull has erupted at least 11 times, and ash from one of these explosions was deposited in the Orkney Islands about 5560 years ago. It most probably experienced two episodes of activity during historical times: the eruption that took place in AD 870, soon after the settlement of the country, and the other in 1477 (Larsen 1984). The outbursts of Torfajökull seem to be set off by, and occur virtually at the same time as, eruptions from the nearby tholeiitic volcanic system of Veidivötn–Bárdarbunga.

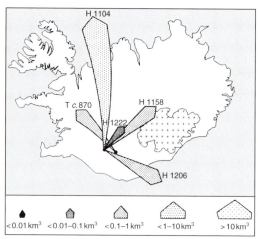

The main axes of silicic layers exploded from Askja (A), Eyjafjallajökull (E), Hekla (H), Öræfajökull (Ö) and Torfajökull (T). Arrow width indicates order of magnitude; length indicates relative volume. (After Larsen et al. 1999.)

ANTHONY NEWTON

Hekla from the southwest along the axis of the Heklugjá fissure.

Öraefajökull

Öraefajökull is the second largest active strato-volcano in Europe after Etna (Thórarinsson 1958). It is also the highest mountain in Iceland, rising to 2119 m, on the southern edge of the Vatna-jökull ice cap. Several distinct peaks surround an elliptical ice-filled caldera, stretching 5 km from north to south and 3 km from east to west. The volcano is older than the last glacial period, for its base is composed not only of basaltic lava flows but also of palagonitic breccias and pillow lavas, as well as fluvioglacial deposits. This mass was then transected by rhyolitic intrusions, one of which forms the Hvannadalshnukur, the highest point of both the volcano and Iceland. Most of the summit cone of Öraefajökull probably formed during the last glacial period when basaltic fragments exploded into the open air above the ice cap. It seems that an eruption from this volcano was responsible for the white pumice found on the 9000-year-old Trandvikan shoreline in Norway. There have been several other explosive but poorly dated episodes, notably one between 2000 and 4000 years ago, and another between 1000 and 2000 years ago.

It was during the eruption in June 1362 that Öraefajökull earned its name, which is derived from Oraefi, "the waste land". This was the largest explosion in Iceland in historical times. It apparently produced 10 km³ of distinctive white rhyolitic pumice, about 2 km³ of which fell on land. Some 4300 km² in the area to the northeast of the volcano was covered in ash 10 cm deep; and ash also fell on Ireland and Scotland. The explosion of hot fragments through the summit glaciers formed vigorous glacier bursts (jökulhlaups) that combined with ashfalls to devastate the adjacent farmlands to such an extent that they were abandoned for over a century. The smaller eruption of andesitic fragments in 1727 produced further jökulhlaups that once again destroyed much of the land that had been re-occupied (Thórarinsson 1958, Scarth 1999).

Hekla

Hekla is the most famous volcano in Iceland. It is also one of the most scientifically informative, because its eruptions spread layers of pale and easily identifiable fragments that are valuable indicator beds in establishing the chronological sequence of many postglacial events in the country. Indeed, Hekla has expelled more than half the total volume of intermediate and silicic materials erupted in Iceland since the end of the most recent ice age (Saemundsson 1979). Hekla rises near the edge of the eastern branch of the active volcanic zone in the centre of a volcanic system that is about 40 km long and reaches a height of 1491 m. It forms an elongated stratovolcano aligned on the 5.5 km-long Heklugjá fissure, which carries three distinct craters that do not always erupt in unison.

Hekla has been growing since the most recent glacial period from a base of intermediate and basaltic palagonites and lava flows, ash and pumice, and fluvioglacial deposits. Five major eruptive episodes were identified, dated and named H5 to H1 (Thórarinsson 1970, Larsen & Thórarinsson 1977), and they have recently been re-investigated. The first (H5) occurred about 6100 years ago, H4 about 3800 years ago, and H3, which produced about 12 km³ of fragments, took place about 2900 years ago. However, the Selsund pumice of the original H2 eruption has now been found below the H3 deposits, and has been dated to 3500 years ago. The old H2 has therefore been renamed the HS eruption. The H1 eruption

The historic eruptions of Hekla

The H1 eruption of Hekla began in the autumn 1104 after a rest of almost 700 years. It was a violent explosion of some 2.5 km^3 of rhyolitic fragments, which has been exceeded in historical times only by the Öraefajökull eruption of 1362. Hekla gave off no lava flows, and the explosive climax was probably brief. But the winds carried fine rhyolitic pumice northwards over half the country, where many farms had to be abandoned, and eventually as far as the British Isles and Scandinavia. Three minor eruptions occurred in 1158, 1206 and 1222. The next eruption began with a Plinian explosion in July 1300 that blanketed the north of the island with fine fragments and dislocated the rural economy so much that 500 people died in the ensuing famine in the Skagafjordur area in the north. Activity lasted for a year and ended with emissions of lava flows. The weak eruption in May 1341 emitted fragments, but the main damage was to the animals asphyxiated by poison gases. The eruption of 1389 saw the development of a new fissure near Skard and a new crater, the Raudoldur, south-southwest of the summit fissure. After more than a century of quiescence, Hekla built up enough energy for a violent Plinian explosion on 25 July 1510. The ashfall caused widespread damage, and volcanic bombs were even thrown 40 km from the volcano. The weak eruption of 1597 was chiefly notable because it lasted six months, and the even more feeble activity beginning on 8 May 1636 lasted a year.

The eruption that continued for seven months from 13 January 1693 was in many ways a repeat of the events of 1300. There was the usual Plinian opening act; then a widespread blanket of pumice combined with emission of poisonous gas to annihilate part of the animal population, and vast lava emissions closed the performance. The longest historical activity of Hekla took place between April 1766 and May 1768. This duration was matched by the exceptional amount of lava, covering 65 km^2, emitted from the central fissure. The ashfalls from the initial Plinian phase caused great destruction on the farms to the north, and a warm jökulhlaup later overwhelmed the Ytri–Ranga Valley. The seven-month eruption that began on 2 September 1845 was very similar. There was much damage from the blanket of ash and pumice, another jökulhlaup invaded the Ytri–Ranga valley, and the events ended with a lava flow. Hekla then rested for a century.

After the second longest period of quiescence since 1104, Hekla produced one of its most powerful eruptions in 1947, lasting for 13 months. The initial blast on 29 March 1947 sent a white column of steam soaring 27 km from the summit fissure within 20 minutes. Soon afterwards, the column settled around a height of 10 km as brown and then black dacitic ash was expelled. The ash covered 70 000 km^2 in Iceland, and within two days it was falling over Finland, 2750 km away. In the early hours of the eruption, the melting ice and hot ash generated many jökulhlaups that flooded as usual into the Ytri–Ranga Valley. Much less usual was the expulsion of lavas on the very first day from the summit fissure, and they covered more than 18 km^2 in 18 hours. Lava emissions became concentrated on Hraungígur, the new vent at the south-western end of the summit fissure. The other two vents on the crest of Hekla, Axlargígur and Toppgígur had frequent explosions for about a month, but then only intermittently for the rest of the active period. On the other hand, the Hraungígur vent displayed 13 months of continuous basaltic-andesite effusions and produced some 800 million m^3 of lava. Cattle were poisoned by carbon dioxide gas emissions, but, most unusually in this sparsely populated area, the eruption caused a fatality when a lava block fell upon a volcanologist.

occurred in AD 1104. All these eruptions apparently began with a major explosion of rhyolite or rhyodacite, and concluded with intermediate or basaltic-andesite emissions. Most of the material erupted since 1104 has been andesitic. Fragments from many of these eruptions have been identified as far afield as Scotland, Ireland, Sweden and northern Germany (Dugmore et al. 1996).

Since 1104, Hekla has usually behaved with a certain regularity. Earthquakes lasting less than two hours commonly provided preliminary warnings. Activity began with a violent Plinian explosion of fine rhyolitic or rhyodacitic fragments, forming an impressive column for several hours, which usually gave off more than two thirds of the total fragmentary output (Thórarinsson 1967b). As this episode concluded, first dacite and then basaltic-andesite flows poured forth, quickly at first, then more slowly after a few days, and then more rapidly again for several weeks. The eruptions ended with the emission of lava flows from the summit Heklugjá fissure for several months. Few eruptions lasted as much as a year, and the more violent outbursts only a few hours. Generally speaking, the more violent eruptions with the higher silica contents follow the longest periods of quiescence. The rate of emission has averaged 10 million m^3 per year. The volcano has produced about 8 km^3 of materials during historical times, probably from a compositionally zoned magma reservoir at about 8 km depth.

To the surprise of Hekla's students, it erupted again in 1970, after an interval of a mere 23 years (Thórarinsson & Sigvaldason 1972). After a 25-minute prelude of earthquakes, the eruption began on 5 May 1970 at 21.23, when three vents rapidly developed lava fountaining and lava

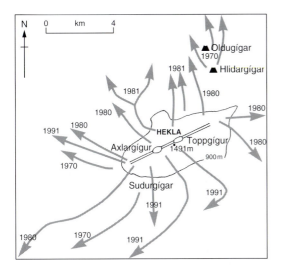

The Heklugjá fissure, with directions and dates of chief lava flows erupted from Hekla between 1970 and 1991.

flows along a fissure, later called Sudurgígar, about 1 km southwest of Axlargígur crater. They were matched by three more vents that developed 7 km northeast of Axlargígur crater and were later called Hlidargígar. The Sudurgígar vents continued to erupt until 10 May 1970, and the Hlidargígar vents continued until 20 May 1970. That same evening a new fissure, Oldugígar, took up the baton about 1 km north of Hlidargígar and gave rise to many lava fountains. Activity on the Oldugígar fissure gradually reduced until only one vent remained to build a cone, 100 m high, before the eruption ended on 5 July 1970. All of these fissures produced basaltic andesites covering 18 km² in area. The vents also gave off much fluorine that killed farm animals in northern Iceland.

The 1970 eruption had several unusual features. Lava flows emerged during the very first stages of the eruption. This was probably because not enough time had elapsed since the 1947 eruption for evolved magma to develop in the reservoir. The 1970 eruption also – and possibly for the same reason – gave off proportionately more lavas and fewer explosive fragments than usual. But the oddest aspect of all was that the summit fissure of Hekla, the Heklugjá, took virtually no part in the eruption, probably for the first time in the present cycle. However, the three newly formed fissures most probably represent lateral eruptions from the Hekla reservoir.

Hekla erupted again at 13.27 on 17 August 1980, without prior warning, after the shortest dormant period since 1104 (Grönvold et al. 1983). The summit fissure, Heklugjá, released a Plinian column 15 km high, which eventually covered 17 000 km² with fluorine-rich ash. At 13.30, as the active summit fissure lengthened to 8 km, six basaltic-andesite flows were expelled and eventually covered 24 km². Activity stopped on 20 August 1980, but not for long. On 9 April 1981 a short-lived modest ash column, 6 km high, exploded from the summit, followed by basaltic-andesite flows that emerged from new flank fissures and soon covered 6 km². This brief episode ceased on 16 April 1981. The 1980 and 1981 eruptions seem to have been part of the same episode because they emitted similar lavas, probably from a magma reservoir about 7–8 km deep.

Hekla resumed its eruptions once again on 17 January 1991. The outburst was preceded, as in 1970, 1980 and 1981, by ground deformations of some 3–4 cm and about 30 minutes of earthquakes (Gudmundsson et al. 1992). It produced mainly basaltic-andesite lavas, which covered about 23 km², and a Plinian column that rose 11.5 km within 10 minutes. But after a day had elapsed, activity was limited to the main fissure, where the main crater then formed. The eruptions came to an end on 11 March 1991.

The agitated phase is continuing. Earthquakes began beneath Hekla at 17.00 on 26 February 2000. Warnings of the impending eruption were announced on the national radio at 18.00, and the volcano duly burst into activity once again at 18.19. The eruption reached its greatest intensity during the next hour. Lava fountains gushed forth over a fissure nearly 7 km long, a column of ash immediately soared 10 km into the air, and a light scattering of ash covered central Iceland. Next day, lava flows advanced down the flanks of Hekla from several craters just south of the summit and, on 28 February, one flow with a snout more than 8 m thick had reached Lambafell, 5 km south of the vents, but it had advanced at a speed of less than 3 m an hour. On 29 February, activity almost ceased, but it revived after noon at the southern end of the fissure. On 1 March, seven craters were in moderate Strombolian activity and the lavas had covered about 17 km² with a volume of about 0.1 km³. The activity declined appreciably thereafter.

SUBGLACIAL VOLCANOES

Volcanoes formed beneath ice sheets are among the most distinctive features of Iceland. Some developed under the extensive glacial ice sheets and were exposed when the ice melted, and others are still active beneath the four remaining ice caps. Subglacial volcanoes owe their special characteristics to the interactions between the ice cover, the molten materials expelled, and the meltwaters that were thereby generated. Morphologically, the now-exposed subglacial volcanoes are distinguished by their relatively flat summits and steep, sharply defined sides. They form table mountains (stapar) and ridges (hryggir), which contrast markedly with the conical crests of most volcanoes formed in the open air (Van Bemmelen & Rutten 1955). Many older shields or stratovolcanoes, such as Hekla, Öraefajökull and Dyngjufjöll, for instance, probably spent much of their early lives erupting beneath ice sheets, although the typical forms have since been masked by the copious products of more recent eruptions in the open air. Grímsvötn (under Vatnajökull) and Katla (under Mýrdalsjökull) provide the finest examples of contemporary subglacial activity (Jones 1969, 1970).

The simplest subglacial features are the flat-topped palagonitic (móberg) ridges formed by subglacial basaltic fissure eruptions, which would have given rise to cone rows and lava-flow vents if they had occurred in the open air. They are exemplified by the Herdubreidartögl ridge in the Ódádahraun desert and the Namáfjall ridge lying east of Lake Mývatn.

Table mountains are more complex than palagonitic ridges, chiefly because they erupted for several decades or more. Had they erupted in the open air, they would probably have developed as lava shields. More than 30 table mountains have been mapped in the central zone of activity. Most of them have a basal diameter of between 2 km and 5 km, and they rise by steep concave slopes to a regular, almost flat, lava-capped plateau. They owe their sharp outlines and steep sides to the confinement of the enclosing ice and the meltwater vault, generated by their eruptions, which stopped the fragments from spreading farther outwards. Their flattish summits were formed when lava flows erupted in the open air after melting had pierced the ice sheet. Below the cap of lava, the bulk of each table mountain is composed of palagonitic tuff, breccia and pillow lavas erupted in a mass of meltwater that was confined beneath the ice sheet (Preusser 1976, Jones 1960, 1970).

Herdubreid, dominating the plains of the Ódádahraun, is one of the most majestic of Icelandic table mountains (Sigvaldason 1968, Moore & Calk 1991). Its summit plateau is 2 km across and rises about 1000 m from its base. The lower 250 m of the mountain is largely composed of pillow lavas piled up close to the vent in the meltwaters confined beneath the ice. They are succeeded by layers of tuffs, about 200 m thick, that were erupted by Surtseyan explosions when the ice sheet had melted to form a shallow open-air lake. In turn, these tuffs are succeeded by a 50 m-thick layer of angular pillow-lava lumps, forming a breccia that was generated as lavas spread out across the lake. It is overlain by a band, 150 m thick, of thin alternating layers of lava flows, broken pillow-lava layers and cinders that were expelled in response to variations in the lake level as eruptions continued. A thicker lava flow forms the summit plateau at 1400 m and its eruption marked the final emergence of the volcano from its glaciolacustrine constraints. This freedom was emphasized when explosions built a cinder-cone that now marks the summit of Herdubreid at 1650 m.

Several Icelandic volcanoes, or volcanic systems, still erupt from time to time beneath the modern ice caps. Grímsvötn, Kverkfjöll, Bárdarbunga and Gjálp lie beneath Vatnajökull, Katla lies under Mýrdalsjökull, and Eyjafjallajökull lies under the ice cap of the same name. Kverkfjöll, which last erupted in 1875, is now marked by strong fumarole activity. Bárdarbunga has experienced many small historical eruptions. Eyjafjallajökull may have produced as many as 17 intermediate lava flows during the past 10 000 years, although it has erupted only twice during historical times, in 1613 and in 1821–3. Both of these were small eruptions from the summit crater and they were made more explosive when water from the melting ice cap interacted with the rising magma. Nevertheless, the fine, dense and dark-grey fragments expelled in 1821–3, for example, attained a volume of only 0.004 km³.

Grímsvötn

Grímsvötn has an ice-covered caldera, about 40 km² in area, that lies beneath Vatnajökull. It has been the source of at least two major prehistoric ash eruptions, as well as about 50 episodes

The Gjálp eruption of 1996 under Vatnajökull

At 10.48 on Sunday 29 September 1996, there was a strong earthquake, followed by 30 hours of smaller tremors. On 30 September, basaltic-andesite lava began erupting at Gjálp, under the 500 m-thick ice cap between the calderas of Grímsvötn and Bárdarbunga. The ice cap quickly began to melt from below and ashy fumes rose from its surface. Soon, all of the ice lying directly above the erupting vents had melted to form lakes. The fragments piled up closer and closer to the surface of the lakes, where the weaker water pressure could no longer stifle the explosions. At 05.18 on 2 October, Surtseyan eruptions were already blasting fine tuffs and black ash into the air. Meanwhile, meltwater was pouring into the Grímsvötn caldera and accumulating under the ice cap there. On 4 October, the water beneath the ice exceeded any levels previously recorded. On 13 October, the eruption stopped. It had built a tuff cone that rose 40 m above the meltwater lake. But the ice continued to melt at Grímsvötn. On 2 November, the waters were 60 m higher than the level at which jökulhlaups were usually unleashed. Then, at about 08.30 on 5 November, the

waters finally lifted the dam of ice from the caldera rim, and the expected and overdue jökulhlaup finally cascaded down the Skeidar valley. Over 3 km³ of meltwater, with a discharge of almost 45 000 m³ per second, charged across the coastal plain. It was the largest jökulhlaup since 1938. No lives were lost in this sparsely populated area of the southeastern coastlands, but electricity cables, a main road bridge and 10 km of the coastal ringroad were ripped away.

Grímsvötn erupted again at 09.20 on 18 December 1998 and an eruption column soon extended 10 km above the southern edge of the caldera. Slightly less vigorous activity continued the next day, when the column was reduced to a height of about 7 km and ash was blown southeastwards. On 20 December, more ash was expelled and the eruptive column collapsed at least once and formed a nuée ardente. Thereafter, the eruption's power diminished, the column developed only intermittently, a lake grew up northeast of the vent, and a tuff ring was formed partly on the ice and partly on the adjacent exposed bedrock. Activity ceased on 28 December 1998.

of activity since AD 1200. These eruptions have been basaltic, with insignificant quantities of silicic materials (Jakobsson 1979). Grímsvötn has given off between 3 km³ and 5 km³ of material during historical times; its eruption in 1938, for instance, was probably the third largest in Iceland in the twentieth century. Many of these eruptions have been accompanied by powerful jökulhlaups, which commonly have discharges as vigorous as the world's major rivers. Grímsvötn is also the site of the largest field of active fumaroles in Iceland. Their heat maintains a lake that is trapped in the caldera beneath about 200 m of ice. When the volcano erupts, the waters

increase greatly in volume, lift up part of the ice cap, and then drain rapidly over the rim of the caldera as jökulhlaups that then rush seawards for several days.

Katla

The Katla volcanic system lies beneath the Mýrdalsjökull ice cap in the eastern branch of the active volcanic zone, and the caldera in its centre is 700 m deep and covers an area of about 110 km². One of the most active volcanoes in Iceland, Katla has probably had more than 150

The southern flanks of Katla and Mýrdalsjökull ice cap.

prehistoric eruptions, and about 20 have been recorded since the settlement in AD 874 (Thórarinsson 1975, Sigvaldason 1983). The active phases are commonly preceded by a few hours of strong earthquakes, but eruptions normally last only two or three weeks. The main active episodes have usually expelled about 0.5 km³ of fine tuffs, all of which have a similar alkali basaltic composition. The most important of these, the great Eldgjá eruption of AD 935, occurred on the northeastern arm of the fissure swarm.

A shallow magma reservoir existed at least between 6600 and 1700 years ago when 12 silicic, largely dacitic, eruptions occurred, each concluding with basaltic emissions. The last five of these occurred 3600, 3139, 2975, 2660 and 1676 years ago. However, they were not nearly so large as the main eruptions of Hekla or Öraefajökull, although pumice fragments that floated from Katla's explosions have been identified in Scandinavia and Scotland. During historical times, basaltic eruptions have often occurred at intervals of fewer than 70 years, and the most important took place in about AD 935, 1485, 1625, 1660, 1721 and 1755–6. The eruption between 17 October 1755 and 13 February 1756 was not only the longest but also the most vigorous in historical times and it gave off some 1.5 km³ of fragments. The latest eruption took place in 1918, when about 0.64 km³ of basaltic fragments were expelled.

The jökulhlaups of Katla are its most dangerous features. They have periodically devastated the nearby coastal plain, which is in fact essentially composed of jökulhlaup debris. Indeed, they have extended seawards by some 4 km during the past thousand years. Their discharge during the last major eruption on 12 October 1918 approached 200 000 m³ per second for two days, which is a thousand times that of the River Thjorsa, Iceland's largest river, and comparable to the discharge of the River Amazon.

These jökulhlaups make Grímsvötn and Katla the most dangerous volcanoes in Iceland. Fortunately, most of the regions invaded by jökulhlaups are sparsely populated, otherwise they would be notable killers.

HYDROTHERMAL ERUPTIONS

Iceland is riddled with **hydrothermal** features, which have contributed much to the local toponymy. Their basic cause is abundant precipitation that falls on fissured ground, which is heated by shallow magma. The lavas provide a maze of conduits that facilitate the downward penetration, lateral distribution and ascent of the waters. The varying geometry of the conduits also contributes to the different speeds of water heating, mineral contamination and circulation. But the sine qua non of the hydrothermal system is the presence of high-level magma, which heats the adjacent rocks to high temperatures. The temperature of the rocks increases rapidly with depth and the average rise in the hydrothermal zones is about 1°C for every 10 m from the surface downwards, but in some areas it can exceed ten times that amount. The water circulation and the predominant temperatures also contribute to the chemical alteration of the enclosing lavas, and thus to the amount and type of material eventually carried by the waters towards the surface. Generally speaking, the areas with the higher temperatures give rise to solfatara fields, which have relatively low water discharges, much chemically altered rock, and a predominance of steam and sulphurous gases. The lower-temperature areas, on the other hand, tend to produce geysers and hot springs (where the waters carry small proportions of altered rock), and mudpots (where the proportion of altered rock is higher).

The hydrothermal areas are concentrated in the axial zone and alongside its southwestern branch. The largest high-temperature areas, for instance, are located at Torfajökull, covering 100 km², at Henzill, covering 50 km², as well as under the Grímsvötn ice cap. The areas of lower temperatures are almost all confined to fissure zones in southwestern Iceland, where geysers and warm or hot springs at less than 100°C are generated. They are especially common in Hveravellir ("hot field") and Kelingarfjoll, for example. The western end of the Reykjanes Peninsula has saline mudpots and hot springs resulting from the rather unusual infiltration of sea waters into the heated substratum.

Iceland is perhaps more famous for its geysers than for any of its magmatic eruptions and has, of course, given the word to the international vocabulary. Iceland now has only about a dozen geysers which, in fact, would operate only sporadically without artificial assistance. The much-fissured area around Geysir is the chief low-temperature thermal centre in Iceland, where the springs reach the surface at 80–100°C. The Grand Geyser itself is waning now after about 8000 years of activity. It erupted every half hour 200 years ago; a century ago it still managed once every three weeks; now it bursts out only after long irregular intervals unless it is stimulated by soap powder. Nevertheless, earthquakes in June 2000 seemed to give Grand Geyser a new lease of life. However, Strokkur, nearby, has been sustained artificially on a bore hole life-support machine and still gushes 20 m high every few minutes so as not to disappoint the customers. Like all hydrothermal systems, geysers are so fragile that they can be altered or destroyed by earth tremors or by natural or human-induced blockages. In compensation, other conduits may be opened by the same token, such as the Niyhver springs that developed after an earthquake in the Reykjanes Peninsula.

Geysers form when water and steam are suddenly released to the surface, but most of the heated waters bubble out more calmly as hot springs that often form distinctly warm streams and rivers, or bubbling pools, such as Bolutler in Hveravellir. Many hot springs are naturally contaminated by dissolved chemical precipitates. They then give rise to mudpots, which bubble threateningly, such as, for example, Krísuvík in the Reykjanes Peninsula and those in Kelingarfjoll. Other hot springs precipitate highly coloured deposits of carbonate or sulphur compounds around their vents, which, for instance, add much to the beauty of Kelingarfjoll. In other cases, dissolved minerals or algae give the colour, and Hveravellir, for example, has both a blue and a green basin. However, many minerals are often precipitated around the vents and conduits near the surface and, as a result, the systems can be blocked by natural means. This is especially noticeable where the geysers deposit mounds of silica, called geyserite, around their vents. In contrast, earthquakes could also widen fissures and rejuvenate the whole process.

So many of the Icelandic hot springs have been exploited that more than three quarters of the population use geothermal heat. Boreholes up to 2000 m deep have been sunk to bring up hot water to provide cheap, clean, ecologically sound heat for homes, industrial and commercial buildings, open-air warm swimming-pools, and greenhouses. All of Reykjavík is heated in this way; Krafla has a steam-powered electricity

Hydrothermal eruptions at Krísuvík.

plant; and at Hveragerdi, the geothermally heated greenhouses, covering 10 ha have become world famous because bananas, as well as less exotic fruits such as tomatoes, have been ripened almost within reach of the Arctic Circle.

ISLAND VOLCANOES

Iceland originated out of submarine eruptions on the Mid-Atlantic Ridge. It has also grown since the most recent glacial period along the south-western branch of the active volcanic zone on the Reykjanes Peninsula. Nine reliably recorded eruptions, and probably at least three more, have also occurred off Reykjanes since the settlement of Iceland, and temporary islands were formed in 1211, 1422 and 1783 (Thórarinsson 1966). Where the active volcanic zone extends off northern Iceland, too, at least one eruption in 1867, and probably three more, have taken place in historical times. But no eruptions occurred on its main southeastern branch towards the Vestmannae-yar (the "-aeyar" suffix meaning "islands") from the outbursts of Helgafell on Heimaey more than 5000 years ago until fires were reported south-east of Geirfuglasker in 1896. Thus, the eruptions of Surtsey in 1963 and on Heimaey in 1973 testify to the new vitality of activity here. At present the Vestmannaeyar rise from a platform that is often less than 130 m deep. No doubt, within a short span of **geological time** all of these islands will be joined together by fissure eruptions to form a large peninsula, projecting southwest-wards into the Atlantic Ocean.

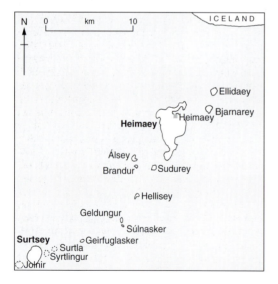

Surtsey, Heimaey and their companions in the Vestman-naeyar.

The eruption of Surtsey 1963–7

The eruption of Surtsey set the standard for what became known as the Surtseyan style of activity all over the world (Moore 1985). The first phase, which may have lasted a week, was the completely submarine eruption of basaltic pillow lavas and then fine tuffs, which built up a volcanic mound rising steeply from the submarine platform at about 130 m depth (Kokelaar 1983). This phase passed largely unnoticed except that the sea was 2°C warmer in the area on 13 November, and there was a distinct smell of sulphur on Heimaey and at Vik on the mainland on 12 November.

The second, truly Surtseyan, phase then began when the eruption expelled its first column of steam and lava fragments into the air, which was first sighted about 33 km from the mainland at 07.15 on 14 November 1963 (Thórarinsson 1964, 1966). Within four hours, the eruption had developed the full fury that it was to maintain with little respite until 4 April 1964. Every few minutes, sea water rushed into the vent, forming a slurry of ash, tuff and water, and met the hot lava and exploding gases rising to the surface. The gases and steam repeatedly shattered the lava into black jets and plumes of fine fragments that were expelled 500 m into the air every second, as if they had been fired from a gun, and the larger fragments formed bombs that whistled outwards 1 km or more from the vent. A billowing column of white steam, often riddled with lightning and laden with fine dark ash, rose 8 km high in calm weather, but gales often blew the column sideways so that it rarely exceeded 2 km in height.

At first the eruptions occurred from two, three or four vents along a northeasterly fissure, but they gradually concentrated on one vent, called Surtur I. They formed a large horseshoe cone that reached 45 m on 16 November, 100 m on 23 November, 130 m on 30 December 1963 and 174 m on 5 February 1964. At the same time, the waves had cut a 100 m-wide marine platform around the cone of unconsolidated tuffs. Yet the eruptions were intense enough to keep pace with such an onslaught. On 16 December, the first geologists to land on the island saw that Surtsey was producing hot olivine basalt, at a temperature of 1150°C, which was very similar to lavas erupted by Katla on the nearby mainland, and they deduced that the eruptions were being supplied by a deepseated magma reservoir.

On 29 December 1963, a new eruption, called Surtla, broke from the sea 2.4 km northeast of Surtsey. However, the expected new island never appeared and it remained as a shoal 6 m below the waves, because activity stopped on 6 January 1964 (Kokelaar & Durant 1983). Activity on Surtsey had waned during this episode, but it increased again when Surtla stopped, thus demonstrating that Surtla was only derived from a lateral vent of Surtsey.

Another, more permanent, change took place on 1 February 1964, when Surtsey developed a new crater called Surtur Junior on the northwestern flanks of the first cone. The original vent, Surtur I, never resumed activity, but this new vent maintained the Surtseyan effort unabated. By the end of March 1964, the island was 1750 m long and 180 m high. Surtur Junior had built a cone as high as its predecessor's and was erupting fragments so quickly that the sea could scarcely penetrate the vent. Thus, the typical Surtseyan explosions seemed doomed.

The third effusive and final phase duly began on 4 April 1964. A lava lake welled up the vent, lava fountains rose 50 m high, and the first lava flows formed a barrier along the beach. The emissions discharged at such a rate that, by the end of April 1964, one third of Surtsey was covered with lava, and the island had the armour plating needed for survival. Surtsey gradually grew into a lava shield.

As Surtsey continued its less spectacular effusions, two further submarine eruptions occurred, most probably from lateral vents on the same fissure as Surtsey itself. On 5 June 1965, the first eruption gave rise to the islet of Syrtlingur, 500 m northeast of Surtsey. But its explosions were never so continuous, nor were its accumulations so voluminous as on Surtsey. Thus, Syrtlingur intermittently grew by Surtseyan explosions and was destroyed by the ocean throughout the summer of 1965. It reached its maximum height of 65 m on 15 September 1965, and was last seen on 17 October, at the start of a week of bad weather. On 24 October, Syrtlingur had stopped erupting and the waves had eroded away the 2 million m³ of fragments that it had expelled above sea level.

The second submarine eruption, which began on 28 December 1965, formed the islet called Jolnir, about 500 m southwest of Surtsey. Again it had neither the explosive persistence nor the volume of fragments to survive for long. Like Syrtlingur, Jolnir appeared after every major emission of fragments and disappeared again when marine erosion gained the upper hand. Once, on 10 August 1966, Jolnir reached 60 m high, but it disappeared altogether on 20 September 1966. In the meantime, on 19 August 1966, Surtsey had begun its last episode of effusive activity, which continued at a decreasing rate until the eruption stopped in July 1967.

By then, more than half of the island's area of 2.5 km² was covered with lava flows, with a total volume of almost 300 million m³. However, the other eruptions formed shoals, which in time will make the supporting pillars for new eruptions on nearby fissures. Surtsey's eruption generated just over 1 km³ of volcanic materials, of which only 10 per cent formed the island above sea level. Surtsey demonstrated that shield volcanoes could be formed within a single eruptive episode, but it also confirmed that no clearcut distinctions could be made between fissure eruptions and central eruptions, for Surtsey evolved from the first to the second, and eventually formed a circular shield. However, throughout the eruption the type of magma, its chemical composition and its gas content remained the same (Thórarinsson 1966).

ANTHONY NEWTON

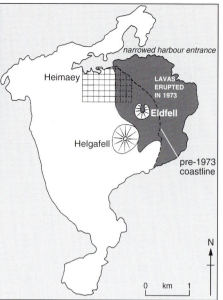

Above A house overwhelmed by lava flows on Heimaey in 1973.

Right Eldfell and the eruption on Heimaey in 1973 with the new land added to the island.

Below Entrance to Heimaey harbour, narrowed on the right by the lava flow erupted from Eldfell in 1973.

ANTHONY NEWTON

The eruption on Heimaey in 1973

Covering an area of only 16 km^2, Heimaey is the largest and also the oldest of the Vestmann Islands, or Vestmannaeyar, and Surtsey is now the second largest. But, in the shallow waters around them, there are many submarine volcanoes that either did not erupt enough material to reach sea level or were too fragile to withstand marine attack. The northern part of Heimaey forms a long volcanic ridge, but the rest of the island is a low plain, where the cinder cones of Saefell and Helgafell erupted in the south about 5000 years ago. The population of 5300 is concentrated in the town of Heimaey, which is situated 1 km due north of Helgafell.

The eruption in 1973 was preceded by some 30 hours of tremors of increasing intensity and frequency (Thórarinsson et al. 1973, Einarsson 1974, Self et al. 1974). At 01.55 on 23 January 1973, lava fountains burst more than 200 m into the air from a fissure, 1800 m long, on the eastern slopes of Helgafell. On 24 January, the lava fountains concentrated upon the northern end of the fissure, but they subsided when the first ash, lapilli and bombs were expelled and olivine basalts started to flow eastwards to the sea. Next day, explosions increased at the northern end of the fissure and rapidly built a cone and showered ash on the town. On 26 January, hot ash set fire to some houses and crushed others under its weight in its eastern parts. The fissure lengthened to 3 km, enabling the lavas to flow in a more threatening northward direction towards the town. Both flows and fragments were discharged at high rates. On 30 January, the cone was already 185 m high and the lava flows had added 1 km^2 to the island. By mid-February, 110 million m^3 of lava covered an area of 3 km^2. On 19 February, the northern crater wall collapsed and the breach helped direct yet more lava towards the town and its harbour. It was saved only by human intervention and the end of the eruption in April 1973. In March, the discharge rates of the lava fell to only 10 m^3 per second, a tenth of the original rates of emission.

The disaster was well handled because the best use was made of contingency planning and the resources available for evacuation. This was the kind of eruption where human initiative was likely to be most successful, because the emissions were limited to a moderate spread of fragments and to rather slowly moving lavas. A seven-person committee was set up to direct operations. The whole population of the island was evacuated to temporary accommodation on the mainland within 7 hours, using the 77 fishing boats sheltering in the harbour from a storm that, luckily, had just abated. The sick and the old were moved by air. The cattle and the cars, US$1 million worth of deep-frozen fish that had been stored in the harbour, the money from the bank and municipal documents were transferred mainly to Reykjavík. The national bank released cash for the evacuees, the government donated US$2 million and increased taxes to pay for the disaster.

From the onset of the eruption, only police and others with technical ability remained on the island, with suitable vehicles and equipment. They faced three dangers: poisonous gas, falling ash and lava flows. Invisible dense carbon dioxide and also hydrogen sulphide, which can at least be detected by its smell of rotten eggs, lurked in the cellars for days, but, once their presence was known, they could be neutralized by gasmasks. But it was carbon dioxide that caused the only fatality during the eruption.

Falling ash caused the greatest danger during the first weeks of the eruption when the explosions were at their most powerful. Bombs damaged rooftops and broke windows, and hot ash often set fire to the wooden houses. The weight of the ash also caused roofs to collapse and was even thick enough to bury houses in the eastern part of the town, nearest the volcano. Boarding up windows with metal sheets, and shovelling ash from rooftops, could not keep pace with the falls, although volunteers worked well into March. In the end, the ash burned and buried over 80 houses, three quarters of them being lost in the first week.

The lavas started to flow towards Heimaey after the fissure lengthened on 26 January and part of the crater wall collapsed on 19 February. They threatened the eastern part of the town and the entrance to its harbour, as well as the electric cables and the freshwater conduit from the mainland. The threat to the harbour was all the more important because the town accounted for over an eighth of Iceland's fish exports. There were, therefore, important financial and humanitarian reasons for trying to stop the advancing flow. In early February, volunteers began pouring sea water onto the lava's snout to solidify it and slow down its advance. When this was successful, in early March, the government sent out a ship whose pumps dowsed the lavas much more effectively. Powerful water pumps were then also despatched from the USA and installed on land. Soon, 20 000 m^3 of lava were being congealed every hour, stopping the advance of the flow. At the same time, the ash swept from the streets was compacted into a barrier around the perimeter of the flow to stop it from spreading sideways and westwards into the town. Moreover, the lavas that reached the harbour narrowed the entrance to 30 m and thereby provided better protection for the fishing fleet than ever before.

But such methods could only be successful during a relatively docile eruption, where the water supply was plentiful and the lavas slow moving. The financial and human costs of seawater pumping and barrier building were high. In spite of all this endeavour, the electricity supply cable was cut, three of the five fish factories were destroyed and the lava flow claimed over 300 houses. Furthermore, the lava flow was stopped only when the eruption was ending and the rates of discharge and advance had already waned considerably. Most of the town was rebuilt and the population has reached its former total. Seismometers have been installed on the mainland nearby and continuously recording tiltmeters have been set up on the island to warn of future crises.

BIBLIOGRAPHY

Annertz, K., M. Nilsson, G. E. Sigvaldason 1985. *The post-glacial history of Dyngjufjoll*. Report 8503, Nordic Volcanological Institute, Reykjavík.

Backstrom, K. & A. Gudmundsson 1989. *The grabens of Sveiner and Sveinagjá, NE Iceland*. Report 8901, Nordic Volcanological Institute, Reykjavík.

Barth, T. F. W. 1942/43. Craters and fissure eruptions at Mývatn in Iceland. *Norsk Geografisk Tidsskrift* **9** (2), 58–81.

—— 1950. *Volcanic geology, hot springs and geysers of Iceland*. Publication 587, Carnegie Institution of Washington, Washington DC.

Beblo, M. & A. Björnsson 1980. A model of electrical resistivity beneath NE Iceland: correlation with temperature. *Journal of Geophysics* **47**, 184–90.

Bemmelen, R. W. van & M. G. Rutten 1955. *Table mountains of northern Iceland*. Leiden: E. J. Brill.

Björnsson, A. 1985. The dynamics of crustal rifting in NE Iceland. *Journal of Geophysical Research* **90**, 10151–62.

Björnsson, H. & P. Einarsson 1990. Volcanoes beneath Vatnajökull, Iceland: evidence from radio-echo sounding, earthquakes and jökulhlaups. *Jökull* **40**, 147–68.

Björnsson, A., K. Saemundsson, P. Einarsson, E. Trygvason, K. Grönvold 1977. Current rifting episode in north Iceland. *Nature* **266**, 318–23.

Björnsson, A., G. Johnssen, H. S. Sigurdsson, G. Thorbergsson, E. Tryggvason 1979. Rifting of the plate boundary in north Iceland, 1975–1978. *Journal of Geophysical Research* **84**, 3029–3038.

Brandsdóttir, B. 1992. Historical accounts of earthquakes associated with eruptive activity in the Askja volcanic system. *Jökull* **42**, 1–12.

Brandsdóttir, B. & P. Einarsson 1979. Seismic activity associated with the September 1977 deflation of the Krafla central volcano, NE Iceland. *Journal of Volcanology and Geothermal Research* **6**, 197–212.

Decker, R. W., P. Einarsson, P. A. Mohr 1971. Rifting in Iceland: new geodetic data. *Science* **173**, 530–33.

Dugmore, A. J. 1989. Icelandic volcanic ash in Scotland. *Scottish Geographical Magazine* **105**, 168–72.

Dugmore, A. J., A. J. Newton, K. J. Edwards, G. Larsen, J. J. Blackford, G. T. Cook 1996. Long-distance marker horizons from small-scale eruptions in Iceland and Scotland. *Journal of Quaternary Science* **11**, 511–16.

Einarsson, P. & B. Brandsdóttir 1980. Seismological evidence for lateral magma intrusion during the 1978 deflation of the Krafla volcano in NE Iceland. *Journal of Geophysics* **47**, 160–65.

Einarsson, E. H., G. Larsen, S. Thórarinsson 1980. The Solheimar tephra layer and the Katla eruption of c. AD 1357. *Acta Naturalia Islandica* **28**, 1–24.

Einarsson, P., B. Brandsdóttir, M. T. Gudmundsson,

H. Björnsson, K. Grínvold 1997. Center of the Iceland hotspot experiences volcanic unrest. *Eos* **78**, 369–75.

Einarsson, T. 1974. *The Heimaey eruption in words and pictures*. Reykjavík: Heimskringla.

Foulger, G. R. & D. R. Toomey 1989. Structure and evolution of the Hengill–Grensdalur volcanic complex, Iceland: geology, geophysics and seismic tomography. *Journal of Volcanology and Geothermal Research* **46**, 157–80.

Gautneb, H. 1989. *Sheet intrusions associated with the Reykjadalur volcano, western Iceland: structure and composition*. Report 8902, Nordic Volcanological Institute, Reykjavík.

Grönvold, K. 1984. *Mývatn Fires 1724–1729: chemical composition of the lava*. Report 8401, Nordic Volcanological Institute, Reykjavík.

Grönvold, K. & H. Johannesson 1984. Eruption in Grímsvötn 1983. *Jökull* **34**, 1–11.

Grönvold, K., G. Larsen, P. Einarsson, S. Thórarinsson, K. Saemundsson 1983. The Hekla eruption 1980–1981. *Bulletin Volcanologique* **46**(4), 349–63.

Gudmundsson, A. 1984. Tectonic aspects of dykes in northwestern Iceland. *Jökull* **34**, 81–96.

—— 1986a. Formation of crustal magma chambers in Iceland. *Geology* **14**, 164–6.

—— 1986b. Mechanical aspects of postglacial volcanism and tectonics of the Reykjanes Peninsula, SW Iceland. *Journal of Geophysical Research* **91**, 12711–21.

—— 1987. Lateral magma flow, caldera collapse, and a mechanism of large eruptions in Iceland. *Journal of Volcanology and Geothermal Research* **34**, 65–78.

—— 1988. Effect of tensile stress concentration around magma chambers on intrusion and extrusion frequencies. *Journal of Volcanology and Geothermal Research* **35**, 179–94.

—— 1995. Ocean ridge discontinuities in Iceland. *Geological Society of London, Quarterly Journal* **152**, 1011–1015.

Gudmundsson, M. T. & H. Björnsson 1991. Eruptions in Grímsvötn, Vatnajökull, Iceland, 1934–1991. *Jökull* **41**, 21–45.

Gudmundsson, A. and 11 co-authors 1992. The 1991 eruption of Hekla, Iceland. *Bulletin of Volcanology* **54**, 238–46.

Gudmundsson, M. T., F. Sigmundsson, H. Björnsson 1997. Ice–volcano interaction of the 1996 Gjálp subglacial eruption, Vatnajökull, Iceland. *Nature* **389**, 954–7.

Haflidason, H., J. Eirikson, S. van Krefeld 2000. The tephrochronology of Iceland and the North Atlantic during the middle and late Quaternary: a review. *Journal of Quaternary Science* **15**, 3–22.

Hammer, C. U. 1984. Traces of Icelandic eruptions in the Greenland ice sheet. *Jökull* **34**, 51–65.

Hammer, C. U., H. B. Clausen, W. Dansgaard 1980. Greenland ice sheet evidence of post-glacial volcanism and its climatic impact. *Nature* **288**, 230–35.

Imsland, P. 1983. Iceland and the ocean floor: comparison of chemical characteristics of the magmatic rocks and some volcanic features. *Contributions to Mineralogy and Petrology* **83**, 31–7.
—— 1989. Study models for volcanic hazards in Iceland. In *Volcanic hazards: assessment and monitoring*, J. H. Latter (ed.), 36–56. Berlin: Springer.

Jackson, E. L. 1982. The Laki eruption of 1783: impacts on population and settlement in Iceland. *Geography* **67**, 42–50.
Jakobsson, S. P. 1979. Petrology of recent basalts of the eastern volcanic zone, Iceland. *Acta Naturalia Islandica* **26**, 1–103.
Jakobsson, S. & J. G. Moore 1980. Through Surtsey: unique hole shows how volcano grew. *Geotimes* **25**, 14–16.
Jones, J. G. 1969. Interglacial volcanoes of the Laugarvatn region, southwest Iceland – I. *Geological Society of London, Quarterly Journal* **124**, 197–210.
—— 1970. Interglacial volcanoes of the Laugarvatn region, southwest Iceland – II. *Journal of Geology* **78**, 127–40.

Kokelaar, B. P. 1983. The mechanism of Surtseyan volcanism. *Geological Society of London, Quarterly Journal* **140**, 939–44.
—— 1986. Magma–water interactions in subaqueous and emergent basaltic volcanism. *Bulletin of Volcanology* **48**, 275–89.
Kokelaar, B. P. & G. P. Durant 1983. The submarine eruption and erosion of Surtla (Surtsey), Iceland. *Journal of Volcanology and Geothermal Research* **19**, 239–46.
Krafft, M. & F. D. de Larouzière 1999. *Guide des volcans d'Europe et des Canaries*, 3rd edn. Lausanne: Delachaux et Niestlé.
Kristjansson, L. (ed.) 1975. *Geodynamics of Iceland and the North Atlantic area* [proceedings of the NATO Advanced Study Institute, Reykjavík, 1–7 July, 1974]. Dordrecht: Reidel.

Larsen, G. 1984. Recent volcanic history of the Veidivötn fissure swarm, southern Iceland: an approach to volcanic risk assessment. *Journal of Volcanology and Geothermal Research* **22**, 33–58.
Larsen, G. & S. Thórarinsson 1977. H4 and other acid Hekla tephra layers. *Jökull* **27**, 28–45.
Larsen, G., K. Grönvold, S. Thórarinsson 1979. Volcanic eruption through a geothermal borehole at Námafjall, Iceland. *Nature* **278**, 707–710.
Larsen, G., A. J. Dugmore, A. J. Newton 1999. Geochemistry of historical-age silicic tephras in Iceland. *The Holocene* **9**, 463–71.

Marquart, G. & W. Jacoby 1985. On the mechanism of magma injection and plate divergence during the Krafla rifting episode in NE Iceland. *Journal of Geophysical Research* **90**, 10178–92.
Mathews, W. H. 1947. "Tuyas", flat-topped volcanoes in northern British Columbia. *American Journal of Science* **245**, 560–70.
McGarvie, D. W. 1984. Torfajökull: a volcano dominated by magma mixing. *Geology* **12**, 685–8.
McGarvie, D. W., R. MacDonald, H. Pinkerton, R. L. Smith 1990. Petrogenic evolution of the Torfajökull volcanic complex, Iceland II: the role of magma mixing. *Journal of Petrology* **31**, 461–81.
Métrich, N., H. S. Sigurdsson, P. S. Meyer, J. D. Devine 1991. The 1783 Lakagígar eruption in Iceland: geochemistry, CO_2 and sulfur degassing. *Contributions to Mineralogy and Petrology* **107**, 435–47.
Miller, J. 1989. *The 10th-century eruption of Eldgjá, southern Iceland*. Report 8903, Nordic Volcanological Institute, Reykjavík.
Moore, J. G. 1985. Structure and eruptive mechanisms at Surtsey Volcano, Iceland. *Geological Magazine* **122**, 649–61.
Moore, J. G. & L. C. Calk 1991. Degassing and differentiation in subglacial volcanoes, Iceland. *Journal of Volcanology and Geothermal Research* **46**, 157–80.

Óskarsson, N. 1984. Monitoring of fumarole discharge during the 1975–1982 rifting in Krafla volcanic center, north Iceland. *Journal of Volcanology and Geothermal Research* **22**, 97–121.
Óskarsson, N., G. E. Sigveldason, S. Steinthorsson 1982. A dynamic model of rift zone petrogenesis and the regional petrology of Iceland. *Journal of Petrology* **23**, 28–74.

Pálmason, G. 1986. Model of crustal formation in Iceland and application to submarine mid-ocean ridges. In *The geology of North America*, vol. M: *the western North Atlantic region*. P. R. Vogt & B. E. Tucholke (eds), 87–97. Boulder, Colorado: Geological Society of America.
Preusser, H. 1976. *The landscapes of Iceland: types and regions*. The Hague: W. Junk.

Rittmann, A. 1938. Die Vulkane am Mývatn in nordost Island. *Bulletin Volcanologique* **4**, 3–38.
—— 1939. Threngslaborgir: eine islandische Eruptionssplate am Mývatn. *Natur und Volk* **69**(5), 275–89.
Rubin, A. M. 1990. A comparison of rift-zone tectonics in Iceland and Hawaii. *Bulletin of Volcanology* **52**, 302–319.

Saemundsson, K. 1974. Evolution of the axial rifting zone in northern Iceland and the Tjörnes fracture zone. *Geological Society of America, Bulletin* **85**, 495–504.
—— 1978. Fissure swarms and central volcanoes of

the neovolcanic zones of Iceland. *Geological Journal* [special issue] **10**, 415–32.

—— 1979. Outline of the geology of Iceland. *Jökull* **29**, 7–28.

—— 1986. Subaerial volcanism in the western North Atlantic. In *The geology of North America*, vol. M: *The western North Atlantic region*, P. R. Vogt & B. E. Tucholke (eds), 69–86. Boulder, Colorado: Geological Society of America.

Scarth, A. 1999. *Vulcan's fury: man against the volcano*. London: Yale University Press.

Schilling, J. G., P. S. Meyer, R. H. Kingsley 1982. Evolution of the Icelandic hotspot. *Nature* **296**, 313–20.

Self, S. & R. S. J. Sparks (eds) 1981. *Tephra studies*. Dordrecht: Reidel.

Self, S., R. S. J. Sparks, B. Booth, G. P. L. Walker 1974. The Heimaey Strombolian scoria deposit, Iceland. *Geological Magazine* **111**, 539–48.

Sheridan, M. F. & K. H. Wohletz 1981. Hydrovolcanic explosions: the systematics of water–pyroclast equilibriation. *Science* **212**, 1387–9.

Sigbjarnarson, G. 1973. Katla and Askja. *Jökull* **23**, 45–51.

Sigurdsson, H. S. & R. S. J. Sparks 1978. Rifting episodes in north Iceland in 1874–1875 and the eruptions of Askja and Sveinagjá. *Bulletin Volcanologique* **41**, 149–67.

Sigurdsson, H. S. & B. Loebner 1981. Deep-sea record of Cenozoic explosive volcanism in the North Atlantic. In *Tephra studies*, S. Self & R. S. J. Sparks (eds), 289–316. Dordrecht: Reidel.

Sigvaldason, G. E. 1968. Structure and products of subaquatic volcanoes in Iceland. *Contributions to Mineralogy and Petrology* **18**, 1–16.

—— 1983. Volcanic prediction in Iceland. In *Forecasting volcanic events*, H. Tazieff & J. C. Sabroux (eds), 193–213. Amsterdam: Elsevier.

Sigvaldason, G. E., K. Annertz, M. Nilsson 1992. Effect of glacier loading/unloading on volcanism: postglacial volcanic production rate of the Dyngjufjöll area, central Iceland. *Bulletin of Volcanology* **54**, 385–92.

Steinthorsson, S., N. Óskarsson, G. E. Sigvaldason 1985. Origin of alkali basalts in Iceland: a plate tectonic model. *Journal of Geophysical Research* **90**, 10027–10042.

Thórarinsson, S. 1958. The Öraefajokull eruption of 1362. *Acta Naturalia Islandica* II(2), 1–98.

—— 1964. *Surtsey, the new island in the North Atlantic*. New York: Viking.

—— 1966. The Surtsey eruption: course of events and the development of Surtsey and other new islands. *Surtsey Research Progress Report* II, 117–23. Reykjavík: Nordic Volcanological Institute.

—— 1967a. Some problems of volcanism in Iceland. *Geologische Rundschau* **57**, 1–20.

—— 1967b. The eruptions of Hekla in historical times: a tephrochronological study. In *The eruption of Hekla 1947–48*. Reykjavík: Societa Scientifica Islandica.

—— 1968. On the rate of lava and tephra production and the upward migration of magma in four Icelandic eruptions. *Geologische Rundschau* **57**, 705–818.

—— 1970a. The Lakagígar eruption of 1783. *Bulletin Volcanologique* **33**, 910–29.

—— 1970b. *Hekla, a notorious volcano*. Reykjavík: Almenna Bókafélagid.

—— 1981. The application of tephrochronology in Iceland. See Self & Sparks (1981: 109–134).

Thórarinsson, S. & K. Saemundsson 1980. Volcanic activity in historic time. *Jökull* **29**, 29–32.

Thórarinsson, S. & G. E. Sigvaldason 1962. The eruption of Askja 1961. *American Journal of Science* **260**, 641–51.

—— 1972. The Hekla eruption of 1970. *Bulletin Volcanologique* **36**, 270–88.

Thórarinsson, S., T. Einarsson, G. Kjartansson 1960. On the geology and geomorphology of Iceland. *Geografisker Annaler* **41**, 135–69.

Thórarinsson, S., S. Steinthorsson, T. Einarsson, H. Kristmannsdottir, N. Óskarsson 1973. The eruption on Heimaey, Iceland. *Nature* **241**, 372–5.

Thordarson, T. 1995. *Volatile release and atmospheric effects of basaltic fissure eruptions*. PhD Thesis, University of Hawaii, Manoa.

Thordarson, T. & S. Self 1993. The Laki (Skaftár Fires) and Grímsvötn eruptions in 1783–1785. *Bulletin of Volcanology* **55**, 233–63.

Thordarson, T., S. Self, N. Óskarsson 1996. Sulfur, chlorine and fluorine degassing and atmospheric loading by the AD 1783–1784 Laki (Skaftár Fires) eruption in Iceland. *Bulletin of Volcanology* **58**, 205–225.

Thordarson, T., D. J. Miller, G. Larsen, S. Self, H. S. Sigurdsson 2000. New estimates of sulfur degassing and atmospheric mass-loading by the AD ~935 Eldgjá eruption, Iceland. *Journal of Volcanology and Geothermal Research*, in press.

Tryggvason, E. 1980. Subsidence events in the Krafla area, north Iceland, 1975–1979. *Journal of Geophysics* **47**, 141–53.

—— 1984. Widening of the Krafla fissure swarm during the 1975–1981 volcanic–tectonic episode. *Bulletin Volcanologique* **47**, 47–69.

—— 1986. Multiple magma reservoirs in a rift zone volcano: ground deformation and magma transport during the September 1984 eruption of Krafla, Iceland. *Journal of Volcanology and Geothermal Research* **28**, 1–44.

—— 1987. Mývatn Lake level observations 1984–1986 and ground deformation during a Krafla eruption. *Journal of Volcanology and Geothermal Research* **31**, 131–8.

—— 1989. Ground deformation in Askja, Iceland: its source and possible relation to flow in the mantle plume. *Journal of Volcanology and Geothermal Research* **39**, 61–71.

—— 1989. *Measurement of ground deformation in Askja 1966 to 1989*. Report 8904, Nordic Volcanological Institute, Reykjavík.

Tryggvason, K., E. S. Husebye, R. Stefansson 1983. Seismic image of the hypothesized Icelandic hotspot. *Tectonophysics* **100**, 97–118.

Vink, G. E. 1984. A hotspot model for Iceland and the Voring Plateau. *Journal of Geophysical Research* **89**, 9949–59.

Walker, G. P. L. 1963. The Breiddalur central volcano, eastern Iceland. *Geological Society of London, Quarterly Journal* **119**, 29–63.

—— 1966. Acid volcanic rocks of Iceland. *Bulletin Volcanologique* **29**, 375–406.

Werner, R., H. U. Schmincke, G. E. Sigvaldason 1996. A new model for the evolution of table mountains: volcanological and petrological evidence from Herdubreid and Herdubreidartögl volcanoes (Iceland). *Geologische Rundschau* **85**, 390–97.

Wolfe, C. J., I. T. Bjarnason, J. C. VanDecar, S. C. Solomon 1997. Seismic structure of the Icelandic mantle plume. *Nature* **385**, 245–7.

Zielinski, G. A., M. S. Germani, G. Larsen, M. G. L. Baillie, S. Whitlow, M. S. Twickler, K. Taylor 1995. Evidence of the Eldgjá (Iceland) eruption in the GISP2 Greenland ice core: relationship to eruption processes and climatic conditions in the tenth century. *The Holocene* **5**, 129–40.

7 Jan Mayen

The lonely Norwegian island of Jan Mayen lies 650 km northeast of Iceland in the North Atlantic Ocean at 71°N and 8°W. It is wholly volcanic in origin, covers an area of 320 km², runs 54 km from northeast to southwest, and is mostly less than 10 km wide. Historical records extend back about 400 years, but they are incomplete and scanty, for this grim, cold, isolated, inhospitable and often icebound island has supported settlement only rarely, and it now acts as a weather and navigation station. The island has two distinct parts, each about 25 km long, that are offset from each other by an isthmus. Both are dominated by fissures trending from northeast to southwest (Noe-Nygaard 1974). The narrow southwestern area, Sör-Jan, is a hilly range rising to 750 m above sea level, where fissure eruptions have formed a series of aligned trachytic domes and many cinder cones, from which recent lava flows have spread down to the coast. The northeastern part of the island, Nord-Jan, is wider, higher and more spectacular, for it is dominated by the basaltic stratovolcano, Beerenberg, the northernmost active volcano on land in the world. Beerenberg rises from 3000 m below sea level to 2277 m above sea level, its summit crowned by an ice cap that radiates glaciers down towards the coast. The central crater, Sentralkrateret, more than 1 km across and 300 m deep, feeds the most powerful of these glaciers, Weyprechtbreen. Beerenberg is often shrouded in fog or storm clouds, but in clear interludes the glistening ice and snow form a stunning contrast with the ice-free areas, which are often covered by fresh black and red cinder cones and lava flows erupted from fissures on its flanks (Fitch 1964, Sylvester 1975, Imsland 1978). In spite of the presence of the great stratovolcano, it is the fissure eruptions that have dictated the relief and outline of Jan Mayen, and even Beerenberg itself is aligned in the predominant northeast to southwest direction. The same trends are also followed by the volcanic Stimen bank, 15 km long and less than 100 m deep, in the ocean southwest of Sör-Jan. The form of Jan Mayen is thus closely related to major regional tectonic features.

Jan Mayen lies near a major offset of the Mid-Atlantic Ridge. The Kolbeinsey Ridge forms its main mid-ocean spine stretching 650 km north-northeastwards from Iceland. It is then offset 200 km to the east before resuming its course towards Spitzbergen as the Mohns Ridge. The line of the offset is marked by the Jan Mayen fracture zone, generated by a sheer displacement that forms a continuous linear depression more than 2000 m deep across the ocean floor (Saemundsson 1986). Jan Mayen rises where the fracture zone joins the Mohns Ridge, where enough lava could reach the surface to build up an island (Havskov & Atakan 1991). However, the tectonic situation of Jan Mayen is more complex than this basic background to its volcanic activity would, perhaps, imply (Johnson & Heezen 1967). It may also have grown up above a hotspot, probably now centred beneath the Eggvin bank, 150 km west of Jan Mayen, near the northern end of the Kolbeinsey Ridge.

From the scarce evidence available, eruptions on Jan Mayen seem to be separated by about a century of rest. Although eruption rates could have been higher in the past, the rate of volcanic activity in historic time has been low in Jan Mayen, and the average eruption has produced only about 0.07 km³ of lava. Some 5.35 km³ of

PALL IMSLAND

January 7 1985, the second day of an eruption on the northern flanks of Beerenberg, the eruption column turning towards the east.

lava has erupted altogether during the past 10 000 years or so (Imsland 1978).

Many of the Jan Mayen lavas are alkali-olivine basalts of relatively high potash content. **Ankaramites** and magnesium-rich basalts are common on Beerenberg; more evolved basalts, associated with trachytes, are more characteristic of Sör-Jan (Imsland 1986). Sör-Jan is the oldest part of Jan Mayen, forming a range of aligned cinder cones and lava flows composed of ankaramitic basalts accompanied by several trachytic domes. Volcanic activity may have begun in the southwest and shifted northeastwards (Wordie 1922, 1926).

Beerenberg clearly represents the modern volcanic climax in Jan Mayen. The rocks are predominantly alkali-olivine basalts. Beerenberg grew up in four distinct phases, each of which is reflected in its morphological characteristics. Little is known about the first − submarine and longest − phase in its growth. It undoubtedly expelled the most lava, for the plinth on which the volcano stands extends southwestwards beneath the rest of Jan Mayen and probably includes some of the Jan Mayen Ridge as well. This submarine activity continued in a northeasterly direction until a phase of vigorous Surtseyan

eruptions, now represented by basal tuffs, heralded the emergence of the island about 500 000 years ago (Fitch et al. 1965, Imsland 1978).

The second phase of activity formed the broad basal shield of Beerenberg, which has a diameter at sea level of 15–24 km. It is now represented by the various layers of the Kapp Muyen group, where the ankaramitic lavas lie between various glacial beds (Fitch 1964). This second phase had three unequal parts. At first, basaltic emissions from a central vent gradually built up the shield to a height of about 750 m. It was interrupted by an explosive episode that expelled the Havhestberget formation, an ashflow of basaltic pumice that was directed mainly to the southwest, where it now covers glaciated older lavas in a wide plateau leading towards Sör-Jan. The last episode marked a return to effusive conditions that built the rest of the basal shield. Innumerable fluid flows of ankaramitic basalts, usually less than 20 m thick, radiated from the central vent, and together they compose the Nordvestkapp formation (Fitch 1964). This is the most voluminous formation in the whole of Beerenberg, and the hub of the shield eventually rose to a height of about 1500 m. When these eruptions waned,

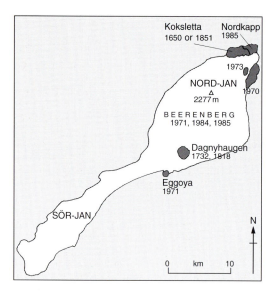

Historical eruptions on Jan Mayen.

the central cone of Beerenberg. The sequence of events has been established in relation to marine platforms around the perimeter of the strato-volcano. About 4000–5000 years ago, the chief fissure eruptions formed the Tromosryggen Ridge of cinder cones and flows, which is 300 m high and 6 km long. The lavas cascaded down the newly formed fault cliff on the northeastern end of Beerenberg and extended in lava deltas at its foot. These features were subsequently eroded by glaciers and cliffed by the waves. A similar episode of fissure eruptions later gave rise both to the Sarskrateret cone, in the far northeast of the island, and to the Koksletta lavas in the north. These lavas flowed down cliffs that had already been carved into the Tromosryggen basalts; they spread out into a broad platform, 4 km long and 1 km wide, at the northern tip of Jan Mayen. They may have erupted 2500–3500 years ago (Fitch 1964). On the other hand, they are so remarkably unweathered that they could have erupted even as recently as 1820–82. For example, the Koks-letta lava platform was mapped in 1882, but it was not marked on Scoresby's quite accurate map in 1820, perhaps for the good reason that it did not then exist (Sylvester 1975). It has been suggested that the Koksletta eruption could have taken place in 1851 (Havskov & Atakan 1991).

This fourth phase of eruptions has certainly continued into historical times, but, even then, isolation, frequent low cloud and the long spells without permanent settlement have combined to make observations of eruptions unreliable, intermittent and scanty. No activity appears to have taken place in Sör-Jan. On Beerenberg, an eruption may have occurred at Koksletta in 1650, but any resulting forms have not been identified. But the results of the more reliably dated erup-tions in 1732 and 1818 can be seen in the fresh unglaciated lavas and cones at Dagnyhaugen, low on the southwestern flanks of Beerenberg. The opposite northeastern flanks, near the Nordkapp, probably saw activity about 1850, and certainly in 1970, 1971, 1973 and 1985

erosion then carved valleys into the flanks of the shield.

The renewal of volcanic activity in the third phase was marked by a significant change. More viscous hawaiites and mugearites rapidly formed the steep-sided lava cone, about 750 m high and 5 km across, that lies like an enormous sand-castle on top of the basal shield. The formation of the main cone of the stratovolcano and its chief crater, Sentralkrateret, was probably completed about 6000–7000 years ago (Fitch 1964). Then followed a period of fluvial, glacial and marine erosion forming barrancos, small high-level glacial corries and steep cliffs. It was probably at the end of this third phase that displacements formed the great fault cliff that bounds the north-eastern flanks of Beerenberg.

The fourth and latest phase of activity on Beer-enberg during the past 6000 years or so has been dominated on its outer flanks by fissure erup-tions that created several small cinder cones and widespread lava flows. These fissures trend along the axis of Jan Mayen and curve through

The eruption in January 1985

Several strong earthquakes approaching magnitude 5.0 on the Richter scale shook the island on 4–6 January 1985. The eruption began in the afternoon of 6 January, and lasted some 35–40 hours before it ended on 8 Jan-uary. It occurred on a small fissure, 1 km long, at the far northeastern corner of Jan Mayen, extending from sea level up to 200 m on the north side of the Sarskrateret cone. Three main craters emitted gas and spatter, and the wind distributed fine ash from a dark-brown column that rose 1 km above the vents. About 7 million m³ of vol-canic material (about one tenth of an average Jan Mayen eruption) was given out. Lavas also gushed vigorously out northwards into the sea and eventually added about 0.25 km² to the land area of Jan Mayen.

(Sylvester 1975, Imsland 1986). All the eruptions developed near sea level on fissures trending northeast–southwest. However, the central crater of Beerenberg has been seen to erupt only twice, briefly emitting steam in March 1984 and April 1985, although bad weather might have hidden other activity. In 1970, for example, a severe storm hid the eruption for three days from the weather station situated only 30 km away.

A fissure curving 6 km from sea level to a height of 1000 m on Beerenberg gave rise to the eruption in 1970. It opened on 18 September 1970 after a magnitude 5.1 earthquake occurred at a depth of 30 km. Activity was first noticed on 20 September from a passing aircraft, when a high ash cloud had developed and the first lavas were probably emitted (Siggerud 1972, Noe-Nygaard 1974, Sylvester 1975). Eruptions eventually concentrated on five vents, just over 1 km apart, which produced tall lava fountains, spatter cones and lava flows cascading seawards. Almost incessant earthquakes of less than magnitude 4.0 (often 800 a day) were registered over the next month. Lava-flow emissions ended in late October 1970, but intermittent explosions of fragments continued until January 1971 (Imsland 1986) or possibly until June 1971 (Siggerud 1972). Some ash and steam may also have been expelled after June 1971 from the summit crater of Beerenberg, and from the Eggoya satellite cinder cone low on its southwestern flanks, and fumaroles were observed in both places in 1974. In all, about 0.5 km^3 of alkali-olivine basalt was emitted, adding about 4 km^2 of new land to northeastern Jan Mayen (Imsland 1986).

The latest eruption was in January 1985 from a fissure prolonging the great fault cliff dominating the relief of the northeastern arm of Jan Mayen. It represents a reactivation of the fault that has already thrown down the lower northern flanks of Beerenberg. Also, the earthquakes accompanying the eruptions were unusually strong and deep, and seem to have been generated by regional tectonic displacements related to strike-slip shear motion in the Jan Mayen fracture zone. Thus, the magma probably rose up a line of weakness where stresses were released on the southern edge of the Jan Mayen fracture zone. Thus, the eruptions were caused by leaking tectonic fractures and were not truly related to the main Jan Mayen system (Imsland 1986). In future, Beerenberg and Jan Mayen can therefore be expected to grow farther to the northeast.

BIBLIOGRAPHY

Bungum, H. & E. S. Husebye 1977. Seismicity of the Norwegian Sea: the Jan Mayen fracture zone. *Tectonophysics* **40**, 351–60.

Dollar, A. T. J. 1966. Genetic aspects of the Jan Mayen fissure volcano group on the mid-oceanic submarine Mohns Ridge, Norwegian Sea. *Bulletin Volcanologique* **29**, 25–6.

Eldholm, O. 1991. Magmatic tectonic evolution of a volcanic rifted margin. *Marine Geology* **102**(1–4), 43–61.

Fitch, F. J. 1964. The development of the Beerenberg volcano, Jan Mayen. *Proceedings of the Geologists' Association* **75**, 133–65.

Fitch, F. J., R. J. Grasty, J. A. Miller 1965. Potassium–argon ages of rocks from Jan Mayen and an outline of its volcanic history. *Nature* **207**, 1349–51.

Hawkins, J. R. W. & B. Roberts 1963. Agglutinate in north Jan Mayen. *Geological Magazine* **100**, 156–63.

Havskov, J. & K. Atakan 1991. Seismicity and volcanism of Jan Mayen island. *Terra Nova* **3**, 517–26.

Imsland, P. 1978. The geology of the volcanic island Jan Mayen, Arctic Ocean. *Nordic Volcanological Institute Report* **7812**, 1–74.

—— 1984. *Petrology, mineralogy and evolution of the Jan Mayen magma system*. Reykjavik: Prentsmidjan Oddi (Vísíndafelag Islendinga 43).

—— 1986. The volcanic eruption of Jan Mayen, January 1985: interaction between a volcanic island and a fracture zone. *Journal of Volcanology and Geothermal Research* **28**, 45–53.

Johnson, G. L. & B. C. Heezen 1967. The morphology and evolution of the Norwegian–Greenland Sea. *Deep-Sea Research* **14**, 755–71.

Morgan, W. J. 1972. Deep mantle convection plumes and plate motions. *American Association of Petrolium Geologists, Bulletin* **56**, 203–9.

Navrestad, T. & A. Sornes 1974. The seismicity around Jan Mayen. *Yearbook: 1972* (pp. 29–40), Norsk Polarinstitutt, Oslo.

Noe-Nygaard, A. 1974. Cenozoic to Recent volcanism in and around the North Atlantic Basin. In *The ocean basins and margins*, vol. 2: *the North Atlantic*, A. E. M. Nairn & F. G. Stehli (eds), 428–30. New York: Plenum.

Saemundsson, K. 1986. Subaerial volcanism in the western North Atlantic. In *The geology of North America*, vol. M: *the western North Atlantic region*, P. R. Vogt & B. E. Tucholke (eds), 69–86. Boulder, Colorado: Geological Society of America.

Siggerud, T. 1972. The volcanic eruption on Jan Mayen, 1970. *Yearbook: 1970* (pp. 5–18), Norsk Polarinstitutt, Oslo.

Sylvester, A. G. 1975. History and surveillance of volcanic activity on Jan Mayen island. *Bulletin Volcanologique* **39**, 313–35.

Wordie, J. M. 1922. Jan Mayen island. *Geographical Journal* **59**, 180–85.

—— 1926. The geology of Jan Mayen. *Royal Society of Edinburgh, Transactions* **104**, 742–5.

PART 4 NORTHERN EUROPE

8 France

No volcano has erupted in France in historical times, but prehistoric peoples almost certainly witnessed some of its latest activity. The most recent act in several million years of eruptions occurred about 6000 years ago when the crater now occupied by Lac Pavin exploded in Auvergne and further eruptions could occur in that province.

During the past 10 million years, eruptions have been concentrated in the Massif Central, and volcanic landforms of great variety dominate the scenery of Auvergne and Velay, in the centre and southeast of this vast upland. Thus, the Chain of Puys has scores of cinder cones and lava flows, with a few majestic domes; the Mont-Dore and Cantal massifs are stratovolcanoes forming the highest peaks in the centre of the region; the Cézallier, Aubrac and Devès all form widespread basaltic plateaux. In Velay, lava plateaux predominate in the west and the extruded sucs in the east. More sporadic activity also occurred in Burgundy and Forez in the north, in Ardèche in the southeast, and in a zone stretching across the Causses to the Escandorgue Chain in Languedoc. There are volcanic lakes everywhere, such as Lac d'Aydat, which have been impounded by lava flows, and maars formed by explosions, such as the Gour de Tazenat and the Lac du Bouchet. Older lava masses have been set in relief by the removal of weaker sediments around them. In the Velay and in the Grande Limagne, erosion has reduced old flows and vents to ridges, buttes or necks that protect ancient villages or form the plinths of monuments that make Le Puy-en-Velay, for instance, one of the most spectacular towns in France. The plateau formed by an ancient lava lake at Gergovie made an ideal site for a Gaullish defensive site; nearby, the Montagne de la Serre points its lava-capped ridge into the Grande Limagne; and, in the far southeast, the Coiron lavas protected a spur like a great paw that juts out from the Massif Central to threaten the Rhône Valley (e.g. Scarth 1967, Bout 1973, Peterlongo 1978, Gèze 1979, Rouire & Rousset 1980, de Goër de Hervé et al. 1991).

However, volcanic activity was not relayed in any regular pattern from one area to another and, for example, some of the oldest and youngest volcanic features in the Massif Central are both found within sight of the Puy de Dôme. The volcanoes all lie upon a basement of faulted blocks and troughs, which are separated by pronounced escarpments. Most of the faulting took place before the eruptions began. But, erosion has incised the area markedly during the past 2 or 3 million years, and streams eroded deep into the volcanoes, both as they grew up and after they became extinct: the central core of Mont-Dore was gutted and deep gorges radiated from the Cantal stratovolcano. They developed a labyrinthine stratigraphy that can be deciphered only by the most sophisticated of modern techniques (Cantagrel & Baubron 1983). The most recent theatre of activity in the area is the Chain of Puys, whose fresh and clearcut features culminate in one of the most distinctive mountains in all of France, the Puy de Dôme.

The eruptive climaxes of the various areas of activity began with the Cantal, and the Aubrac and Cézallier plateaux, between 9 and 7 million years ago. They continued with those in Velay from about 8 to 6 million years ago. The climax of the Mont-Dore occurred about 3 million years

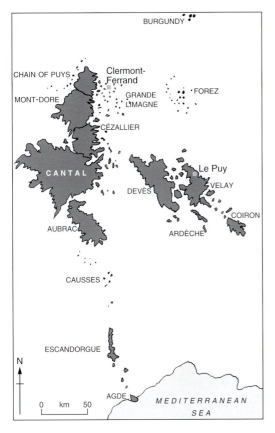

Volcanic areas of France

Puys, rhyolites eventually developed in the Mont-Dore, and trachyrhyolites in the Cantal to form domes and pumice ashflows. Phonolites also developed widely in Velay and from small secondary reservoirs in Mont-Dore.

Aligned vents have produced by far the most numerous and widespread eruptions in the Massif Central. The vast majority are related to fissures, but the major faults have permitted the ascent of magma only very rarely. Thus, for example, in the area near Clermont-Ferrand that is riddled by hundreds of vents, the major faults forming the border between the Grande Limagne and the plinth of the Chain of Puys have produced only two volcanoes (Scarth 1966). Most of the fractures and fissures trend from north to south, northwest to southeast, or north-northeast to south-southwest. An important feature of the alignments is that eruptions occurred upon them from time to time, but not all at once, such as on the fissures in Iceland.

Closely spaced eruptions along such fissures gave rise to the major basaltic plateaux of the Massif Central. The Cézallier, Aubrac, Devès and Coiron plateaux all trend from northwest to southeast, and their lavas show differentiation towards hawaiites and mugearites. These lavas often emerged in such quantities that they swamped the previous relief. But in Devès, in particular, the fissures have not only formed lava flows but also many maars, and line after line of cinder cones, known as gardes. Aligned volcanic activity reached its climax in eastern Velay and especially in the Chain of Puys, where not only basaltic but also more evolved magmas erupted. Thus, in eastern Velay, viscous phonolitic extrusions formed the steep-sided sucs that rise above the basalts; and, in the centre of the Chain of Puys, trachytic domes dominate the scenery among the lines of cinder cones.

The centrally clustered vents of the Massif Central form the smallest category of volcanic features, but also two of its most prominent landforms: the stratovolcanoes of the Mont-Dore and Cantal. However, much smaller stratovolcanoes have been all but hidden in the Boutières of eastern Velay and the Luguet in the Cézallier plateau. These stratovolcanoes developed after an initial phase of more scattered basaltic eruptions. The spatial concentration was associated with magma evolution towards mugearites and benmoreites (here the "trachyandesitic" doreites and sancyites), followed on the one hand by trachytes and phonolites and on the other by less common

ago, but eruptions continued until about 250 000 years ago. The Devès plateau erupted mainly about 2 million years ago, but the apogee of the Chain of Puys took place only around 10 000 years ago. However, the intervals of repose were undoubtedly much longer than the eruptive acts. Thus, the total number of active years was small in comparison to the millions of years encompassed by the whole eruptive sequence.

Most of the lavas erupted in the Massif Central are basanites and alkaline basalts. They occur in the older eruptions but are also very common in the planèzes of Mont-Dore and the Cantal, as well as the lava plateaux of Aubrac and the Cézallier. More differentiated lavas form hawaiites (here called labradorites or trachybasalts), mugearites (called doreites and ordanchites, for instance, in the Mont-Dore), and benmoreites (known as sancyites in the Mont-Dore). In the local terminology, too, the doreites and sancyites were grouped together as "trachyandesites", in both the Mont-Dore and the Cantal (Peterlongo 1978). Trachytes developed from alkaline basalts in the Chain of rhyolites. The central clusters of vents erupted repeatedly over several million years, piling up material into quite gently sloping stratovolcanoes, perhaps 2000 m high, and spreading more than 30 km in diameter. Violent eruptions then ensued, which expelled vast blankets of trachytic or rhyolitic pumice and ash flows that were followed by the collapse of calderas in both the Mont-Dore and the Cantal. Their last phases comprised caldera infilling and especially lower-flank basaltic eruptions forming the planèzes skirting both volcanic massifs. There was thus a late return to the first phase of predominantly effusive activity.

The Massif Central is one of the classic areas of volcanology. It was here that Guettard (1752) first identified the extinct volcanoes of France. It was here too that Desmarest (1806) and Poulett-Scrope (1827) demonstrated that rivers did indeed erode their valleys, by comparing the present valleys with the ancient, higher courses fossilized beneath lava flows on the edges of the Grande Limagne.

THE CHAIN OF PUYS

The Chain of Puys forms the beautiful western skyline to Clermont-Ferrand, the capital of Auvergne. It stretches from north to south for 40 km along the granitic block of the Plateau des Dômes, which rises west of the Grande Limagne, and it has a small annexe on the fringes of the Mont-Dore stratovolcano. Most of the puys are about 100 m high, but the Puy de Dôme, rising 500 m, provides a majestic exception in the centre of the Chain.

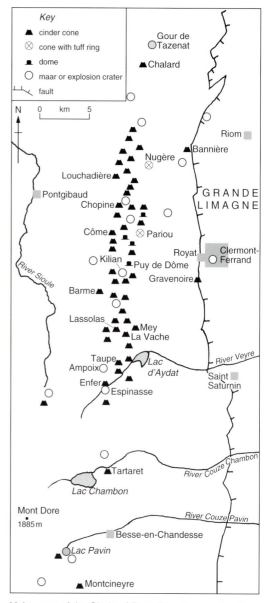

Volcanoes of the Chain of Puys (lava flows omitted).

The northern part of the Chain of Puys from the Puy de Dôme.

The puys, or peaks, arose from almost a hundred vents that provide a museum of cinder cones, lava flows and domes. The eruptions began about 95 000 years ago and the latest occurred no more than about 6000 years ago, and the puys are so well preserved that they almost make up for the absence of an active volcano among them. However, the Chain is probably not extinct, although there have been no credible accounts of activity during the 2000 years since Vercingetorix repulsed Julius Caesar at Gergovia in 52 BC. Most eruptions took place along fissures running from north-northeast to south-southwest, but they rarely occurred simultaneously. Many vents gave off lava flows 10–20 m thick, which often have rugged aa surfaces known locally as cheires, or stoney ground. Each eruption probably continued for a few weeks or a few years. Thus, the eruptions can scarcely have lasted more than a total of 500 years during the whole activity period and there were long episodes of quiescence when the whole Chain would have looked just as calm as it is today.

The eruptions were probably fed by two superposed reservoirs, one about 25–30 km deep, and the other lying 5–15 km below the surface, where the magmas have undergone two periods of evolution, and perhaps the beginnings of a third. In spite of this evolution, 67 per cent of the lavas erupted are alkaline basalts, 20 per cent are hawaiites, 10 per cent are trachytes, and 3 per cent are mugearites and benmoreites

(Camus 1975, de Goër de Hervé & Belin 1985, de Goër de Hervé et al. 1991, Scarth 1994, Camus & de Goër de Hervé 1995).

The growth of the Chain of Puys

The first eruptions began about 95 000 years ago. As the lower reservoir developed, first basalts and then hawaiites were expelled, and the earliest maars exploded. These emissions formed, for instance, the lava flows from the Puy de Chanat, which date from about 90 000 years ago, and the flow in the River Auzon valley, which is about 60 000 years old. A period of calm then ensued. About 45 000 years ago, the hawaiites stored in the shallow reservoir gradually evolved to mugearites, which led to the first eruptive climax in the Chain about 30 000 years ago. The Puy de Lemptégy, the Puy de Laschamp, the Puy de la Taupe and the Puy des Goules, as well as the Petit Puy de Dôme and most of the lava flows directed westwards towards the River Sioule – all belong to this period. Maar-forming explosions apparently preceded the eruption of many cinder cones, possibly aided by the permafrost on the tundra during that glacial period. A second period of repose then followed, which was broken only by rare eruptions, but during this interval an injection of basaltic magma into the lower reservoir started the second episode of evolution as hawaiites began to develop again.

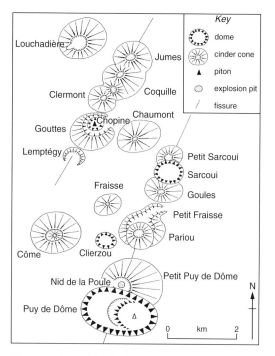

The central part of the Chain of Puys (lava flows omitted).

the start of yet another period of magma evolution. Their finest expression is in the twin breached cones of the Puy de la Vache and the Puy de Lassolas, which together formed the dark grey flow of the Cheire d'Aydat; and it is possible that the basaltic Puy de Côme could also belong to this episode. It might be the youngest volcano in the northern Chain, because it is not covered by fragments thrown from any other volcano, although, in fact, such fragments might not have reached it because of its rather isolated westerly position. It was dated to 7700 years ago on the third edition of the map, *Volcanologie de la Chaîne des Puys* (1991), but it has also been assigned to the period between 15 000 and 12 000 years ago (Camus & De Goër de Hervé 1995).

The latest eruptions broke out about 6000 years ago and were exceptional in two respects: they occurred in a small annexe area, due south of the Chain, on the eastern flanks of the Mont-Dore massif, and they resulted from a very rapid magma evolution that expelled materials ranging from basanite to trachyte in perhaps less than 200 years.

From about 15 000 to about 12 000 years ago, these hawaiites rose into the upper reservoir and occasionally erupted at the surface. During the next few thousand years, benmoreites and trachytes developed in the upper reservoir. The stage was set for the last episode in the growth of the Chain of Puys, which lasted from 12 000 until about 8000 years ago, and has contributed so much to their character. In the central part of the Chain – and for the first time in the area – trachytic magmas extruded the domes of the Puy de Dôme, Sarcoui, Clierzou, the Puy de Vasset and the Puy Chopine. Soon afterwards, trachytic nuées ardentes exploded from the Puy Vasset and the Puy Chopine, and from the Kilian explosion crater, just south of the Puy de Dôme, and covered the neighbourhood with three indicator beds. Eruptions of mugearites and benmoreites were even less common than trachytes, but they were well represented by the Puy de Pariou and the lava flows from the Puy de la Nugère that provided the Volvic stone used to build the striking cathedral in Clermont-Ferrand.

Then, about 8000 years ago, basaltic eruptions resumed. They came from the least evolved magmas in the history of the Chain and seem to have risen straight from the mantle, without halting in either reservoir; they might, therefore, signify

Domes

The domes form a small but distinctive group in the centre of the Chain, which comprises the Puy de Dôme itself, Sarcoui, Clierzou, the Puy Chopine and the Puy de Vasset. They are composed of domite, a pale-grey trachyte rich in silica, which resembles, and is almost as light as, pumice. They formed when the viscous lava extruded in a bulbous mass, although explosions of nuées ardentes sometimes destroyed or interrupted their growth. At least three of the domes – Sarcoui, the Puy Chopine and the Puy de Vasset – extruded from vents that had previously undergone hydrovolcanic explosions.

Sarcoui is the most perfect dome in the Chain of Puys, resembling an upturned cauldron that rises 250 m above a base 800 m across. It is bald and stark. and its pale grey rocks show through its thin grass cover. The remains of a tuff ring on its eastern base show that the eruption began with a hydrovolcanic explosion. The onion-skin layers, formed during the extrusion, can be seen in the quarries excavated in Gallo-Roman times. Nuées ardentes formed arcuate niches where they exploded from its flanks. The largest spread northeastwards for 600 m, where secondary explosions formed a rash of rootless cones. Its

The Puy de Dôme

The Puy de Dôme is, indeed, a volcano that stands out in a crowd. Although this striking individual reaches only 1465 m above sea level, it soars over 200 m above any rival in the Chain, is a landmark for over 100 km around, and is the very symbol of Auvergne. It is, in fact, two volcanoes, now joined together side-by-side, which arose in succession from the same vent about 10 000 years ago (de Goër de Hervé et al. 1991). The younger eastern part nestles within the older western part. Both are composed of trachytic domite. The original dome resembled the Soufrière in Guadeloupe (French West Indies), and was spiked with radiating ridges and pinnacles that gave it a certain ruggedness.

A violent blast blew away its eastern sector, forming a hole like a Greek theatre, with two arms jutting out eastwards. Within these enclosing arms, the second dome extruded from the same vent. By an extraordinary coincidence, it resembled the domes on Montagne Pelée, the other great volcano in the French West Indies. But, it was smoother than its predecessor and also covered by trachytic fragments, so that the eastern side of the Puy de Dôme contrasts clearly with its more rugged western flanks. A distinct atrium would also separate the two domes at the summit, but human interference, ranging from a Roman temple of Mercury to modern tourist installations, has tended to mask the morphology.

The Puy de Dôme, from the southeast, with the smooth newer dome rising on the east above the more rugged older dome on the west.

youthful appearance reveals its age, for Sarcoui formed between 12 000 and 8500 years ago, but it is older than the Puy Chopine, because it is mantled by ash from that eruption.

The steep, darkly wooded pyramid of the Puy Chopine rises only 160 m, but it has one of the most distinctive profiles in the Chain of Puys. Its parent vent produced at least two previous eruptions. The first built the bare red cinder cone of the Puy des Gouttes; the second was a hydro-volcanic explosion that destroyed the northern sector of the cone about 8300 years ago, and scattered an indicator bed of trachytic fragments over the northern parts of the Chain. At once, an almost solid piston of trachyte extruded and formed the Puy Chopine. The lavas dragged

blocks of granite and diorite, 15–20 m across, from the walls of the vent and pushed them upwards, and they are now largely responsible for its distinctive pointed crest.

Monogenetic cones

More than three quarters of the puys are cinder cones, and most of them are about 100 m high and some 400 m across, with craters less than 50 m deep. They erupted from vents along en echelon fissures trending from north-northeast to south-southwest. The younger cones have well defined craters, and their straight, scarcely altered outer flanks usually lie at 30°, akin to the

The crescent of cinders forming the Puy des Gouttes encloses the southern flanks of the Puy Chopine, whose wooded crest rises above the crescent on the left.

angle of rest of 32°–35° of their bedded fragments. Each cone was formed by a brisk phase of Strombolian eruptions lasting for a few weeks, a few months, or possibly a few years. Many fine sections are available at the Puy de la Taupe, Puy Gravenoire and Puy de Lemptégy, for example, because these cones have been quarried for pouzzolane, which is used as road metal and a building component. Indeed, the Puy de Lemptégy had been almost completely sold off before it was taken over as a volcano museum.

The simplest cones result from eruptions of similar intensity from a single vent. One such eruption built the Puy des Goules about 30 000 years ago. It is 150 m high and has a crater 200 m wide and 40 m deep, partly filled by fragments ejected from its younger neighbours. Its sharp crater rim is some 10 m higher on the northeastern side, where the southwesterly winds blew the fragments during the eruption. The cone has no gullies, but creep has already reduced its slopes to an angle of 25°. When Sarcoui erupted some 20 000 years later, its lower northern flanks were destroyed.

The Puy de Côme from the southwest. Most of its slopes are forested, but its nested cone can be discerned on the grass-covered summit.

The twin breached cones of the Puy de Lassolas and the Puy de la Vache. Their lava flows combine to form the now forested Cheire d'Aydat, which flowed away to the southeast on the right.

Multiple eruptions from a single vent often build up larger-than-average cones, but their forms depend upon when the main eruptions took place in the sequence. A large final eruption can bury the earlier cones: thus, the bulky form of the Puy de Mercoeur buries a smaller cone that erupted earlier from the same vent. On the other hand, when the later eruptions give off fewer fragments than their predecessors, a smaller cone then often nestles within the crater of the larger. This sequence happened at the Puy de Côme, where the nested cone is much smaller than the puy as a whole. A circular atrium, about 10 m deep, separates the two crater rims, which are both highest on their northeastern sides, where the southwesterly winds blew slightly more fragments as they erupted. The bulk of the Puy de Côme could have formed in several months, or possibly for several years, because it is the largest cinder cone in the Chain, with a height of 250 m. Then, as activity waned, a short eruption built the cone nested in the larger crater, perhaps in a matter of days. This last phase might have been one of the latest eruptions in the Chain, because it could have ejected the ash in the Etang de Fung, some 6 km away, that was dated, rather questionably, to 3890 years ago.

When vents are closely spaced along a fissure, the fragments often pile up into single elongated cone, such as the Puy de Barme, which rises 4 km southwest of the Puy de Dôme. The two small craters at its summit operated simultaneously to form an elliptical cone about 15 000–12 000 years ago. Subsequently, a third vent erupted on its southwestern flanks and extended the cone down the fissure. More widely spaced vents can sometimes give rise to twin cones if they function at the same time. The Puy de Jumes and the Puy de Coquille are twin cones, set 600 m apart, on the same fissure. Both are composed of basaltic lapilli and cinders and are so close together that their beds intermingled, which confirms that they erupted simultaneously.

Clustered vents form bulky cones, crowned by a group of craters. Four vents built up the Puy de Monchier about 8540 years ago, and four separate craters of similar size are assembled near its summit. On the other hand, eruptions from the central vent built up most of the Puy de la Taupe, but its companion cones are scarcely more than satellites on its flanks, because their vents probably functioned for only a few weeks.

Breached cones

The Chain of Puys has over 60 cones that were breached by lava flows. Generally, a whole sector of the cone is missing down to its very base, but, more rarely, the breach is limited to the crater itself. For instance, after the Puy de Louchadière had formed, a lava flow welled up inside the crater, destroyed its upper western sector, and spread down beyond the base of the cone. The rest of the crater now forms an armchair-like hollow that probably gave Louchadière its name – *La Chaise*, the chair. After the Puy Chalard had formed, the breaching was even more severe, because a flow not only removed a whole sector of the cone, but was powerful enough to push it bodily some 100 m to the northwest, where it now remains as a ridge of cinders.

PUY DE DÔME

PUY DE CÔME

PUY DE LOUCHADIERE   PUY DE BARME

The northern part of the Chain of Puys from the southwest.

However, most breaches develop while the cones are erupting and they are almost always caused by lavas that leave the vent at the same time. This is how the Puy de la Vache and the Puy de Lassolas attained the characteristic shapes that have made them among the best known of all the cones in the Chain. Their reddish cinders form two horseshoe masses rising 200 m above the lavas that they both erupted to form the Cheire d'Aydat about 8000 years ago. The lavas carried away the cinders that fell upon them as if they were on an escalator, and the sectors where each flow emerged were therefore never constructed. The volcanoes have remained crescent shape, like a French croissant, ever since the eruption began.

Cones breached solely by explosions are rare in the Chain of Puys because moderate eruptions prevailed. However, when these eruptions encountered water, more vigorous hydrovolcanic explosions destroyed sectors of adjacent cinder cones. The northern sector of the Puy des Gouttes, for instance, was obliterated by the explosions from the same vent, which led to the growth of the Puy Chopine.

Polygenetic cones

The Puy de Pariou grew up in several stages. The first eruption formed a cinder cone about 9000

years ago, but it remains as only a ridge of basaltic cinders at the western base of the volcano, because it was largely destroyed by hydrovolcanic explosions that formed a maar and a mugearitic tuff ring. About 8500 years ago, mugearitic lavas surged up the vent, invaded the maar and formed a lava lake, pushed away the northern sector of the tuff ring, left it stranded near the Puy des Goules, and then flowed out eastwards in a long tongue that reached the suburbs of Clermont-Ferrand. The final eruptions then built the 160 m-high uppermost cone of the Puy de Pariou. Its pristine crater, which is 90 m deep and enclosed within the sharpest rim in the Chain, betrays its youth. It is about 8000 years old.

The Puy de Tartaret is composed of two superimposed cinder cones that stand about 90 m above the Couze Chambon stream. About 27 000 years ago, rising magma met the waters of the Couze, and the resulting hydrovolcanic explosions formed a maar and tuff ring that now emerges from the southern base of the volcano. When the water could no longer gain access to the vent, Strombolian eruptions built a cinder cone, and a basaltic flow invaded the Couze Chambon valley. Together, they impounded the river and formed Lac Chambon. About 13 000 years ago, magma rose again, and the presence of the lake induced an initial Surtseyan eruption that formed another maar and a tuff ring, whose yellowish beds, mixed with lacustrine silts,

The southern part of the Chain of Puys.

The Puy de Pariou and the Puy de Dôme from the air.

The latest eruptions in France

The latest eruptions in France took place beyond the southern end of the Chaîne des Puys, on the edge of the Mont-Dore stratovolcano, about 10 km southwest of Besse-en-Chandesse. They all took place within the space of about a couple of centuries some 6000 years ago. The episode began with the eruption of the breached basanitic cinder cone of the Puy de Montcineyre that held back the Lac de Montcineyre. Thus, as the cone grew up, Surtseyan explosions interrupted the Strombolian eruptions whenever the water invaded the vent. Soon afterwards, a shallow hydrovolcanic explosion burst out 4 km to the north and formed the maar of Estivadoux and its thin ring of tuffs and fragments of country rock. A little later, 1 km to the west of Estivadoux, another Strombolian eruption built up the cinder cone of the Puy de Montchal and also expelled three hawaiitic flows. The first two flows were small, but the third was 12 km long. It filled the maar of Estivadoux and sent out a thin tongue, with a fine blocky surface, that invaded the upper reaches of the Couze Pavin valley. During the short interval that ensued, magmatic differentiation continued apace, for the next and latest eruption was a trachytic explosion on the lower flanks of the Puy de Montchal. Nuées ardentes blasted out 15 km to the southeast, and formed the hole that was then filled by Lac Pavin – L'épouvantable ("the fearsome") – a name that would have been appropriate for the eruption itself, but in fact it describes the 92 m-deep lake and the surrounding dark wooded hollow.

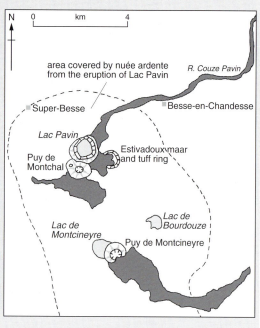

The latest eruptions in France.

outcrop on the northern base of the volcano. Then, when water could no longer reach the vent once again, a second cinder cone erupted. This new Puy de Tartaret covered much of the first and erupted one of the finest basaltic flows in Auvergne, which moulds the floor of the Couze Chambon valley for 22 km, as far as the Grande Limagne. When the molten lavas flowed into the stream, about 30 rootless conelets, from 5 m to 30 m high, developed over the first 2 km of the flow. The Puy de Tartaret and its lavas used to hold back the waters of Lac Chambon, but the cone does not now reach its shores. A great landslide later avalanched from the northern slopes of the lake, came to rest between the cone and the waters, and left behind a scar 200 m high, called the Saut de la Pucelle ("the virgin's leap").

Hydrovolcanic eruptions

Most of the hydrovolcanic eruptions in the Chain of Puys formed maars, which are sometimes known as gours or, when they are filled by sediments, as narses. Eruptions into periglacial permafrost probably encouraged much hydrovolcanic activity, and many older maars have now been identified. Many maars are encircled by tuff rings of exploded fragments of country rock and shattered magma. In several cases, also, such as at the Puy de Gravenoire and the Puy de Tartaret, hydrovolcanic eruptions were a prelude to the formation of cinder cones, and quarries in the Puy de la Taupe, for instance, reveal alternations of hydrovolcanic and Strombolian eruptions before the latter eventually predominated.

The most famous maar in Auvergne is the Gour de Tazenat, in the north, which is 66 m deep and 900 m across. It probably formed at about 30 000, or perhaps as much as 90 000, years ago, when water from the Rochegude stream seeped down a fault and met the rising basaltic magma. The consequent explosions occurred at increasing intervals for a few days and piled up thin layers of fine fragments of lava and country rock around the northeastern edge of the Gour. The Narse d'Espinasse, in the south, started life as a maar, when the waters of the River Veyre infiltrated a fissure, and caused hydrovolcanic explosions at the same time as Strombolian eruptions were building the Puy de l'Enfer from a dry vent nearby. When the eruptions stopped, the Puy de l'Enfer formed a crescent of cinders around the maar. The River Veyre continued its course through the maar, filled it almost to the brim with sediments, and formed the flat-floored Narse d'Espinasse, which unfortunately has not been accurately dated. A smaller narse formed about 10 700 years ago on the same fissure at Ampoix. Cores taken from its sediments have established a biostratigraphic chronology for the area (Juvigné & Gwelt 1987).

Trachytic explosion craters are as rare as trachytic eruptions in the Chain of Puys. The Kilian crater, 1 km south of the Puy de Dôme, was formed by one such explosion. It was one of the latest widespread explosions in the centre of the Chain of Puys and it provided a valuable indicator bed about 8500 years ago.

The lava flows of the Chain of Puys

In spite of the tendencies towards magmatic differentiation in the Chain of Puys, nearly 90 per cent of the lavas in the flows are basaltic. The flows were also fluid enough to enter the valleys draining to the River Sioule and the Grande Limagne. The lavas piled up more thickly in the west. The Cheire de Côme, for instance, has seven superposed flows with a total thickness of 135 m, although not all of them necessarily came from the Puy de Côme. Indeed, the upper reaches of many flows are so thickly covered with ash that it is often hard to distinguish their sources. Thus, for instance, the fine lava delta known as the Cheire de Mercoeur came, in fact, from the Puy de Mey. However, farther away from the cones the lavas almost always develop the rugged maze of pits and pinnacles, flow channels,

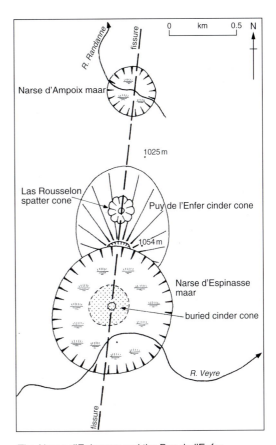

The Narse d'Epinasse and the Puy de l'Enfer.

The Gour de Tazenat.

pressure ridges and lava levees typified by the aa surfaces of the Cheire de Côme, the Cheire d'Aydat, and the mugearitic flow extending from the Puy de la Nugère to Volvic. The upper parts of many flows descending westwards are blistered by tumuli and occur most clearly on the lavas emitted from the Puy de Pourcharet and the Puy de Combegrasse. Near Nébouzat, the lavas from the Puy de Pourcharet invaded a marshy area, and the steam explosions developed some small rootless cones.

Elsewhere, the streams have continued to flow underneath or alongside the invading flows. The streams often rapidly eroded the ribbon of lava. Thus, for example, the basaltic flow that threaded into the River Auzon valley some 60 000 years ago has already been cut into several pieces; and, above Royat, the River Tiretaine has already laid bare the central structures of the 40 000-year-old basalts from the Petit Puy de Dôme.

In some cases the lava flows were voluminous enough to impound the drainage of the valleys that they entered. The temporary damming of the

Rootless cones formed in the lava flow near Nébouzat.

River Sioule at Pontgibaud by the hawaiites of the Cheire de Côme has been famous ever since the eminent geomorphologist W. M. Davis honoured it with an illustration in 1912. On a smaller scale, 10 km to the south, the Cheire d'Aumone blocked the Mazayes stream about 14 000 years ago. Again, the lake drained and left behind the marshy zone that is called the Etang de Fung.

The most interesting of all the lava flows in the region is the Cheire d'Aydat. The fluid basalts swept in four quick pulsations from the Puy de la Vache and the Puy de Lassolas down into the River Veyre valley and thence into the Grande Limagne. On its northern side, it impounded a small stream, surged 500 m up its valley, and blocked the Lac de Cassière. On its southern side, it held back the Lac d'Aydat when it invaded the River Veyre valley. The river was forced to flow first alongside, and then underneath, the Cheire d'Aydat, before it emerged once again alongside the lavas. The River Veyre is now eroding the weaker sands of the Grande Limagne at Saint Saturnin. They have thus started volcanic inversion of relief, which is so well represented in the ridge of the three-million-year-old Montagne de la Serre, just to the north (Scarth 1967). In another three million years, the Cheire d'Aydat could be in the same state as its neighbour is at present.

BIBLIOGRAPHY

Bout, P. 1973. *Les volcans du Velay*. Brioude, France: Watel.

—— 1978. *Problèmes du volcanisme en Auvergne et Velay*. Brioude, France: Wattel.

Brousse, R. & C. Lefevre 1990. *Les volcans en France*. Paris: Masson.

Brousse, R., R. Maury, J. P. Santoire 1976. L'âge de la coulée du Tartaret. *Académie des Sciences à Paris, Comptes Rendus* **D282**, 531–2.

Camus, G. 1975. *La Chaîne des Puys: étude structurale et volcanologique*. Thèse d'état, Université de Clermont-Ferrand II. [Also published as *Annales de l'Université de Clermont-Ferrand* 28, 1–328.]

Cantagrel, J. M. & J. C. Baubron 1983. *Géochronologie du massif des Monts Dore: implications volcanologiques*. Bulletin 2, Bureau de Recherches Géologiques et Minières, Orleans.

Desmarest, N. 1806. *Sur la détermination de trois époques de la nature par les produits des volcans, et sur l'usage qu'on peut faire de ces époques dans l'étude de ces volcans*. Mémoires de l'Institut Tome VI, 1ère Prairial: an 12 (published in 1806).

Gèze, B. 1979. *Languedoc, Méditerranéen, Montagne Noire*. Paris: Masson.

Goër de Hervé, A. de & J. M. Belin 1985. *Le volcanisme en Auvergne*. Clermont-Ferrand: Centre de Recherches et Documentation Pédagogiques.

Goër de Hervé, A. de and 6 co-authors 1991. *Volcanologie de la Chaîne des Puys* (3rd edn). Aydat, Puy-de-Dôme: Parc Naturel Régional des Volcans d'Auvergne.

—— 1995. *Volcanisme et volcans de l'Auvergne.* Aydat, Puy-de-Dôme: Parc Naturel Régional des Volcans d'Auvergne.

Guérin, G., P. Y. Gillot, M. J. Le Garrec, R. Brousse 1981. Age subactuel des dernières manifestations éruptives du Mont-Dore et du Cézallier. *Académie des Sciences à Paris, Comptes Rendus* **D292**, 855–7.

Guettard, J. E. 1752. *Mémoire sur quelques montagnes de la France qui ont été des volcans.* Mémoire 27 de l'Académie des Sciences de Paris pour 1752 (published in 1756).

Jung, J. 1946. *Géologie de l'Auvergne et de ses confins bourbonnais et limousins.* Mémoire 11, Service de la Carte Géologique de France, Paris.

Juvigné, E. & M. Gwelt 1987. La narse d'Ampoix comme tephrostratotype dans la Chaîne des Puys méridionale (France). *Bulletin de l'Association française pour l'étude du Quaternaire* **1987**(1), 37–49.

Krafft, M. & F. D. de Larouzière 1999. *Guide des volcans d'Europe et des Canaries*, 3rd edn. Lausanne: Delachaux et Niestlé.

Lorenz, V. 1974. On the formation of maars. *Bulletin Volcanologique* **37**(2), 183–204.

Maury, R. C., R. Brousse, B. Villemant, J. L. Joron, H. Jaffrezic, M. Treuil 1980. Cristallisation fractionnée d'un magma basaltique alcalin: la série de la Chaîne des Puys (Massif Central France) I: pétrologie. *Bulletin Minéralogique* **103**(2), 250–66.

de Montlosier, F. 1788. *Essai sur la théorie des volcans d'Auvergne.* Riom.

Pelletier, H., G. Delibrias, J. Labeyrie, M. T. Perquis, R. Rudel 1959. Mesure de l'âge de l'une des coulées volcaniques issues du Puy de la Vache (Puy-de-Dôme) par la méthode du carbone 14. *Académie des Sciences à Paris, Comptes Rendus* **D214**, 2221.

Peterlongo, J. M. 1978. *Massif Central.* Paris: Masson.

Poulett-Scrope, G. 1827. *The geology and extinct volcanoes of central France.* London: John Murray.

Rouire, J. & C. Rousset 1980. *Causses, Cévennes, Aubrac.* Paris: Masson.

Scarth, A. 1966. The physiography of the fault scarp between the Grande Limagne and the Plateaux des Dômes, Massif Central. *Institute of British Geographers, Transactions* **38**, 25–40.

—— 1967. The Montagne de la Serre. *Geographical Journal* **133**(1), 42–8.

—— 1994. *Volcanoes.* London: UCL Press.

9 Germany

The volcanic eruptions in Germany were concentrated in a belt, about 600 km long, following the valley of the River Rhine, which occurred when a rift developed across Europe as the Atlantic Ocean opened. Most of the eruptions took place along major faults trending mainly from north to south or sometimes from northwest to southeast. Successive episodes of rifting were reflected in the activity that reached its climax in areas such as the Odenwald, Spessart, Taunus and Westerwald, Swabia and the Kaiserstuhl more than 15 million years ago.

It was only after a long interval that eruptions resumed in the Eifel Massif, which forms an extensive plateau, about 500 m high, in the area between the River Rhine and its tributary the River Moselle. The West Eifel zone, near Daun, runs for 50 km from northwest to southeast in a band 20 km wide. The East Eifel zone, centred on the Laacher See, near Mayen, is an area with some 50 vents that stretches 35 km from the River Rhine in a belt about 25 km broad. The West Eifel massif has long been famous for its maars, the circular lakes reaching up to more than 2 km in diameter, for which the region has become the type locality. But these beautiful relics of hydro-volcanic eruptions are also accompanied by tuff rings and many cinder cones, lava flows and domes, as well as extensive blankets of pumice. The younger flows, cinders and pumice have been widely quarried, whereas the older materials have weathered to provide a mixture of rich arable and meadow lands and woods surrounding prosperous rural towns.

Volcanic activity began in the Eifel region about 800 000 years ago and reached its distinct late climax between about 12 000 and 10 000 years ago. It was also some 11 000 years ago that the great eruptions of Laacher See distributed indicator beds as far afield as Switzerland, northern Italy and southern Sweden (Bogaard & Schmincke 1985). One or perhaps two magma reservoirs erupted alkaline basalts, basanites, tephrites, phonolites and some trachytes. Basalts, tephrites and basanites were responsible for most of the lava flows, cinder cones, maars and tuff rings, whereas the domes and most of the layers of pumice are chiefly phonolitic. Some domes and layers of tuff are also composed of selbergite, a local phonolite rich in leucite. Although the region has been quiet for several thousand years, emissions of carbon dioxide from Laacher See, for instance, indicate that the magma reservoir has cooled only to about 400°C.

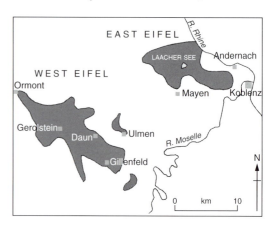

Volcanic zones of the Eifel Massif.

The growth of Rothenberg cone

Rothenberg, which rises on the low ridge separating the Rieden depression from Laacher See is, perhaps, typical of the rather complex evolution of the basanitic–tephritic cinder cones in the East Eifel (Houghton & Schmincke 1986, 1989). It was emitted from six vents aligned on a north-northeast to south-southwest fissure, about 600 m long. The present much-quarried summit of Rothenberg now lies 60 m and originally probably rose about 120 m above its base. The bulk of the volcano is composed of two coalescent cinder cones, although Rothenberg was constructed in six phases of activity. The eruption probably started about 12 000 years ago in an area blanketed by more than 20 m of phonolitic nuée ardente fragments that had exploded from the Rieden depression. These deposits constitute a major water-holding layer and were primarily responsible for the hydrovolcanic explosions from the Rothenberg vents. The first hydrovolcanic eruption formed the small tuff ring, 200 m across, of thin and fine tephritic beds. However, a rapid increase in magma discharge soon excluded any groundwater from the vent and built the first tephritic cinder cone, 65 m high, which eventually buried the tuff ring. Then, when the discharge decreased again, the magma could no longer rise up the main northern vent and could reach the surface only to the south. Here, abundant groundwater was still available in the Rieden layers, and hydrovolcanic explosions immediately generated another set of tuffs. The last two phases occurred when basanitic magma erupted. Activity was only weak on the northern cone, but a new vent formed farther south along the fissure. Hydrovolcanic eruption again began the sequence on the new vent. But once again, as magma discharge increased and eliminated the groundwater, Strombolian eruptions expelled more cinders that formed the southern basanitic cinder cone of Rothenberg, which soon coalesced with its northern predecessor.

EAST EIFEL

The volcanic activity of the East Eifel region was helped by the development of a zone of fault troughs, running west-southwest across the plateau from the River Rhine near Andernach. The eruptions of alkali basalts, tephrites and basanites often formed cinder cones and lava flows, but violent widespread phonolitic explosions of pumice were sometimes followed by volcano-tectonic foundering or by extrusions of phonolitic domes. Although hydrovolcanic activity was both powerful and frequent, voluminous lavas were emitted fast enough to prevent the formation of maars here; and the volcanotectonic depressions, such as those at Wehr, Rieden and the Laacher See are at least twice the size of the maars farther west. Two dozen or so basanitic cinder cones predominate in the fault-trough

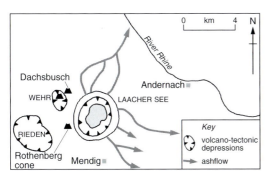

Volcanic depressions of the East Eifel massif and the main directions taken by the ashflows emitted from the Laacher See.

zone east of Laacher See and a similar number of leucitic or leucitic phonolite cones and domes mark the upthrown blocks west of Laacher See (Houghton & Schmincke 1989). Many of the parent vents in both areas are aligned on regional fissures trending from northwest to southeast or northeast to southwest.

The volcanic depressions of Wehr, Rieden and Laacher See seem too large to be maars and too small to be calderas. They are clearly associated with volcanotectonic collapse as well as with violent explosions that often had a distinct hydrovolcanic component. These larger depressions were, perhaps, simply caused by more violent explosions of phonolitic magma in eruptions that reached Plinian proportions.

The Rieden depression is the largest in the East Eifel area and is 2 km broad, stretching 4 km from northwest to southeast. Its edges are sometimes scalloped, which suggests that the depression collapsed intermittently. It contains 15 volcanic vents, which were the likely sources of the thick blanket of pale phonolitic (selbergitic) fragments that filled the depression and radiated more than 5 km from it (Bogaard et al. 1987, Houghton & Schmincke 1989). The fragments were expelled in nuées ardentes, which most probably accompanied the collapse of the Rieden depression, which began 470 000 years ago and climaxed about 410 000 years ago.

A smaller and simpler collapse then occurred at Wehr some 2 km to the northwest. The Wehr depression stretches almost 2 km from northwest to southeast. The phonolitic-trachyte

Laacher See

Situated 40 km south of Bonn, Laacher See is the largest lake and the most famous volcanic formation in the Eifel Massif. It is 2.5 km in diameter and 270 m deep, and lies in the midst of a swarm of tephritic and basanitic cinder cones that rise from a plateau blanketed with thick layers of white pumice (Schmincke & Mertes 1979). Laacher See seems to be a complex ashflow caldera formed by a combination of vigorous explosions and volcano-tectonic foundering. There is no doubt about the explosions. About 11 000 years ago, in the space of perhaps little more than a week, and almost certainly less than a year, Laacher See was the site of a Plinian eruption of over 16 km³ of fragments and flows of phonolitic ash and pumice that are still over 50 m thick near the vent. The debris covers many adjacent cinder cones and forms a clear indicator bed in deposits far beyond the confines of the region (Bogaard & Schmincke 1984, 1985).

The eruptions were generated by a phonolitic magma in a reservoir situated between 3 km and 6 km below the surface, which had probably evolved from an original basanitic magma that had already erupted the surrounding swarm of cinder cones about 270 000 years ago. The most differentiated, highly alkaline and gas-rich phonolites from the uppermost parts of the reservoir were ejected first, followed by less differentiated crystal-rich phonolites, at temperatures between 800°C and 880°C, as the eruptions concluded. The eruption may have been precipitated a few hours after fresh basanitic magma invaded the reservoir, although it is possible, on the other hand, that the new magma arose only when the old had been ejected from the reservoir (Wörner & Wright 1984).

The eruptions proceeded at a very rapid pace and most of them had marked hydrovolcanic components. The breccias expelled during the initial vent clearing were soon superseded by the Plinian columns of phonolitic fragments. The wind winnowed pumice from the columns and spread it widely, and nuées ardentes surged outwards whenever the columns collapsed (Fisher et al. 1983, Schumacher & Schmincke 1990). The nuées ardentes rushed between the cinder cones and swamped the valleys with as much as 10 m of pumice for up to 3 km around the vent. The finer material spread in a blanket of better-sorted pumice that reached 1 m deep over a wide area of the East Eifel. And the finest materials covered some 700 000 km² and reached as far afield as Stockholm, Berlin and Turin (Bogard & Schmincke 1984, 1985). This eruption brought the activity of the East Eifel area to a spectacular, but perhaps temporary, conclusion. However, as new magma apparently invaded the magma reservoir about 11 000 years ago, and fumaroles are still present at Alte Burg, the Laacher See volcano should not be considered extinct.

fragments erupted about 213 000 years ago and, no doubt, mark the initiation of the depression. It has a simple outline and encloses only one vent, although the Dachsbusch cone of alkali basalt also rises on its northern rim. The later eruptions from Wehr covered the Dachsbusch basalts and formed four white layers of phonolitic pumice, dated to 60 000, 52 000, 32 000 and 25 000–12 000 years ago, which were themselves interspersed with basaltic eruptions. The Wehr depression is not entirely extinct, for Welschmiesenmuhle, at its northern end, still manifests fumarole activity.

WEST EIFEL

The volcanic activity of the West Eifel area spread densely over two areas of the plateau centred on Daun, and most of the vents are aligned along en echelon fissures. The basaltic, tephritic and basanitic magmas erupted at intervals that began about 400 000 years ago and ended about 12 000 years ago in the northwest, but lasted only from about 60 000 years until 10 000 years ago in the southeast. The major differences in both these volcanic areas are determined by the presence or absence of groundwater. In the northwestern part of the region, Strombolian eruptions took place along the fissures without water interference and formed tephritic and basanitic cinder cones, such as the Dohm, Hoher List and Radersberg, for instance, and they were often accompanied by small lava flows. Most of these cones have already lost their steepness and sharp outlines after many millennia of weathering.

In the southeast, the eruptions were not only generally more recent, but also were greatly altered by water interference. Although a dozen or so ordinary cinder cones have been formed, the area is dominated by the famous maars (Ollier 1967, Lorenz 1973). These are rounded

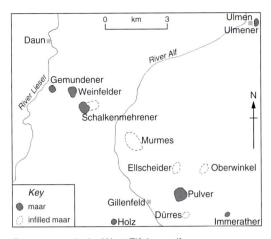

Some maars in the West Eifel massif.

hollows of varying depth and they derive their name from the lakes or meers that now occupy the enclosed depressions, simply because rainwater cannot completely drain away from them. Most maars are 500–1000 m across and 10–200 m deep. Pulvermaar is perhaps the finest example, 900 m across, 120 m deep and encompassed by a regular basaltic tuff ring over 10 m high. Some maars, such as the older Schalkenmehrenermaar, the Dürresmaar and the Oberwinkelermaar, have already been filled by sedimentary or volcanic accumulations, but others, like the Meerfeldermaar, are still over 200 m deep. The maars were caused by shallow hydrovolcanic explosions that were accompanied by calderalike subsidence down ring faults (Lorenz 1973). The repeated explosions formed a low tuff ring of volcanic and lithic fragments. No less than 28 of the 29 maars in this area were formed in valleys where streams provided the water that infiltrated down fissures, met the rising magma, suddenly generated steam in a confined space, and triggered off violent explosions. The Weinfeldermaar, where copious groundwater was available anyway, provides the only exception. Most of the maars were formed during the waning phases of the most recent glaciation, when melting permafrost would also provide great quantities of additional water to generate further explosions. The maars have characteristically rounded outlines; many, like Pulvermaar, are circular; some have slightly scalloped edges, like Gemundenermaar and Weinfeldermaar; and others, such as Immerathermaar and Meerfeldermaar, are elliptical. The surrounding tuff rings often have steep inward-facing slopes of about 25° and gentler outer slopes that rarely exceed

12°. These tuff rings are often only between 10 m and 30 m high and contain many thin beds of fine basanitic or tephritic tuffs and varying amounts of fragments of the country rock. However, there are differences of detail in the maars, which are related, for example, to variations in the amount of magma supplied and to the depth of the explosion. The greater amount of magma erupted at Gemundenermaar, for instance, was probably responsible for the construction of its 100 m-high tuff ring. The 200 m-deep Meerfeldermaar, on the other hand, is surrounded by a ring that is chiefly made up of fragments of the country rock, which indicates that little magma participated in most of its eruptions. At Ulmenermaar, too, similar explosions dominated, because the ring is also composed largely of country-rock fragments.

It was during the period of the waning phases of the most recent glaciation, from 12 500 to 10 000 years ago, that the present maars were formed. For example, one of the oldest, Holzmaar, formed 12 500 years ago and it was followed about 11 000 years ago by Mosbruchmaar and the largest in the area, Meerfeldermaar, which is 1480 m long and 1200 m wide. At about the same time the pair of maars were formed that soon coalesced into the Schalkenmehrenermaar, where the products of the younger western eruption choked the eastern maar so completely that it is now dry. Between 11 000 and 10 000 years ago came Pulvermaar, Gemundenermaar and Weinfeldermaar, and the explosion of Dürresmaar made a cavity in the side of the adjacent Römerberg cinder cone, and Sprinkermaar cut into the southern flanks of the Wartgesberg cinder cone. These are but examples of many that mark the most recent volcanic activity in the West Eifel region. They owe their survival to their relative youth and it is likely that maars also exploded during earlier periods. But the older specimens have probably been eroded or weathered away, or masked beneath a cover of younger volcanic materials.

BIBLIOGRAPHY

Bogaard, P. van den & H-U. Schmincke 1984. The eruptive center of the late Quaternary Laacher See tephra. *Geologische Rundschau* **73**, 935–82.

—— 1985. Laacher See tephra: a widespread isochronous Late Quaternary tephra layer in central and northern Europe. *Geological Society of America, Bulletin* **96**, 1554–71.

Bogaard, P. van den, C. M. Hall, H-U. Schmincke, D. York 1987. ^{40}Ar/^{39}Ar laser dating of single grains: ages of Quaternary tephra from the East Eifel volcanic field, FRG. *Geophysical Research Letters* **14**(12), 1211–14.

Dunda, A. & H-U. Schmincke 1978. Quaternary basanites, melilite nephelinites and tephrites from the Laacher See area, Germany. *Neues Jahrbuch für Mineralogie Abhandlung* **132**, 1–33.

Fisher, R. V. & A. C. Waters 1970. Base surge bed forms in maar volcanoes. *American Journal of Science* **268**, 157–80.

Fisher, R. V., H-U. Schmincke, P. van den Bogaard 1983. Origin and emplacement of a pyroclastic flow and surge unit at Laacher See, Germany. *Journal of Volcanology and Geothermal Research* **17**, 375–92.

Frechen, J. 1976. *Exkursionen im Siebengebirge am Rhein, Laacher Vulkangebiet und Maargebiet der Westeifel-vulkanologisch-petrographische Exkursionen*. Berlin: Borntraeger.

Fuhrmann, U. & H. J. Lippolt 1986. Excess argon and dating of Quaternary Eifel volcanism, II: phonolitic and fioditic rocks near Rieden, East Eifel, FRG. *Neues Jahrbuch für Geologie und Palaeontologie Abhandlung* **172**, 1–19.

Houghton, B. F. & H-U. Schmincke 1986. Mixed deposits of simultaneous Strombolian and phreatomagmatic volcanism: Rothenberg volcano, East Eifel volcanic field. *Journal of Volcanology and Geothermal Research* **30**, 117–30.

—— 1989. Rothenberg scoria cone, East Eifel: a complex Strombolian and phreatomagmatic volcano. *Bulletin of Volcanology* **51**, 28–48.

Krafft, M. & F. D. de Larouzière 1999. *Guide des volcans d'Europe et des Canaries*, 3rd edn. Lausanne: Delachaux et Niestlé.

Lippolt, H. J., U. Fuhrmann, H. Hradestsky 1986. ^{40}Ar/^{39}Ar age determinations on sanidines of the Eifel volcanic field (Federal Republic of Germany): constraints on age and duration of a middle Pleistocene cold period. *Chemical Geology* **59**, 187–204.

Lorenz, V. 1973. On the formation of maars. *Bulletin Volcanologique* **37**, 183–203.

Lorenz, V. 1986. On the growth of maars and diatremes and its relevance on the formation of tuff rings. *Bulletin of Volcanology* **48**, 265–74.

Moore, J. G. 1967. Base surge in recent volcanic eruptions. *Bulletin Volcanologique* **30**, 337–63.

Nairn, A. E. M. & J. Negendank 1973. Palaeomagnetic investigations of the Tertiary and Quaternary igneous rocks, VII: the Tertiary rocks of southwest Germany. *Geologische Rundschau* **62**, 126–37.

Ollier, C. D. 1967. Maars, their characteristics, varieties and definition. *Bulletin Volcanologique* **31**, 45–73.

Schmincke, H-U. 1990. *Die quartären Vulkanefelder der Eifel*. Stuttgart: Schweizerbartsche.

Schmincke, H-U. & H. Mertes 1979. Pliocene and Quaternary volcanic phases in the Eifel volcanic fields. *Naturwissenschaften* **65**, 614–15.

Schumacher, R. & H-U. Schmincke 1990. The lateral facies of ignimbrites at Laacher See volcano. *Bulletin of Volcanology* **52**, 271–85.

Wörner, G. & T. L. Wright 1984. Evidence for magma mixing within the Laacher See magma chamber (East Eifel, Germany). *Journal of Volcanology and Geothermal Research* **22**, 301–327.

Glossary

Many terms are described more fully in the text. All percentages are expressed by weight.

aa A Hawaiian word describing lava flows with a very rough, broken, angular surface.

alkali basalt Basalt that is richer in alkalis and lower in silica than the more common tholeiitic basalt.

andesite A greyish volcanic rock, intermediate in composition between basalt and rhyolite, and containing about 53–62 per cent of silica. Emitted at temperatures between 1100°C and 900°C, it is often viscous and occurs in aa flows, domes and fragments, and is a common constituent of stratovolcanoes in subduction zones. Its fragments often result from violent explosions.

ankaramite A type of basalt that is especially rich in large pyroxene and olivine crystals.

ash Pulverized volcanic rock exploded violently from a vent in fragments less than 2 mm in size, often composed of silicic magma. It can form widespread blankets on land, and the finest particles may remain in the stratosphere for years.

ashflow A nuée ardente or pyroclastic flow consisting essentially of ash. It is a turbulent mixture of volcanic ash and gas, which is expelled at high temperatures and great speed from a vent, and flows down slope and covers vast areas.

barranco A deep gully, or ravine incised by a stream into any steep slope, and often particularly well developed where unconsolidated fragments clothe a volcano.

basalt A dark pasty-grey volcanic rock, with only about 40–52 per cent of silica, but relatively rich in iron, calcium and magnesium. By far the most common volcanic rock, forming the bulk of the ocean floors, and on land it occurs in many lava flows, cinder cones, shield volcanoes and volcanic plateaux. It is usually emitted hot and fluid (1100–1200°C) and flows are commonly 10 km long. It is chiefly erupted in effusive emissions without violent explosions, from fissures as well as single vents. *See also* tholei-

itic basalt and alkali basalt.

basanite A type of very alkaline basalt that is rich in ferromagnesian minerals.

benmoreite An intermediate alkaline volcanic rock, similar to trachyte, but containing less silica (55–60 per cent), that is usually emitted at temperatures of about 1000°C.

block A large solid angular fragment, often of old lava material, expelled during an eruption.

block lava A lava flow that has solidified with a surface of angular blocks.

bomb A large fragment of molten lava, often 1 m across, expelled during an eruption, which develops a rounded or almond shape when twirled through the air. Some bombs have a characteristic breadcrust surface, others resemble cauliflowers or cowpats, depending on the way that they solidify.

bradyseism Small, oft-repeated upward and downward movements of the Earth's crust, which bring about slow and prolonged vertical displacements. They are often caused by oscillations in the levels of subterranean magma.

caldera A large, almost circular or horseshoe shape hollow, several kilometres across, formed mainly by violent explosions and collapse. Usually bounded by steep enclosing walls and formed most often on stratovolcanoes. The term is derived from the Spanish word for cauldron. The Portuguese term caldeira is sometimes used.

caliche A buff or pale ochre calcrete in the Canary Islands, that was formed by chemical precipitation after the capillary rise of water rich in calcium carbonate during the wetter episodes of the Glacial Period.

cinders Fragments of lava, commonly a few tens of centimetres across, that are expelled by moderate explosions, which often form cones. They are usually light and riddled with gas holes. They are most often composed of basalt or andesite and are frequently interbedded with lapilli. They are also commonly known by their Italian name of scoria.

cinder cone A steep conical hill, usually less than 250 m high, with straight slopes that are initially at the angle of rest of the loose materials composing the cone. Formed above a vent when moderate explosions accumulate layers of cinders, lapilli and ash.

crust The solid outer layers of the Earth forming both the continents and the ocean floors.

dacite A pale volcanic rock, rich in silica (63–68 per cent), which is emitted at temperatures usually about 800°C to 900°C and is viscous and slow moving. A common constituent of domes, it is often also involved in violently explosive eruptions.

debris avalanche A landslide composed of a mixture of fragments of all sizes, and often predominantly of old material, which occurs when a sector of a volcano collapses and is deposited at its base.

dense rock equivalent (DRE) The fragments expelled during an eruption are riddled with holes that can double or triple their volume compared with that of a compact rock such as a lava flow. Thus, in order to compare the sizes of different eruptions, the measurements of erupted volumes taken in the field are changed into their equivalent in dense rock. These dense rock volumes correspond broadly with the volumes of magma expelled.

dome A rounded convex-sided mass of volcanic rock, which is usually silicic and too viscous to flow far from the vent. Often formed on stratovolcanoes towards the end of an eruption. Frequently composed of dacite, phonolite, trachyte or rhyolite.

dyke A vertical or steeply inclined sheet of magma injected along, and solidifying in, fractures in the Earth's crust. Volcanoes are fed by deep dykes, and other dykes, in turn, branch or radiate from them.

effusion An eruption of lava that takes place with little or no gaseous explosions and thus most commonly gives rise to lava flows.

eruption The way in which gases, liquids and solids are expelled onto the Earth's surface by volcanic action ranging from violently explosive outbursts to noiseless, effusive or hydrothermal outflow.

eustatic The term used to describe worldwide changes in sea level, which are themselves caused by changes in the volume of water in the oceans.

fajã A relatively isolated area in the Azores, which has been formed at the base of cliffs, either when rubble has fallen from them, or where lava has flowed into the sea in a broad delta.

fissure A crack, fault or cluster of joints, cutting deep into the Earth's crust, which may allow magma to reach the surface. A fissure usually gives rise to effusive emissions, which may be accompanied by rather more explosive eruptions forming cones of cinders or spatter.

fragments Ash, bombs, cinders, lapilli or pumice shattered by explosions during an eruption. They are the main constituents of cinder cones and many stratovolcanoes. Also called pyroclasts and tephra.

fumarole A small vent giving off gases or steam, and often surrounded by fragile precipitated crystals. They are a major aspect of hydrothermal activity, along with solfataras and geysers.

geological time The whole history of the Earth, extending back about 4.6 thousand million years.

gravitational collapse The collapse of part of a volcano often caused by magmatic pressure from beneath. Collapse also occurs when a volcano is built up so quickly that it becomes gravitationally unstable, or when a sector of the volcano is undermined by erosion of its lower slopes. Also called sector collapse or flank collapse.

hawaiite A grey alkaline volcanic rock, with about 50 per cent of silica, that forms fragments and fairly fluid lava flows. Sometimes called trachybasalt.

historical time The timespan during which events have been recorded, in however fragmentary a fashion, by observers. In the Mediterranean area it may reach back 3000 years, whereas in the New World it can be less than 200 years.

hornito A small cone or mound of spatter, usually less than 10 m, but occasionally 100 m in height. They are notably rough and steep-sided, whatever their size. The name is derived from the Spanish word for little oven, which smaller hornitos resemble.

hotspot A stationary plume of convectively rising mantle coming from a deep source, perhaps even from the boundary between the core and the mantle. A hotspot generates chains of volcanoes when the plates move over it.

hydrothermal eruption A term used to describe eruptions of gases, steam, and hot water, without magma.

hydrovolcanic eruption A term used to describe violently explosive eruptions in which both water and magma play a significant role. Such eruptions are termed Surtseyan in shallow sea water and lakes, and hydrovolcanic or hydromagmatic on land (formerly known as phreatomagmatic eruptions).

Icelandic eruption Eruptions that are usually basaltic in character which take place notably along fissures caused basically by plate divergence in Iceland. They form long rows of relatively small cones but often give out vast lava flows. Sometimes they also develop into large volcanic systems.

ignimbrite Voluminous deposits, often covering more than 1 km³, of pumice, broken crystals, and elongated pieces of glass ("fiamme") in a matrix of ash, laid down by large nuées ardentes. They can be welded when they are deposited at high temperatures.

island arc A gently curving chain of volcanic islands rising above sea level from the ocean floor, formed when an oceanic plate is subducted beneath another. Volcanic chains are their equivalent on land.

jökulhlaup A glacier burst in Iceland, generated when an eruption melts part of the ice lodged in a caldera, forms a lake that eventually lifts the ice cap, and bursts over the rim of the caldera and rushes down slope at great speed.

lapilli Small pea or nut-size volcanic fragments, usually between 2 mm and 64 mm, that are expelled in a molten, or nearly solid hot state during moderate eruptions, and accumulating in layers within cones. The term derives from the Italian word for little stones.

latite A silicic lava in which potassium and sodium occur in similar volumes. It is a term that is now little used, and is considered to be a potassic variety of trachyandesite.

lava Molten rock or magma which reaches the surface and solidifies on cooling. Lava occurs as flows, domes, fragments within cones and as pillows formed on the ocean floors.

lava tunnel A linear hollow formed when molten lava flows away beneath an already solidified outer crust. Smaller forms are sometimes called lava tubes.

Maar A German word used to describe an almost circular crater, often about 1 km across, formed mainly by hydrovolcanic eruptions. They may or may not be bordered by a ring or crescent of fine fragments, sloping gently outwards from a low crest overlooking the crater. The crater is usually filled with rainwater, forming small lakes from which the German name is derived.

magma Hot mobile rock material, mainly formed by partial melting of the mantle, commonly at depths between 70 km and 200 km. It is composed of hot viscous liquid material containing hot, but still solid, crystals or rock fragments and small proportions of included gases. It is less dense than the materials surrounding it and is thus able to rise slowly towards the Earth's surface by buoyancy. If it overcomes the pressure and resistance of the rocks of the Earth's crust, it erupts in a fluid state, releasing its contained gases with varying degrees of explosive violence and emits lava in flows or fragments.

magma reservoir A large zone of ill defined fissures and cavities in the lithosphere where rising magma halts for varying lengths of time. Reservoirs are most often a few cubic kilometres in volume and are situated usually between 2 km and 50 km in depth.

magmatic differentiation A very complex process that takes place when crystals separate out from magma as it cools. The first crystals to separate out, made up of dark, basic and dense silicates of magnesium and iron, sink towards the base of the reservoir and accumulate. Thus, the remaining liquid magma becomes increasingly silicic and concentrated in the upper reaches of the reservoir, thereby stratifying the cooled reservoir. Eruptions that take place from the upper reaches of the reservoir are therefore often violently explosive and give off fine silicic fragments, whereas those from the lower reaches are usually characterized by relatively gentle effusions of lava.

mantle The hot but not wholly mobile layer of the Earth situated below the crust and which encloses the core.

mid-ocean ridge A ridge on the ocean floor where volcanic eruptions generate new oceanic crust and where two adjacent plates diverge.

mudflow A current of water and a great proportion of fragments of all sizes and types, which is commonly formed when an eruption melts part of an ice cap or disturbs a crater lake. Mudflows travel down valley at high speed and often cause much damage. Often known by the Indonesian term, lahar.

mugearite A grey alkaline intermediate volcanic rock, containing about 52–58 per cent of silica, and rich in potassium. It is often also called trachyandesite.

nuée ardente A French term used to describe an incandescent cloud, or glowing avalanche, of scorching hot gas and fragments of all sizes, including ash, pumice and rock debris in an aerosol-like emulsion expelled by explosive eruptions, which travels across the ground at very high speeds and gives off glowing billowing clouds. It is, perhaps, the most dangerous of all the forms of volcanic eruptions. Its deposits are often known as pyroclastic flows, where the content is dense, and pyroclastic surges, if formed from the accompanying cloud of ash and dust.

obsidian Volcanic glass. A dense, shiny black or brown, glassy and rare form of rhyolite, which rises and cools rapidly, and is usually too viscous to flow far from the vent. It forms domes, mounds and short rugged lava flows.

pahoehoe A Hawaiian word used to describe smooth gently undulating or ropy lava flows often having a typical sheen. Formed most often by basalts emitted in a hot fluid state.

palagonite Yellowish-brown angular fragments commonly formed when lavas erupt beneath water or ice.

pantellerite Rhyolite that is relatively rich in alkalis, and thus also known as peralkaline rhyolite.

Peléan eruption A violent eruption that gives rise to blasts and nuées ardentes like those expelled by Montagne Pelée on 8 May 1902.

peralkaline A volcanic rock in which the combined molecular proportions of sodium and potassium exceed those of aluminium oxides.

phonolite A pale volcanic rock rich in sodium and potassium but relatively poor in silica (55–60 per cent), which is usually emitted at temperatures of less than 1000°C. It is derived from very alkaline basalts such as basanite or nephelinite. It is viscous and it occurs in domes and rugged lava flows. It is so named because it often breaks into plates that give a characteristic ringing sound when struck.

picón The cover of black ash and lapilli expelled during the historical eruptions on Lanzarote.

pillow lava Lava that erupts at depth under water in the form of piles of pillows or cushions.

plate Usually rigid slabs into which the lithosphere is divided. Their edges constantly diverge, or converge and plunge beneath each other. All are composed of oceanic crust and some also carry continental crust. Between 10 and 15 major plates, and a similar number of micro plates are generally recognized.

Plinian eruption A powerful explosion of gas, steam, ash, and pumice, lasting for several minutes, or hours, which rise in vertical columns to 30 km or more and often branch out in the form of a Mediterranean pine, like that at Vesuvius in AD 79. Whenever the vertical impetus wanes, the part of the column might collapse and form nuées ardentes. They are generated mainly by silicic magmas, although they can develop from hydrovolcanic eruptions of basalt. The gas and dust expelled to the stratosphere often form an acidic aerosol that can modify the weather over large tracts of the Earth.

pumice Very pale volcanic fragments riddled with gas holes, formed by the expansion of contained gases as the magma reaches the surface and explodes very violently over vast areas during an eruption. Most pumice floats on water and sometimes forms ephemeral floating islands after eruptions at sea. It varies from small fibrous chips to knobbly lumps and often resembles solidified foam. It is commonly expelled in eruptions of rhyolite, dacite, trachyte or phonolite.

pyroclastic flow or **pyroclastic surge** *See* nuée ardente.

rhyolite A pale volcanic rock very rich in silica (69–75 per cent), which is usually emitted at temperatures about 700–800°C and commonly forms extensive pumice and ashflows when expelled as fragments, but it also gives rise to viscous lavas forming domes and stubby flows.

rootless cone When molten lavas invade boggy land or very shallow surface water, their heat can suddenly convert the water into steam. The resultant explosions shatter the lava into fine fragments that accumulate in small cones on the flow. They are termed rootless because, unlike cinder cones, they have no vents extending down into the terrain beneath the lava flows.

satellite cone A small cinder cone erupted on the flanks of a stratovolcano or shield. They can occur singly, in swarms, or along fissures radiating from the summit. They are also called lateral, adventive or parasite cones.

seamount A volcanic mountain found below sea level, and often rising 1000 m from the ocean floor. Active seamounts could eventually form new volcanic islands. They may also be extinct submerged remains of old volcanoes.

shield A large gently sloping volcano, composed mainly of fluid basaltic lava flows with relatively few fragmented layers, emitted from clustered vents.

shoshonitic lava A lava with a relatively high potassium content that occurs in subduction zones.

silica The molecule, formed of silicon and oxygen (SiO_2), that is a fundamental component of volcanic rocks, and is the most important factor controlling the fluidity of magma. Other things being equal, the higher the silica content of a magma, the greater its viscosity.

silicic magma Magma rich in silica (60–75 per cent). Also called acid magma.

solfatara An Italian word used to describe the emission of sulphurous gases from a fumarole.

spatter Lava fragments of cinder size, often emitted as clots in lava fountains. They are still molten when they return to the ground and thus flatten out and form "cowpats". Spatter often welds together to form steep-sided cones and ramparts, as well as hornitos.

stratovolcano A large and fairly steep-sided volcanic cone composed of stratified bedded layers of lava fragments and flows, as well as many other volcanic products. Also known as composite volcanoes or stratocones.

Strombolian eruption Repeated moderately explosive activity, expelling fragments of lava, that produces cones of ash, lapilli and cinders, and copious lava flows over a period of several months or a few years. The most common kind of eruption on land.

subduction zone A zone where two plates converge and one plunges beneath the other into the mantle. The subducted slab releases volatiles that stimulate melting in the wedge of mantle above it, which helps form volcanoes. The plunging action of the slab also generates deep-seated violent earthquakes.

subglacial eruption Any eruption that takes place under ice, but most commonly associated with activity beneath the ice caps of Iceland.

sub-Plinian eruption A less violent form of Plinian eruption.

Surtseyan eruption An eruption that takes place in shallow lakes or seas where water can enter the vent, mix with the rising magma, and repeatedly form steam that shatters the magma into fine fragments often called tuffs. Without such water interference, most of these eruptions would probably be Strombolian in nature. One of the major types of hydrovolcanic eruption.

tephrite A volcanic rock that is relatively poor in silica (45–50 per cent) and rich in alkalis. After crystallization, some of the latter help to form feldspathoids, which are like feldspars but with less silica. Nepheline is a sodic feldspathoid; leucite is a potassic feldspathoid. The leucite tephrite of Vesuvius often contains large rounded white crystals of leucite.

tholeiitic basalt Basalt containing relatively high amounts of silica, with relatively small proportions of sodium and potassium.

trachyandesite A term comprising both mugearite and benmoreite.

trachybasalt More or less synonymous with hawaiite.

trachyte A pale greyish acidic volcanic rock relatively rich in silica (60–65 per cent) and in sodium and potassium, which is usually emitted at temperatures of about 1000°C. It is viscous, and can be involved in violent explosions. It also forms rugged lava flows and domes.

tsunami A Japanese term used to describe huge rapidly moving sea waves generated by violent eruptions or earthquakes. They increase in size and speed as they reach shallow water and often cause much damage and death on nearby coasts.

tuff cone A steep squat conical hill, usually less than 300 m high, composed of innumerable thin layers of fine fragments with a deep wide crater, formed above a vent by Surtseyan eruptions.

tuff ring A broad circular accumulation of fine fragments, often 1 km or more in diameter, surrounding a broad shallow crater. Both the outer and craterward-facing slopes are relatively gentle compared with those of a cinder cone, and the crater is much wider. It has the approximate proportions of a doughnut or motorcar tyre.

vent The usually vertical conduit or pipe up which volcanic material travels from the magma source to the Earth's surface.

volcanic gas The volatile component of magma, mainly including steam, carbon dioxide, sulphur dioxide and smaller amounts of chlorine and fluorine. As the magma approaches the Earth's surface, the gases are exsolved and can become the chief factor in the violence of eruptions.

volcano A hill or mountain formed around and above a vent by accumulations of erupted materials such as ash, pumice, cinders or lava flows. The term refers both to the vent itself and to the often cone-shape accumulation above it.

Vulcanian eruption A short violent explosion expelling fragments of usually silicic lava, commonly in the form of ash, as well as bombs with a characteristic surface resembling a bread crust.

Vocabulary

This vocabulary is not intended to be exhaustive, but the selection of descriptive terms may help the reader to visualize the appearance of many volcanoes presented in the text.

F – French, **G** – Greek, **Ger** – German, **It** – Italian, **Ice** – Icelandic, **P** – Portuguese, **S** – Spanish

achada	**P**	abutting onto	couze	**F**	river in Auvergne
ajuda	**P, S**	help	cuchillo	**S**	narrow ridge
alto	**P**	high	cumbre	**S**	ridge
ancien	**F**	old	dyngja	**Ice**	lava shield
antico	**It**	old	echeide	**S**	inferno (a Guanche word)
antigua	**S**	old	égueulé	**F**	breached
apalhraun	**Ice**	aa lava flow	eldfjell	**Ice**	volcano
areia	**P**	sand, ash	eldgjá	**Ice**	fire fissure
arena	**It, S**	sand, ash	eldur	**Ice**	fire, eruption
askja	**Ice**	box	enfer	**F**	hell
áspros	**G**	white	enxofre	**P**	sulphur, fumarole
azufrado	**S**	sulphur	fjal	**Ice**	mountain
bagacina	**P**	cinders	fljót	**Ice**	river
baixo	**P**	low, lower	fogo	**P**	fire
bajo	**S**	low, lower	fossa	**It**	crater
barranco	**S**	deep gully, ravine	fuego	**S**	fire
bermejo	**S**	red	fuencaliente	**S**	hot spring
bianco	**It**	white	fuoco	**It**	fire
biscoitos	**P**	rugged lava flow	furna	**P**	oven, cavern, hot spring
blanc	**F**	white	gemelos	**S**	twins
blanco	**S**	white	geysir	**Ice**	hot water fountain
boca	**S**	vent, or mouth	gjá	**Ice**	fissure
bocca	**It**	vent, or mouth	gordo	**P, S**	large, fat
borg	**Ice**	rocky hill	gour	**F**	round lake, or maar
branco	**P**	white	hághios	**G**	holy
butte	**F**	small mesa	helluhraun	**Ice**	smooth pahoehoe lava flow
cabeço	**P**	head, cone	hnúk	**Ice**	peak
caldeira	**P**	cauldron, large crater	hornito	**S**	small oven
caldera	**S**	cauldron, large crater	hryggur	**Ice**	ridge
cancela	**P**	sheltered place	hver	**Ice**	hot spring
carvão	**P**	burnt, charcoal	jameo	**S**	lava tunnel in the Canary Islands
cendres	**F**	ash, cinders	jökulhlaup	**Ice**	glacier burst
cenere	**It**	ash, cinders	jökull	**Ice**	glacier
cheire	**F**	rugged lava flow in Auvergne	kaménos	**G**	burnt
chinyero	**S**	ashy	kókkinos	**G**	red
cima	**It, P**	summit	lac	**F**	lake
colorado	**S**	coloured, red	lago	**It**	lake

lagoa	**P**	lake		rajada	**S**	split
lajes	**P**	stoney, stones, flagstones		raudur	**Ice**	red
lombo	**P, S**	ridge		redondo	**P**	round
loutrá	**G**	hot springs		reventado	**S**	breached
Maar	**Ger**	round lake or meer		revento	**S**	breached
malpaís	**S**	area of rugged lava flows		rhaun	**Ice**	lava flows
mávros	**G**	black		ribeira	**P**	river
megálos	**G**	big		rio	**S**	river
mesa	**S**	tableland, or flat-topped hill		rivière	**F**	river
micrós	**G**	small		rojo	**S**	red
mistério	**P**	lava flow in the Azores		rosso	**I**	red
mont	**F**	mountain		sarcoui	**F**	cauldron
montaña	**S**	mountain		sciara	**It**	rugged lava flow
monte	**It, S**	mountain		secco	**It**	dry
narices	**S**	nostrils		seco	**P, S**	dry
narse	**F**	infilled maar in Auvergne		serra	**It, P**	ridge
negro	**P, S**	black		serre	**F**	ridge
neós	**G**	new		solfatara	**It**	sulphur, sulphurous fumarole
nero	**It**	black		soufrière	**F**	sulphurous place
nuée ardente	**F**	incandescent cloud		stéfanos	**G**	crown
paleós	**G**	old		suc	**F**	narrow pointed lava dome
partido	**S**	cloven		timão	**P**	arched yoke
pavin	**F**	fearsome		vatn	**Ice**	water, lake
pic	**F**	peak		vecchio	**It**	old
pico	**P, S**	peak		vermelho	**P**	vermilion, red
pietre	**It**	stones		vermelo	**S**	vermilion, red
puy	**F**	hill in Auvergne		vieux	**F**	old
queimado	**P**	burnt		víti	**Ice**	Hell, explosion crater
quemado	**S**	burnt		voragine	**It**	chasm
quente	**P**	hot		vounó	**G**	hill

Eruptions in Europe in historical times

A question mark indicates doubt about:
- after the date: the date of the eruption
- before the date: the occurrence of the eruption itself
- after the place name: the location of the eruption.

Many eruptions cited without question marks are not necessarily certain to have taken place, but the reference often represents what seems to be the most likely location and date of the events. Many of the eruptions listed are discussed more fully in the text and in the relevant articles to which reference is made. Data come mainly from historical written documents; some information, however, is inferred as follows from dating systems: A archaeological, AM archaeomagnetic, ^{14}C radiocarbon, Ra radium disequilibria, K–Ar potassium–argon

Vesuvius

A	persistent mild activity
IE	intermediate eruption
FE	final eruption
SPE	sub-Plinian eruption
PE	Plinian eruption, see text

N.B. The distinction between IE and FE, or SP and P, is often not so clear as it may seem in this list.

1944 18–29 March, strong FE, lava flows to the NW at Massa di Somma and San Sebastiano, lava fountains more than 2 km in height, hydrovolcanic explosions and small nuées ardentes, 28 deaths

1929 13 July to 17 March 1944, A including 4 IE

1929 3–8 June, strong IE with lava fountains 500 m high and lava overflows to the E base of the cone

1913 5 July to 2 June 1929, increasing A fills the 1906 crater

1906 4–22 April, very strong FE with lava fountains 3 km high, flows to the SE destroying Boscotrecase, ballistic projectiles causing heavy damage at Ottaviano, 218 deaths

1875 18 December to 3 April 1906, A including 6 IE, of which that of Colle Margherita and Colle Umberto

1872 26–30 April, FE with lava fountains and flows to the NW, at least 12 deaths

1870 2 November to 25 April 1872, A including one IE

1868 15–26 November, FE

1865 10 February to 14 November 1868, A with one IE

1861 8–10 December, FE through a fissure extending to the sea on the SW side

1855 19 December to 7 December 1861, A including one IE

1855 1–27 May, FE (?) with lava flows to the NW reaching Cercola

1850 6–16 February, FE

1841 20 September to 5 February 1850, A including one IE

1839 1–5 January, FE

1835 13 March to 31 December 1838, A

1834 23 August to 2 September, FE

1824 2 July to 22 August 1834, A including 3 IE

1822 21–25? October, strongly explosive FE

1806 27 January to 20 October 1822, A including 4 IE

1805 14–16 October?, FE

226

1796   15 January to 13 October 1805, A
       including 3 IE
1794   15 June to 7? July, large FE, fissure on the
       SW flank and lava flow to the sea
       destroying Torre del Greco, at least 18
       deaths
1783   18 August to 14 June 1794, A including
       2 IE
1779   3–15 August, strong FE, lava fountains
       4 km high
1770   16 February to 2 August 1779, A
       including 2 IE
1767   19–27 October, FE
1764   1 July to 18 October 1767, A with one IE
1760   23 December to 5 January 1761, FE with
       fissure low on the S flank
1744   1 July to 22 December 1760, A including
       3 IE
1737   20–30? May, strong FE
1732   25 December to 19 May 1737, A
       including one IE
1730   17–23 March, FE (?)
1712   5 February to 16 March 1730, A
       including 10 IE
1707   28 July to 15? August, strongly explosive
       FE
1699   22 April to 27 July 1707, A
1698   20–31 May? FE (?)
1638?–98   A including 5 (?) IE
1631   16–19 December, strong SPE with nuées
       ardentes (AM, ^{14}C), most of the villages S
       and W of the volcano destroyed, more
       than 4000 deaths
1500 c. "fire" in the crater (hot gases?)
?1347–50   Strombolian activity?
1139   29 May to 5 June, strongly explosive
       eruption and flows (AM, ^{14}C)
1037   January–February, extensive lava flows
       to the sea (AM)
1006–1007   explosive eruption, blocks hurled "3
       miles from the crater"
999   cinders and lava flows (AM)
991   "flames and ashes"
968   lava fountains and flows to the sea (AM)
787   October-December, lava fountains and
       flows (AM, ^{14}C)
685   February–March, huge eruption
       column, lava or pyroclastic flows to the
       sea (^{14}C)
536   Strombolian activity
?512   probable confusion with 472
472   5–6 November, large SPE with nuées
       ardentes to the NW (AM, ^{14}C), fallout to
       the NE
379–95   Strombolian activity and probable lava
       flows
222–35   persistent Strombolian activity
203   strong explosions and ash deposits
       (SPE?, ^{14}C)

172   Strombolian activity
AD 79   24–26 August, large PE burying
       Herculaneum (AM), Pompeii, Stabiae
?217 BC   possible SPE
c. 700 BC   (A), SPE
c. 1000 BC   (A), SPE
?   . . . at least 3 undated explosive
       eruptions (SPE?)
c. 1800 BC   (^{14}C), "Avellino" PE
?. . .   at least 2 undated SPE
c. 6000 BC   (^{14}C), "Mercato" PE

Phlegraean Fields and Ischia

1538   29 September to 6 October, Monte
       Nuovo, Phlegraean Fields
1302   18 January to March, Arso, Ischia
c. 670–890   Fiaiano, Ischia
c. 430   Molara, Vateliero and Cava Nocelle,
       Ischia
c. AD 60   Montagnone–Maschiatta, Ischia
c. 19 BC?   Rotaro II?
c. 350 BC   Porto d'Ischia, Ischia?
c. 470 BC   Porto d'Ischia, Ischia?
c. 600 BC   Rotaro I, Ischia

Aeolian Islands

Stromboli
Almost continuously active since at least 1788;
only major or unusual events reported here.

1999   26 August, strong explosion, tourists
       injured
1998   16 January, 23 August, 8 September,
       stronger explosions
1996   1 and 6 June, 4 September, incandescent
       material on vegetation
1994   21–22 August, continuous lava
       fountaining
1993   10 February, 16 and 23 October, strong
       explosions, tourists injured
1990   15 April, lithic fallout on village
1985   December to April 1986, lava flow to the
       sea
1975   4–24 November, ashfall, lava flow to the
       sea
1972   December, explosions, lithic fallout on
       village
1967   19 April to 13 August, lava flow to the
       sea
1959   19 May, 11 July, strong explosions
1956   January–March, lava flows
1955   28 February to 22 March, flank flow at
       the foot of Sciara del Fuoco
1954   1 February, paroxysm, ashfall, hot
       avalanche, tsunami

1952 7–22 June, explosion, intermittent lava flows

1949 6–9 June, ash and block fall, fire on vegetation, lava flow

1944 20 August, 2000 m plume, hot avalanche, one house destroyed

1943 December to October 1944, intermittent flows to the sea

1943 3 December, ash and block fall, fires, houses damaged

1938–9 lava fountains 1 km high, several flows to the sea

1934 2 February, block fall near village, air shocks damage houses

1932 3 June to 1 February 1934, repose period

1931 7 July to May 1932, repose period

1931 23 April, 7 July, strong explosions

1930 22 October to 2 December, intermittent lava fountains and flows

1930 11 September, major paroxysm, blocks onto villages, nuée ardente to Vallonazzo trough, tsunami, flows, 6 deaths

1930 3 and 6 February, ashfall, lava flows

1919 22 May, major outburst, bombs on villages, ashfalls to Sicily, tsunami, 4 deaths

1916 30–31 June, major explosions, flows

1915 18 June to 20 December, explosions and flows

1912 June–November, intermittent lava fountains, block falls, ash to Calabria and Sicily

1907 June to May 1910, repose period

1907 January–April, intermittent ashfalls and lava flows to the sea

1905 February–December, repeated lava flows and ashfalls

1900 April–October, frequent explosions and ashfalls, one hot avalanche

1891 24 July and 31 August, two paroxysms, block falls, fire on vegetation

1889 June to 1890, repose period

1888 October to June 1889, lava fountains and flows

1887 31 March and 18 November, pumice falls

1882 17–30 November, paroxysm, opening of new vents, block falls, landslides

1879 February and August, lava fountains, ashfalls, fire on vegetation

1874 June, block fall on village

1857 repose period

1855 3–4 October, lava fountains and scoria fallout

1850 ash and block falls, houses damaged

1822 22 October, ashfall, hot avalanche, destruction of land

1768–70 explosions, lava flows, submarine eruption

1558? destruction of land . . .

Lìpari

c. 1200 (AM) Rocche Rosse obsidian flow

AD 729? Mt Pilato Plinian eruption

c. AD 350 Forgia Vecchia flow

Vulcano

1888 3 August to 22 March 1890, series of powerful explosions

1886 11 January to 31 March, intermittent explosions, new vents

1881–3 strong fumaroles, incandescent gases

1879 6–13 January, detonations heard at Lìpari, "flames,"

1877 September–October, strong fumaroles and (?) weak explosions

1876 29 July (?), eruption with ashfall to Salina

1873 22 July to June 1875, fumaroles, intermittent explosions

1822–5 greater fumarole activity

1786 12 January to February, explosive eruption

1783 February, explosions heard in Calabria

1776–86 fumaroles larger than usual

1731–1739 eruptive period ending with Pietre Cotte obsidian flow

1727 Forgia Vecchia satellite crater?

1688 June, explosions

1626 March–April, greater activity?

1540–50? Erupted material links Vulcano to Vulcanello

1444 5 February, powerful explosions with incandescent material

c. 1300–40 "great fire", detonations

c. 1250 eruptions from both Vulcano and Vulcanello?

1184 December, possible confusion with Lìpari

? 900–50 stronger activity?

? 729–87 possible confusion with Lìpari

c. 500–80? Explosions?

c. 200–250 activity

? 144 probably no eruption

c. AD 25–100 activity?

c. 29–19 BC activity

91 BC both Vulcano and Vulcanello active?

126 BC submarine eruption at or near Vulcanello

?186–183 BC doubtful reference to Vulcano or Vulcanello

c. 350 BC activity?

c. 425 BC "fire" from Vulcano Island

? 475 BC probable confusion with 425 BC

c. 1000–800 BC to(AM) lava flow from the W basement of Vulcanello

c. 2600 BC (K–Ar), major eruptive cycle

c. 3500 BC (K–Ar), beginning of La Fossa activity

Etna

CC	Central Crater
V	Voragine
BN	Bocca Nuova
NEC	Northeast Crater
SEC	Southeast Crater
GM	geological map, 1979

Updated from Tanguy & Patanè (1996), lava volume indicated when measured (Murray 1990), Ra dating from Condomines et al. (1995).

2000   26 January. SEC eruptions (lava fountains and flows, more that 60 events as of 24 June)

1999   October, explosions and overflows from BN

1999   4 September, lava fountains from V, explosions at BN

1999   4 February to 6 November, fissure and flows at the SE base of SEC

1998   September to January 1999, SEC eruptions (explosions and overflows)

1998   22 July, 5 August, lava fountains at V

1998   27 March, lava fountains at NEC

1996   November to July 1998, explosions BN, V, SEC, NEC, lava overflows at SEC

1996   July–August, explosions and flows at NEC

1995–6   intermittent lava fountains at NEC

1991   14 December to 31 March 1993, SE flank eruption (231×10^{6}m^{3} of lava)

1990   January, intermittent lava fountains and flows at SEC

1989   10 September to 9 October, lava fountains at V and SEC, then NE flank eruption (~20×10^{6}m^{3})

1986   30 October to 25 February 1987, SEC and NE flank eruption (~50×10^{6}m^{3}), Mt Rittmann

1986   24 September, lava fountain NEC

1985   25–31 December, E flank eruption

1985   10 March to 13 July, S flank eruption

1984   29 April to 16 October, SEC explosions and lava flows

1983   28 March to 6 August, S flank eruption (79×10^{6}m^{3})

1981   17–23 March, NNW flank eruption (18×10^{6}m^{3})

1980   September, and February 1981, NEC lava fountains

1979   3–9 August, SEC and SE, W, ENE flank eruptions

1978   23–29 November, SEC and SE flank eruption

1978   23–29 August, SEC and SE, ENE flank eruption

1978   29 April to 5 June, SEC and SE flank eruption

1977   July to March 1978, NEC intermittent lava fountains and flows

1975   24 February to 8 January 1977, NEC and N flank

1974   10 October . . ., NEC explosions and flows

1974   30 January to 17 February and 11–29 March, W flank eruptions, Mt De Fiore

1971   5 April to 12 June, S, SE and ENE flank eruptions (40×10^{6}m^{3})

1966–71   March, NEC continuous explosions and flows

1964   February–July, intermittent explosions and flows from CC

1960   17 July, powerful lava fountain at V, minor explosions on following weeks

1955–64   continuous explosions and flows at NEC (250×10^{6}m^{3}), various episodes at V

1950   25 November to 2 December 1951, E flank eruption (124×10^{6}m^{3})

1949   1–4 December, S, N and NW flank (4×10^{6}m^{3})

1947   24 February to 10 March, N flank (7×10^{6}m^{3})

1942   30 June, SSW flank and CC (3×10^{6}m^{3})

1940   16 March, powerful lava fountain at CC

1928   2–20 November, NEC and ENE flank (26×10^{6}m^{3}), Mascali overwhelmed

1923   17 June to 18 July, NNE flank (48×10^{6}m^{3})

1918   29 November?, NW flank (11×10^{6}m^{3})

1917   24 June, lava fountain at NEC

1911   10–21 September, NNE flank (27×10^{6}m^{3})

1910   23 March to 18 April, S flank (37×10^{6}m^{3})

1908   28 December, Messina tectonic earthquake

1908   29–30 April, SE flank (2×10^{6}m^{3})

1899   19 July and 5 August, powerful explosions at CC

1892   9 July to 28 December, S flank (145×10^{6}m^{3}), Mt Silvestri

1886   19 May to 7 June, S flank (51×10^{6}m^{3}), Mt Gemmellaro

1883   22–24 March, S flank, Mt Leone

1879   26 May to 7 June, SSW and NNE flanks (45×10^{6}m^{3}), Mt Umberto–Margherita

1874   29 August, NNE flank (2×10^{6}m^{3})

1869   26 September, E base of central cone (3×10^{6}m^{3})

1868   27 November, 8 December, CC lava fountains 2 km high

1865   30 January to 28 June, NE flank, Mt Sartorius

1852   20 August to 27 May 1853, ESE flank, Mt Centenari

1843   17–28 November, W flank

1838/39/42   CC explosions and overflows

1832   1–22 November, W flank, Mt D'Ognissanti or Nunziata

1819   27 May to 1 August, SE flank, "La Padella" crater

1811   27 October to 24 April 1812, E flank, Mt Simone

1809   27 March to 9 April, NNE flank

1802   15–16 November, E flank

1798–1809   CC strong intermittent explosions

1792   12 May to May? 1793, SSE flank, "La Cisternazza" collapse pit

1787   17, 18, 19 July, CC lava fountains 3 km high and overflows

1780   18–31 May, SSW flank

1766   28 April to 7 November, S flank, Mt Calcarazzi

1763   19 June to 10 September, S flank, "La Montagnola" cone

1763   6 February to 15 March, W flank, Mt Nuovo

1755   10–15 March, E flank

1723–24   CC recorded persistent activity

1732–65   CC recorded persistent activity

1702   8 March to 8 May, SE flank

1693   9 and 11 January, powerful tectonic earthquakes

1689   March, E flank

1682   September?, upper E flank

1669   11 March to 11 July, S flank large eruption (600–800×10^{6}m^3), about 15 villages and part of Catania overwhelmed, strong explosions at CC on 25 March

1651?   . . . (AM), lava flow towards Macchia di Giarre on the E flank

1651   17 January to 1653, W flank large eruption destroying part of Bronte

1646   20 November to 17 January 1647, NNE flank, northern Mt Nero

1634   18 December to June? 1636, SSE flank

1614   1 July to 1624, N flank voluminous eruption, "Due Pizzi" or "Fratelli Pii" hornitos

1610   6 February to 15 August, SW flank, Grotta degli Archi crater row

1607   June, NW flank, Pomiciaro cinder cone

1603–1610   CC strong persistent activity with frequent overflows

1579   September, SE flank

1566   November–December? N flank

1537   10 May to July, S flank

1536–37   CC strong persistent activity

1536   22 March to 8 April? CC overflows, then S and possibly SW flanks

1493–1500c.   CC persistent activity

1446   September, E flank

?1444 . . .   SSE flank? Possible confusion with 1408

1408   8–25 November, SSE flank, partial destruction of Pedara village (AM, Ra)

?1381   5 August, SSE flank, possible confusion with 1329

?1334   confusion with 1329

1329   28 June to August . . ., E and lower SE flanks, Mt Rosso di Fleri (AM, Ra), flows near Acireale

1284–5   E or SE flank

c. 1250   (AM), S flank, flow near Serra La Nave (GM: indicated as 1536)

c. 1200   (AM), SW flank, upper Gallo Bianco flow (GM: indicated as 1595)

c. 1180   (AM, Ra), NE flank, flow to Linguaglossa village (GM: indicated as 1566)

1169   4 February, destructive tectonic earthquake, doubtful Etna eruption (see below)

c. 1160   (AM), SSE flank, flow to the sea (GM: indicated as 1381)

1062–1064   W flank, lower Gallo Bianco flow? (see below)

c. 1060   (AM), WSW flank, lower Gallo Bianco flow (GM: indicated as 1595)

c. 1050   (AM), E flank, Mt Ilice cone and extensive flow to the sea (GM: indicated as 1329)

c. 1020   (AM), NE flank, Scorcia Vacca flow (GM: indicated as 1651)

c. 1000   (AM), SSW flank, Mt Sona and flow to Paterno town (GM: indicated as 812/1169)

c. 970   (AM), N flank, Mt Pizzillo cone

c. 950   (AM, Ra), NW flank, lower Pomiciaro flow (GM: indicated as 1537)

c. 800   (AM), SE flank, flow north and east of Trecastagni (GM: indicated as 1408)

c. 700   (AM), SE flank, Mt Solfizio spatter rampart and flow

c. 650   (AM), W flank, Mt Lepre

c. 500   (AM), S flank, Mt Ciacca lava flow

c. 450   (AM), W flank, flow South of Bronte

c. 300   (AM), S flank, Mompeloso cinder cone and flow

252   1–9 February, SSE flank, Cibali flow in Catania (AM)

c. 200   (AM), SSE flank, Carvana flow in Catania

c. 100   (AM), SSE flank, Mt Pizzuta Calvarina

AD 38–40   large explosive eruption heard in Messina

c. AD 0–20   persistent activity within summit caldera

32 BC   lava flow

36 BC   N or NW flank

44 BC   tephra fall to Reggio Calabria

49 BC   large summit plume, then flow on W flank

122 BC   (^{14}C) large Plinian eruption, flows near Catania

126 BC   earthquakes, summit and (?) flank eruptions

135 BC	gas plume and lava flows
140 BC	"fires"
396 BC	spring, E or SE flank, lava flow to the sea
425 BC	March–April, S flank, lava flow near Catania
?475 BC	probable confusion with 479
479 BC	August, SE flank, lava flow to the sea
?695 BC	. . .? . . .
c. 1100 BC	(AM, Ra), SW flank Mt Arso and flow toward Licodia
c. 1400 BC	large eruptions force the Sicanians to emigrate
c. 3300 BC	(AM, Ra), Fortino Vecchio flow in Catania (GM: indicated as 693 BC)

Straits of Sicily

?2000	Graham Bank
1891	Foerstner Bank
?1863	
?1845	
1831	Graham Bank
?1707	
?1632	

Greece

1950	Neá Kaméni, Santoríni
1939–41	Neá Kaméni, Santoríni
1928	Neá Kaméni, Santoríni
1925–6	Neá Kaméni, Santoríni
1887	Caldera, Nísyros
1873	Caldera, Nísyros
1871	Caldera, Nísyros
1866–70	Neá Kaméni, Santoríni
1707–11	Neá Kaméni, Santoríni
1650	Kolómbos Bank, off Santoríni
1570	Mikrá Kaméni, Santoríni
726	Paleá Kaméni, Santoríni
AD 46	Paleá Kaméni, Santoríni
197 BC	Hierá (Bankos Bank), Santoríni
c. 250 BC	Kaméni Xorió, Méthana

Canary Islands

1971	Teneguía, La Palma
1949	San Juan, La Palma
1909	Chinyero, Tenerife
1824	Tao–Nuevo–Tinguatón, Lanzarote
1798	Narices de Teide, Tenerife
1793?	Lomo Negro, El Hierro?
1730–36	Montañas del Fuego, Lanzarote
1712	El Charco, La Palma
1706	Garachico, Tenerife
1705	Arafo, Tenerife

1705	Fasnia, Tenerife
1704–1705	Siete Fuentes, Tenerife
1677–8	"San Antonio"– Fuencaliente, La Palma
1646	Martín, La Palma
1585	Tahuya, La Palma
1492	Teide, Tenerife
1478–80?	Teide, Tenerife
1470–90 or 1430–40	Montaña Quemada, La Palma
1455?	Teide? Tenerife
1430?	Orotava Valley, Tenerife
1393? or 1399?	Tenerife

Azores

1998–9	off Serreta, Terceira
1981	Monaco Bank, S of São Miguel
1964	off Rosais, São Jorge
1911	Monaco Bank, S of São Miguel
1907	Monaco Bank, S of São Miguel
1902	off Pico-Terceira
1867	off Serreta, Terceira
1811	Sabrina, off Ferraria, São Miguel
1808	Urzelina, São Jorge
1800	off Serreta, Terceira
1761	Negro & Fogo, Terceira
1720	Don João de Castro Bank
1682	off Ferraria, São Miguel
1652	Fogo I & II, São Miguel
1638	off Ferraria, São Miguel
1630	Furnas, São Miguel
1580	Queimada, São Jorge
1564	Lagoa do Fogo, São Miguel
1563	Sapateiro–Queimado, São Miguel
1563	Lagoa do Fogo, São Miguel
1562–4	Praínha, Pico
1445	Furnas, São Miguel
1439	Sete Cidades, São Miguel

Iceland

2000	Hekla
1998	Grímsvötn
1996	Gjálp–Grímsvötn
1991	Hekla
1983	Grímsvötn
1981	Hekla
1980	Hekla
1975–84	Krafla
1973	Heimaey
1970	Hekla
1963	Surtsey
1961	Askja
1947	Hekla
1934	Grímsvötn

1922	Grímsvötn		1416	Katla
1918	Katla		1410	Bárdarbunga–Veidivötn
1903	Grímsvötn		1389	Hekla
1892	Grímsvötn		1362	Öraefajökull
1883	Grímsvötn		1357	Katla
1875	Askja		1354	Grímsvötn
1862–4	Bárdarbunga–Veidivötn		1341	Hekla
1860	Katla	c.	1340	Bárdarbunga–Veidivötn
1845	Hekla		1300	Hekla
1838	Grímsvötn		1262	Katla
1823	Grímsvötn		1245	Katla
1821–3	Eyjafjallajökull		1231	Reykjanes
1783–5	Grímsvötn		1226	Reykjanes
1783–4	Laki		1222	Hekla
1774	Grímsvötn		1206	Hekla
1766–8	Hekla		1179	Katla
1766	Bárdarbunga–Veidivötn		1159	Bárdarbunga–Veidivötn
1755	Katla		1158	Hekla
1739	Bárdarbunga–Veidivötn		1104	Hekla
1724–9	Krafla	c.	935	Eldgjá–Katla
1727	Öraefajökull	c.	920	Katla
1721	Katla		870	Bárdarbunga–Veidivötn–Torfajökull
1717	Bárdarbunga–Veidivötn			
1711	Bárdarbunga–Veidivötn			
1706	Bárdarbunga–Veidivötn			
1697	Bárdarbunga–Veidivötn			
1693	Hekla			

Jan Mayen

1985	Nordkapp
1984–5	Beerenberg?
1975	Nordkapp
1973	Nordkapp
1970–71	Nordkapp
1851?	Koksletta
1850?	Nordkapp?
1818	Dagnyhaugen–Eggoya?
1732	Dagnyhaugen–Eggoya?
1650	Koksletta?

1660	Katla
1659	Grímsvötn
1636	Hekla
1625	Katla
1619	Grímsvötn
1612	Katla
1597	Hekla
1580	Katla
1510	Hekla
c. 1500	Katla
1477	Bárdarbunga–Veidivötn

Index

PLACES, FEATURES AND PERSONAL NAMES

Page numbers in *italics* denote illustrations; those in **bold** denote sections of text.

Aci Castello   57
Acireale   58, 64
Acquacalda   40, 45, 48, 49, *49*, 50
Adrano   57
Aegean plate   81
Aegean Sea   2, 3, 81
Aegina   81, 82
Aeolian Islands   3, 7, 9, **33–51**, *33*
African plate   2, 3, 4, 9, 33, 55, 56, 81, 100, 132, 133
Agnano   24
Água de Pau stratovolcano   134, **137–8**, *137*, 139
Ajuda   150
Akrotíri   83, 84, 87
Alban Hills   *see* Colli Albani
Alcantara   67
Alferes   140
Algar do Carvão   146
Alicudi   33, 34, 42
Anaga peninsula   103
Angra do Heroísmo   133, 144, 146, 148, *148*
Antímilos   92
Apennines   2, 7, 8, 30
Arafo   109
Arbol de Piedra   *see* Roque Cinchado
Ariccia crater   8
Arrecife   120
Arso   30, 32
Askja   167, 170, **172–4**, *173*, 177, *178*
Aspronísi   82, *88*
Astroni   24
Atalaya de Femés   113
Atlantic Ocean   2, 3, 101, 102, 104, 112, 122, 128, 132, 134, 135, 143, 144, 151, 158, 161, 172, 173, 186, 194, 214
Atrio del Cavallo   10, 14, 15, 18
Aubrac   200, 201
Auvergne   200, 202, 205, 210
Avellino Pumice   13, 17
Averno   24, 28, 29, 30
Axarfjordur   174
Axlargígur   180, 181
Azores   2, 4, 132, *132*, 133, 134, 138, 139, 141, 142, 143, *143*, 144, 145, 149, 151, 153, 154, 155, 156, 157, 162

Bagacina   146
Baia   22, 24
Bandama   123
Bárdarbunga   178, 182, 183
Basiluzzo   33
Beerenberg   2, 4, 194, 195, *195*, 196, 197
Bejanedo volcano   125
Belpasso   66
Betancuria   121
Biancavilla   57
Biscoitos   146, *147*, 154, 160
Boca de Tauce   104
Bocca Grande   27
Bocca Nuova   53, 55, 59, *60*, *61*, 66, 67, 69
Bolsena   7
Bolutler   185
Boscotrecase   12, 15
Bracciano   7
Breiddalur   168
Bronte   62, 65

Cabeço
    do Canto   159
    do Capelo   159
    do Fogo   154, 157, 160
    Gordo   158
    do Pacheco   159
    Redondo   158
    Verde   159
Cachorro   154
Caesar, Julius   64, 203
Cala dell'Altura   70
Caldeira (Graciosa)   133, 140, 141, 149, *150*, 156, 157
Caldeira do Alferes   140
Caldeira Seca   140, 141
Caldeirão   159
Caldera
    Blanca   113
    de las Cañadas   100, 102, 103, 104, *104*, 105, *105*, 106, *106*, 107, 108, 110, 111
    Fuencaliente   114
    del Piano (Etna)   *52*, 58, 64, *67*
    del Piano (Vulcano)   45, *45*, 46
    de Taburiente   104, 125
Calogero   42
Caloura   138
Camaldoli della Torre   10
Campania   7, 9, 10, 15, 17, 18, 20, 21, 23, 32
Campotese   31
Cañadas Volcano   102, 103, 104, 105, 107, 108
Canal do Faial   156, 157, 158
Canary Islands   2, 4, 100, *100*, 101, 102, 103, 109, 110, 112, 114, 117, 120, 121, 124, 126, 127, 128, 132

Cangueiro 158
Canneto 41
Cantal 200, 201, 202
Cape Riva 83
Capelas volcano 141
Capelinhos 133, 150, 156, 157, 158, 159, *160*, 161, *161*, *162*
Capelo Peninsula *157*, **158–9**, *160*
Capo di Bove flow 8
Capo Graziano 43
Capri 12, 20, 32
Carneiro 158
Casamicciola 30, 32
Castel Savelli 8
Castelo Branco 158
Catania 51, 55, 58, 59, 63, 64, *64*, 65, 69
Cerrado das Sete 145
Cerrado Novo 138
Cézallier 200, 201
Chahorra de Teide 106, *107*, 110; *see also* Narices de Teide, Pico Viejo
Chain of Puys 3, 200, 201, **202–212**, *202*, *203*, *204*
  cones
    breached **207–208**
    monogenetic **205–207**
    polygenetic **208–210**
  domes **204–205**
  growth **203–204**
  hydrovolcanism **210–211**
  lava flows **211–12**
Cheire
  d'Aydat 204, 208, 211, 212
  de Côme 211, 212
  de Mercoeur 211
Chiancone 57
Chinyero 102, 111, *111*
Chupadero 116
Cinque Denti 70
Cinquo Picos **144**
Cinta Esterna caldera 8
Cinta Interna 8
Cisternazza 61, 65
Clermont-Ferrand 201, 202, 204, 208
Coiron lavas 200
Coiron plateaux 201, *201*
Colle
  Margherita 11
  Umberto 11
Colli Albani 7, 8
Columbus, Christopher 109
Commenda 46, 47, 48
Contrada Diana 41
Corazoncillo 115, *117*
Corvo 43, 132, 136, **162**
Costa

d'Agosto 40
da Nau 159, *160*
Sparaina 32
Couze
  Chambon 208, 210
  Pavin 210
Cova da Burra *132*, 136
Crete 51, 83, 84, 87
Cuddia
  Attalora 70
  Khamma 70
  Randazzo 71
Cuevas del Diablo 121
Cuma 23
Cumbre
  Nueva 125, 127
  Vieja 125, 126, 127, *Plate 7b*
Curbelo, Andrés-Lorenzo 114, 116
Cuvigghuni 57

Dáfni dome 90
Dagnyhaugen 196
Devès 200, 201
Didýme 43
Diodorus Siculus 17, 64
Don João de Castro Bank 133, 142, 143
Due Pizzi 63, 65
Dürresmaar 217
Dyngjufjöll **172–4**, *173*, 182

Eggoya 197
Eggvin bank 194
Eifel Massif 3, 214, *214*
  East **215–16**, *215*
  West **216–17**, *217*
El Charco 127
El Duraznero 126, 127, *Plate 7b*
El Golfo 114, *119*, 124
El Hierro 100, 101, *124*, *124*, 128
El Pitón 102, 105, 106, 107, 109, *Plate 7a*
El Portillo 104
El Quemado 115
El Tabonal Negro 107
El Tiñor 124
Eldfell (Heimaey) *188*
Eldgjá 171, 172, *172*, 184
Ellittico caldera **57–8**
Empedocles 53, 58
Encantada 122
Eolo and Enarete seamounts *33*, 44
Eolus 34
Estivadoux 210
Etna 2, 3, 6, 9, 11, **51–69**, *52*, *53*, *55*, *59*, *63*, *64*, *67*, *72*, 102, 109, 179, *Plate 2*, *Plate 4a*

ancient **56–7**
central cone **58–9**
geological setting **55–6**
historical eruptions **64–9**
hydrovolcanic eruptions **59–61**
lateral eruptions **61–3**, *Plate 4a*
Northeast Crater 53, **59**, 61, *61*, 66, 67, 69
Southeast Crater 53, 56, 58, **59**, 61, *61*, 67, 68, 69, *Plate 2*, *Plate 3*
Euonymos 51
Eyjafjallajökull *178*, 182

Faial 132, 133, 150, **156–61**, *157*
  growth **157–8**
  historical eruptions **160–61**
Famara Plateau 112, 114
Faraglione **48–50**
Faraglioni 33, 49
Fasnia volcano 109
Ferdinandea 72
Feteiras 139
Fiaiano Pumice 28
Fili di
  Baraona 38, 39
  Sciacca 43
Filicudi 33, 34, 41, 43, *43*
Filo del
  Banco 43
  Fuoco 38, 39
Flores 132, **162**
Foerstner Bank **72**
Fogo 134, 137, 138, 139, 146, 154, 157, 160
Fogo-A pumice 138
Fogo caldera 138, 139
Forgia Vecchia 41, 46, 47, 48
Fornazzo 55, 59, 66, 67
Forum Vulcani 22, 27
Fossa
  cone 3, 11, 24, 33, 43, 44, *44*, 45, 46, *46*, 47, 48, 49, *Plate 5a*
  Felci (Filicudi) 43, *43*
  delle Felci (Salina) 43
  Monaca 11
Fouqué dome 90
Franklin, Benjamin 171
Freiras 141
Fuencaliente 126, 127
Fuerteventura 100, 101, **121–2**, *121*
Funda caldera 162
Furnas
  Achada das F. **137**
  caldera 133, 135, 136, *136*, 137
  do Enxofre 144, 145
  stratovolcano 134, *134*, **135–7**, *136*, 137, 144, 145
Furno Ruim 159

Gáldar 123
Gañañías 109
Garachico 110, *110*, 111
Garonjay 128
Gauro 24
Geirfuglasker 186
Gemundenermaar 217
Geórgios (dome) 88, 90, *90*, *95*
Gergovie 200
Geysir 185
Giarre 59
Ginostra 35, 37
Gjálp 182, 183
Goree, Father 89
Gour de Tazenat 200, 210, *211*
Graciosa 114, 132, 133, **149–50**, *149*, 151
Graham Bank **72**
Graham Island 9, 72
Gran
    Canaria 100, 109, 115, 116, **123**, *123*
    Cono 10, 15
    Cratere *44*, 47, 48
Grande Limagne 200, 201, 202, 210, 211, 212
Grímsvötn 4, 171, 172, **182–3**, 184
Guilherme Moniz strato- volcano **144–5**, 146
Güímar 108, 109

Hághios
    Elías 94, 95
    Geórgios 95
    Ioánnis 95
Halepá dome 92
Havhestberget ashflow 195
Heimaey 168, 186, *186*, 187, *188*, 189
Hekla 4, 167, 173, *178*, **179–81**, *179*, *181*, 182, 184
Heklugjá fissure 179, *179*, 180, 181, *181*
Helgafell 186, 189
Hellenic arc 3, *81*
Héphaistos 45
Herculaneum 6, 10, 16, 17, 18, *20*, 19, *Plate 1b*
Herdubreid 182
Herdubreidartögl ridge 182
Hesiod 64
Hierá 45, 86
Hlidargígar 181
Hofsjökull 167, 178
Holuhraun 173
Homer 34, 64
Horta 133, 156, 158, *158*
Hoyo Negro (explosion crater) 127, *Plate 7b*

Hraungígur 180
Hrossaldur cone 175
Hvannadalshnukur 179
Hveragerdi 185
Hveravellir 184, 185
Hverfisfljót gorge 171
Hverfjall 174

Iálysos *81*, 87
Iblean Plateau 56
Iceland
    active volcanic zone **169–72**
        northern **172–7**
    hotspot 167, *169*
    island volcanoes **186–9**
    Plinian eruptions **178–81**
    shield volcanoes **177**
    submarine volcanoes **181–4**
Icod 110
Ilhéus dos
    Capelinhos 159, 161
    Mosteiros 133, 143
Immerathermaar 217
Ionian Sea 55, 56, 57, 58, 59
Ischia 9, **30–32**, *31*
Islote de Hilario *113*, *114*, 121

Jan Mayen
    fracture zone 194, 197
    historical eruptions *195*
    ridge 195
Jandía Peninsula 122
Jolnir 187
Julán 124
Julia 72

Kaiserstuhl 214
Kaméni Islands (Santoríni) 82, 83, 84, 86, 89, 90, *90*, *91*
Kaméni (Methana)
    Vounó 93
    Xorió 93, *93*
Kapp Muyen group 195
Katla 171, 172, 182, **183–4**, *183*, 187
Káto Zákro *81*
Kéfalos peninsula 92
Kelingarfjoll 184, 185
Ketildyngja 177
Kilian 204, 211
Kímolos 92
Knossós 83, 87
Koksletta eruption 196
Kolbeinsey Ridge *4*, 167, 194
Kollóttadyngja 177
Kolómbos Bank 88
Kós 81, 82, **92**, *92*
Krafla 168, 170, 172, **174–7**, *174*, 185

Krísuvík 185, *185*
Ktenás dome 90
Kverkfjöll 182

La Bocaina straits 122
La Gomera 100, 101, 109, **127**, 128
La Isleta 123
La Orotava 109, 111
La Palma 100, 101, 104, **125–7**, *125*, *Plate 7b*
La Quemada 114
La Rambleta 105, 106
La Solfatara 22, 24, **26–8**, *26*, 27, 29
La Sommata 46
La Vecchia caldera 70
La Ventosilla 122
Laacher See 214, 215, *215*, 216
Lac
    Chambon 208, 210
    d'Aydat 200, 212
    Pavin 200, 210
Lacroix 33, 36, 152
Ladeira do Moro *150*
Lago
    Albano 7, 8, *8*
    Averno *29*
    Nemi 7, 8
Lagoa
    Azul 139, 140, *140*, 141
    das Furnas 135, *135*, 136
    de Santiago 140, 141
    do Canário 141
    do Congro 137
    do Fogo 137, *137*, 138
    do Negro 146, *146*
    do Pilar 141
    Rasa 140, 141
    Verde 139, 140, *140*
Lakagígar 170, 171, 172
Lake Mývatn 171, 174, *174*, 175, *176*, 177, 182
Lakes Massif 141
Laki 170, *170*, 171, 172, *Plate 8*
Langjökull 167, 178
Lanzarote 100, 101, **112–21**, *112*, *113*, *115*, 122
Larderello 7
Las Arenas 109
Las Montañetas 123
Las Nueces 115
Las Palmas 123
Las Piños 123
Latium 7
Laurenço Nuñes 154
Laxá 174, 177
Lazio 7
Le Puy-en-Velay 200
Leirhnjúkur fissure 174, 175

Leone 57
Liátsikas 90
Limagne 200, 212
Linguaglossa 63
Linosa 9
Lìpari 9, 33, 34, 39, *39*, 40, 41, 42,
  42, 43, 44, *44*, 47, 48, 49, 50, *50*, 51
Lisbon 132
Lisca Blanca 33
Los Ajaches Plateau 112
Los Azulejos 105
Los Cuervos 115
Los Frailes 109
Los Gemelos 106
Los Lanos 126
Los Marteles 123
Lucretius 64
Ludentsborgir 171, 174

Maciço das Lagoas 141
Madalena *153*, 154, *154*, 155
Mafra 141, 143
Mállia *81*
Manadas complex 151
Mancha Blanca 116
Mandráki 94
Maretas 116
Marsili 44
Martinique 147, 152
Mascali 55, 62, 66
Maschiatta 32
Massif Central 200, 201, 202
Mata das Feiticeiras 138
Matias Simão *147*
Mato do Leal 139
Mazo 115, 116
Mediterranean Sea **3**, 33
Meerfeldermaar 217
Megálo Vounó 83
Mercalli 15, 33, 36
Mercato Pumice 13
Méthana 81, 82, **93**, *93*
Miccio, Scipione 28
Mid-Atlantic Ridge 2, **3–4**, *4*, 132,
  132, 133, 156, 157, 162, 167, 169,
  186, 194
Mikrá Kaméni 88, 89, 90
Mílos 81, 82, **92**
Minoan
  Aegean *81*
  caldera 84
  civilization 83, 84, 87
  deposits 84
  eruption 81, **83–6**
Misenum 17, 18, 19, 20, 21, 22
Molara 32
Mohns Ridge *4*, 194
Monastero caldera 70
Mongibello 52, 53, 57, **58**

Montagna Grande 70, 71
Montagne Pelée 147, 205
Montagne de la Serre 200, 212
Montagnola 43, *55*, 61, 65
Montagnone 32
Montaña
  Abejera 107
  Blanca 107, *107*, 108, 113
  Chamuscada 124
  Colmenar 108
  Colorada 114, 115
  Corona 114
  de Arucas 123
  de Bilma 111
  de la Cruz de Tea 108
  de los Helechos 114
  Entremontañas 124
  Guajara 103
  Majua 107
  Mareta 107
  Quemada 126
  Rajada (Lanzarote) 114, 115
  Rajada (Tenerife) 107, *107*, 108
  Reventada 108
  Rodeo 116
  Roja 113
  Roja del Fuego 114
  San Andreas 122
Montañas
  de las Lajes 107
  del Señalo 115, 116, *118*
  Quemadas 115, 116
Montcineyre 210
Mont-Dore 200, 201, 202, 204, 210
Monte
  Amiata 7
  Areia 136
  Aria 44
  Brasil 144, 148, *148*
  Calanna 57
  Carneiro 158
  Chirica 40
  das Mocas 158
  dei Porri 43, *43*
  Epomeo 30, 31, 32
  Escuro 138
  Gelfiser 71
  Gelkhamar 71
  Gemmellaro 65
  Giardina 40, 41
  Gibele 70, *71*
  Guardia 40, 41, *41*
  Guia 156, 158, *159*
  Lentia 45, 46
  Leone 62, 65
  Mazzacaruso 40
  Nuovo 22, 24, 26, **28–30**, 29, *29*
  Pilato 33, 39, *40*, 41, 42
  Queimada 152

Queimado 139, 158, *158*
Rosso 64
San Leo 68
Sant'Angelo 40
Somma 10, 13, 15, 16, 17, 39,
  Plate 1a
Spina 24
Tabor 32
Vulture 7, 9
Montefascione 7
Monterosa peninsula 40
Monti
  Cimini 7
  Iblei 9
  Rivi 43
  Rossi 65
  Sabatini 7
  Silvestri 54, *54*, 66
  Vulsini 7
Morro Grande 162
Moselle, River 214
Mosteiros *142*, 143
Mourgue de Mentredon 171
Mycénae 87
Mýrdalsjökull 167, 171, 182, 183,
  183

Námafjall 175
Naples 7, 10, 12, 15, 16, 17, 19, 21,
  22, 23, 28, 29, 30, 72
Narices de Teide *105*, 106, *107*,
  110; *see also* Chahorra, Pico Viejo
Narse
  d'Ampoix 211
  d'Espinasse 210, 211, *211*
Neá Kaméni 88, *88*, *89*, 90
Negrão 145
Nicolosi 65, 66
Nífios 95
Níki dome 90
Nísyros 81, 82, **94–5**, *95*, *Plate 5b*
Niyhver springs 185
Nordeste *134*, 135
Nord-Jan 194
Nordvestkapp formation 195
Norse settlement of Iceland (AD
  874) 168, 171, 184
North American plate 3, 4, 132,
  162, 166

Oberwinkelermaar 217
Ódádahraun 172, 177, 182
Odenwald 214
Odysseus 22
Oldugígar 181
Olibano 24, 27
Oplontis 17, 18, 19
Öraefajökull 2, 167, 178, *178*, **179**,
  180, 182, 184

Orchilla 124
Orosius 64
Orotava Valley 108
Osorio 123

Padella 61
Pajara cones 122
Palaeá Kaméni 86, 88, *88*, 89
Palizzi 46, 47
Panarea 33, 34, 41, 43, 44, 51
Pantelleria 7, 9, **70–71**, *70*, *71*, 72
Pappalardo, Diego 65
Paternò 65
Pedara 64
Peloritani Range 56
Perret, Frank A. 12, 14, 15
Phaestós 87
Phíra 84, *85*, *89*, *91*, *Plate 6*
Phlegraean Fields (Campi Flegrei)
  9, **22–30**, *22*, 31, 32
Piano
  del Lago 53
  Sterile 26, 27
Picarito 157, 160
Pico (island) 132, *132*, 133, 147,
  151, **153–6**, *153*, *154*, *156*, 161
  Eastern **155–6**
Pico
  Alto (Terceira, stratovolcano)
    145, *145*, 146
  Alto (Santa Maria) 143
  da Barrosa 137
  Cabras 107
  das Camarinhas 141
  da Candela 145
  do Carvão 141
  da Catarina Vieira 145
  do Cedro 139
  Cruz 139
  del Cuchillo 113
  de Eguas 141
  dos Enes 145
  do Enforçado 139
  da Esperança 151, *151*
  do Ferro *135*, 136
  do Fogo 146
  dos Fragosos 150, *150*
  do Gaspar 136, 146, *146*
  Gordo 146
  Grande 139
  da Ladeira do Moro 149
  das Marcondas 136
  das Mos 138
  del Nambroque 127
  dos Padres 145
  Partido 114, 115, 116, *117*, *118*,
    119, 120
  do Pico 2, 132, 133, **153–5**, *153*,
    154

Rachado 145
Sapateiro 139
de Teide 2, 100, 101, 102, 103,
  104, 105, 106, 107, *107*, 108,
  109, 110, *Plate 7a*
Timao 150
das Tres Lagoas 137
da Vara 134, 135
Viejo 102, 104, 105, 106, *106*,
  110; *see also* Chahorra de
  Teide, Narices de Teide
de la Zorza 121
Piedra Sal 122
Piedras Arrancadas 108
Pietre Cotte 34, 45, 46, 48
Pindar 30, 64
Pingarotes 159
Piquinho *154*, 155
Pirrera 41
Pizzi Deneri 55, 58
Pizzo sopra la Fossa 35, 38, *Plate
  4b*
Plákes dome 92
Plateau des Dômes 202
Pliny
  Elder 18, 19, 21, 30, 47, 51
  Younger 6, 10, 18, 21
Pollara 43
Pollena eruption 11, 16, 17
Polybius 34, 45
Polýegos 92
Polyphemus 6, 64
Pomici di Base 13
Pomiciazzo 39, 41
Pompeii 6, 10, 13, 17, 18, 19, *20*,
  21, *Plate 1a*
Ponta
  da Ferraria 141
  da Varadouro 159
  de São João 154
  Delgada 136, 142
  do Escavaldo 140
  dos Biscoitos 146, 155, *156*
  do Mistério *145*, 154
  Garça 136
Ponza Islands 9
Póros 81, 82
Porto di Levante 46, 48, 49
Posidonius 51
Posillipo 24
Povoação 134, **135**, 136
Pozo de las Nieves 123
Pozzuoli 9, 22, 23, 24, **25–6**, *25*,
  26, 28, 29
Praia 149, *150*
Praia do Norte 157, 160
Praínha 154, 155, 156
Procida 9, 23, 30, 32
Profítis Elías 82, 83

Pulvermaar 217
Punta Lucia 58
Punte Nere 46, 47
Puy
  de Barme 207, *208*
  Chalard 207, *211*
  de Chanat 203
  de Chaumont *203*
  Chopine *203*, 204, 205, *206*,
    208
  de Clierzou *203*, 204
  de Côme *203*, 204, *206*, 207,
    208, 211
  de la Coquille *203*, 207
  de Dôme 200, 202, 204, 205,
    205, 207, *208*, *209*, 211
  de l'Enfer 211, *211*
  de Fraisse *203*, 204
  des Goules 203, *206*, 208
  des Gouttes *203*, 205, *206*, 208
  de Gravenoire 210
  de Jumes *203*, 207
  de Laschamp 203
  de Lassolas 204, *207*, 208, *209*,
    212
  de Lemptégy 203, 206
  de Louchadière *203*, 207, *208*
  de Mercoeur 207, *209*
  de Mey 211
  de Montchal 210
  de Montcineyre 210
  de la Nugère *203*, 204, 211
  de Pariou *203*, 204, *205*, 208,
    209
  de Pourcharet *209*, 211
  de Sarcoui *203*, 204–205, 206
  de Tartaret 208, 210
  de la Taupe 203, *206*, 207, 210
  de la Vache 204, *207*, 208, *209*,
    212
  Petit P. de Dôme 203, *203*, *204*,
    205, *209*, 211
  du Petit Fraisse *204*, *209*

Quattrocchi 41, *41*, *42*
Quattropani 40
Queimada (São Jorge) 152
Queimadas (Faial) 158
Queimado (São Miguel) 139, 158
Quemadas (Lanzarote) 114, 116
Quemado de Orzola (Lanzarote)
  114
Quitadouro 150

Randazzo 67
Raudoldur 180
Rebenoda 122
Rectina 19
Redondo 150

Região dos Picos   134, **138–9**, *139*
Resina   16, *20*
Reykjanes   167, 169, 184, 185, 186
   Ridge   *4*
Reykjarfjördur   *168*
Reykjavík   170, 185, 189
Rhine, River   214, 215
   Rift Valley   *3*
Rhodes   87
Ribeira
   do Cabo   160
   Chà   137
   Grande   138
   Quente   136
   Seca   138, 139, 151
      faults   151
Rieden   215
Rinquim   158
Riviera dei Ciclopi   57
Roccamonfina   7, 9
Rocche Rosse   33, 34, 40, *40*, 42
Rodeo   116
Römerberg   217
Roque
   Cinchado   104, *Plate 7b*
   del Conde   103
   Imoque   103
   de las Muchados   125
   Nublo   123
   Vento   103
Roques
   Blancos   106
   de García   104, 105
Rosais complex   151
Rotaro   32
Rothenberg   215
Royat   211
Rubicón   113, 114

St Agatha   64
St Calogero   42
St Vincent   147
St Willibald   42
Salina   33, 34, 41, 43, *43*
San Antonio cone   126, 127
San Bartolo   35, 117
San Bartolomé   117
San Giovanni di Galermo   65
San Juan eruption   127
San Vincenzo   35
Sandiago volcanic field   108
Santa Bárbara stratovolcano   **145**
Santa Catalina   115, 116
Santa Cruz das Ribeiras   155, *156*
Santa Luzia   154
Santa Maria   132, 133, **143**
Santoríni   3, 81, *81*, **82–91**, *85*, *91*, 100, *Plate 6*
   Cape Skaros   *88*

eruptions in historical times **86–90**
   growth   **83**
   Minoan eruption   **83–6**
São João   154, 155
São Jorge   132, 133, **151–2**, *151*
São Lourenço volcano   143
São Miguel   132, 133, **134–43**, *134*
São Roque   *153*, 155
Sarskrateret cone   196
Saut de la Pucelle   210
Schalkenmehrenermaar   217
Sciacca   72
Sciara del Fuoco   35, 36, 37, 38, 39, *39*
Scorcia Vacca flow   63
Seara   140
Senga   24
Sentralkrateret   194, 196
Serra
   Branca   149, 150
   do Cume   144
   Dormida   149, 150
   das Fontes   149, 150
   Gorda   138
   da Ribeirinha   144
   do Topo   151
Serreta   145, 147
Sete Cidades stratovolcano   134, *134*, 138, **139–41**, *140*
   caldera   138, 139, 141
Skaftá, River   171
Sicily   6, 7, 9, 33, 41, 47, 51, 55, 57, 64, 65, 70, 72
Sída   171
Siete Fuentes   108, 109
Simeto   57
Sioule, River   203, 211, 212
Sisifo   44
Skagafjordur   180
Skáros   83
Skjaldbreidur   177
Snaefellsnes Peninsula   178
Sobaco   116
Soldão   154
Somma   10, *10*, *11*, 12, 13, *13*, *17*
Sóo   113
Sör-Jan (Jan Mayen)   194, 195, 196
Soufrière (Guadeloupe)   147, 205
Spallanzani   35, 36
Spessart   214
Spitzbergen   194
Stabiae   17, 18, 19, 21
Starza   24, 25, 28
Stéfanos   94, *Plate 5b*
Strabo   17, 30, 34, 45, 51, 86, 93
Strokkur   185
Stromboli   3, 6, 9, 33, 34, **35–9**, *35*, *37*, *38*, *39*, 44, *Plate 4b*

   growth   **36–9**
Strombolicchio   33, 35, 36, *37*, 38
Sudurgígar   181
Surtla   187
Surtsey   168, 186, *186*, 187, 189
Surtur   187
Surtur Junior   187
Sveinagjá lava flow   173
Swabia   214
Syrtlingur   187

Taburiente   125, 127
Tacitus   10, 18
Taco   108
Tahiche   113
Tahuya (or Teguso) cinder cone   126
Tanque   110
Taunus   214
Teide   *see* Pico de Teide
Tejeda   123
Temejereque   122
Temple of Serapis   25, 29
Teneguía   126, 127
Tenerife   100, 101, **102–111**, *102*, 120, 124, 126
   historical eruptions   **109–111**
      1730–36   **114–17**
      1824   **117–21**
Teno peninsula   103
Terceira   132, 133, 142, **144–8**, *144*, *147*, 149
   Rift   **146**
   submarine eruptions   **146–8**
Theistareykir   172
Théra   82, 83, 84, 86, 87, 88
Therasía   82, 83, 84, 86
Thingvallavatn   170
Thingvellir   170, *170*, 177
Threngslaborgir   171, 174
Thucydides   47, 64
Timanfaya   114, 115, 116
Timão   150
Timpa   58
Timpone
   Carrubbo   40
   del Fuoco   35, 38
   Ospedale   40
   Pataso   40
Tiretaine   211
Titus, Emperor   51
Tjörnes fracture zone   167
Toppgígur   180
Torfajökull   *178*, *178*, 184
Torre
   Annunziata   11, 16
   del Filosofo   58, *60*
   del Greco   14, 16
Trifoglietto   57

Tripergole 28, 29
Tríton dome 90
Tromosryggen ridge 196
Turkey 2, 81, 86, 87
Tyrrhenian Sea 7, 9, 12, 23, 24, 31,
    33, 35, 37, 39, 42, 43, 44, 55

Ulmenermaar 217
Urzelina 152
Ustica 9

Valle
    del Bove 53, 56, 57, 58, 62, 63,
        65, 66, *66*, 68, *68*, *61*
    di Calanna 68
    dell'Inferno 10, 12
Vallonazzo 37
Vancori 35, 38
Várzea 141
Vateliero 32
Vatnajökull 167, 171, 172, 178,
    179, 182, 183
Vavalaci 57
Vega de Ugo (Lanzarote) 116
Veidivötn 178
Velas 152
Velay 200, 201
Vermelho 150
Vestmannaeyar (Vestmann
    Islands) 168, 169, 186, *186*, 189
Vesuvius 2, 3, 6, 9, **10–21**, *10*, *11*,
    13, *15*, *17*, *18*, *20*, 26, 39, 100,
    Plate 1a
    growth of **11–14**
    historical eruptions **11**
    Plinian eruptions
        large scale **17–21**
        sub-Plinian **16–17**
    semi-persistent activity **14–16**
Veyre 210, 211, 212
Vico 7
Vigna Vecchia 35
Vikrahraun 173
Vila do Porto 143
Virgil 45, 47, 64
Víti
    crater (Askja) 173
    maar (Krafla) 174, *176*
Vitruvius 17
Viulo 11
Volcán
    del Clerigo Duarte 117
    Martín 126, 127
    Negro 114
    Nuevo del Fuego 120
    de Tao 117
    de Tinguatón 120, 121
Volvic 204, 211
Voragine 53, 58, 59, *61*, 66, 67, 69

Vulcan 6, 33, 47
Vulcanello 33, 34, 41, *44*, 45, 46,
    46, 47, 48, 49, *49*, 50, *50*, 51
Vulcano 3, 6, 9, 33, 34, 40, 41, 42,
    42, 44, 45, *45*, 46, 47, 48, 49, 50, 51,
    56, *Plate 5a*

Wartgesberg 217
Wehr 215, 216
Weinfeldermaar 217
Westerwald 214

Yaiza 114, 116, 117
Yalí 81, 92, 94

Zafferana 68
Zara 32
Zonzomas 113
Zucco Grande 43

TOPICS AND THEMES

Page numbers in italics denote illustrations; those in bold denote sections of text.

aa lava 86, 105, 106, 112, 120, 133, 154, 167; *see also* apalhraun, biscoitos, malpais
acidity peak 84, 86, 172
aligned
  cone 153
  crater 175
amphibole 57
andesite 4, 33, 38, 40, 42, 43, 70, 82, 83, 94, 168; *see also* intermediate eruption
apalhraun 167, 178; *see also* aa lava, biscoitos, malpais
archaeomagnetism 11, 42, 47, 51, 63
  dating 63, *63*
ash *passim*
ashfall 138
ashflow 23, 31, 41, 195, 216
  ashflow tuff, brown 41
atrio 6

bagacina 133, 146
barrancos 112, 121, 123, 153, 196
barrier of ash and cinders 68
basalt 2, 4, 6, 38, 42, 43, 44, 45, 46, 52, 56, 57, 70, 82, 83, 93, 101, 103, 109, 110, 112, 120, 121, 122, 123, 127, 135, 139, 141, 143, 144, 145, 146, 150, 155, 157, 158, 159, 161, 162, 166, 167, 168, 169, 170, 172, 174, 177, 195, 196, 197, 201, 203, 211, 214, 216
  alkali 9, 72, 103, 134, 147, 151, 152, 153, 154, 155, 157, 172, 178, 184, 215, 216
  fluid 101, 108, 110, 127, 154, 212
  olivine 109, 111, 112, 114, 117, 122, 138, 144, 146, 167, 187, 189
  plateau (Iceland) **168**
  tholeiitic 9, 33, 56, 167, 168, 170, 171, 173, 174, 177
basaltic
  andesite 33, 38, 82, 93, 94, 181
  cone 108
  dome 149
  magma 56, 203, 210
  plateau 103, 153, 169, 200, 201
  shield 4, 168
basanite 141, 204
benmoreite 9, 56, 57, 124, 144, 145, 157, 201, 203, 204; *see also* sancyite, trachyandesite

biotite 38
biscoitos 146, 154, 155, 160; *see also* aa lava, apalhraun, malpais
blast 16, 17, 19, 46, 51, 152, 180, 205
block 9, 30, 31, 36, 37, 45, 47, 48, 68, 103, 106, 107, 108, 114, 120, 133, 147, 149, 157, 174, 180, 200, 202, 205, 210, 215
block lava 120
bomb 15, 36, 48, 54, 59, 69, 72, 89, *111*, 120, 161, 180, 187, 189
  tortoise 161
bradyseismic movement 22, 23, 24, **25–6**, *25*
breached cone 150, 204
breadcrust bomb 84, 106

cabeço 133, 154, 155, 157, 158, 159, 160
calc-alkaline (magma) 9, 33, 38, 44, 82
caldera *passim*
  collapse 8, 10, 12, 19, 17, 21, 23, 31, 40, 46, 56, 57, 58, 70, 82, 83, 84, 87, 92, 95, 101, 103, 106, 123, 125, 135, 136, 137, 138, 139, 157, 170, 173, 174, 202, 215
caliche 112, 113, 121, 122
Campanian Ignimbrite 9, 12, **23**, 24
carbon dioxide 18, 27, 50, 51, 55, 116, 127, 180, 189, 214
central vent 24, 42, 124, 195, 207
cheire 204, 208, 211, 212
cicirara lava 57
cinder cone 3, 6, 7, 8, 14, 36, 46, 51, 52, 53, 54, 59, 61, 62, 65, 66, 67, 70, 83, 90, 100, 102, 104, 108, 109, 110, 111, 112, 113, 120, 121, 122, 123, 126, 127, 133, 134, 137, 138, 139, 140, 141, 143, 144, 145, 146, 149, 150, 151, 152, 153, 154, 155, 156, 158, 159, 162, 172, 173, 189, 194, 195, 196, 197, 200, 201, 203, 205, 207, 208, 210, 214, 215, 216, 217
cinders *passim*; *see also* bagacina
Campanian Ignimbrite 23, 24
Canary Island hotspot 124
collision 2, 3, 4, 6, 7, 55
cone sheets 23, 24
country rock 210, 217
crater *passim*

crust 2, 3, 6, 7, 14, 28, 36, 42, 56, 57, 94, 120, 170, 171, 173
  accretion 166
  extension 3, 6
crystal
  content 56
  fractionation 14; *see also* differentiation, stratified reservoir
cuchillo 103, 122

dacite 43, 82, 83, 84, 88, 90, 92, 93, 95, 170, 174, 180, 184
death *see* fatalities
debris avalanche 39, 56, 57, 58
deflation (crustal) 175, 177
dendrochronology 86
diatremes 8
differentiation (of magma) 6, 13, 14, 31, 57, 82, 167, 201, 210, 211; *see also* crystal fractionation, stratified reservoir
diverging plate 166
diverted lava 68, 116
dome 3, 6, 7, 22, 23, 24, 27, 31, 32, 33, 34, 40, 41, 42, 43, 46, 70, 71, 82, 83, 86, 88, 90, 92, 93, 94, 95, 103, 106, 107, 108, 123, 135, 136, 137, 139, 140, 144, 145, 146, 149, 150, 158, 168, 194, 195, 200, 201, 203, 204, 205, 214, 215
  extruded 145
domite 204, 205; *see also* trachyte
doreite 201; *see also* mugearite, ordanchite, trachyandesite
dust 16, 86
dyke 3, 45, 57, 83, 101, 103, 121, 124, 125, 128, 144, 162, 171, 175

earthquake 6, 8, 12, 15, 16, 17, 18, 19, 20, 21, 26, 27, 28, 29, 30, 56, 62, 64, 65, 66, 69, 72, 84, 86, 87, 89, 94, 109, 110, 111, 115, 116, 117, 124, 126, 127, 133, 136, 138, 141, 142, 143, 144, 146, 147, 149, 152, 154, 156, 157, 160, 161, 167, 170, 171, 173, 174, 175, 180, 181, 183, 184, 185, 189, 196, 197
effusion 8, 53, 56, 62, 63, 64, 66, 67, 68, 70, 71, 88, 90, 92, 93, 110, 126, 168, 170, 171, 174, 180, 187
  rate 53, 62, 63, 68, 90
effusive 3, 9, 11, 12, 17, 23, 24, 32, 40, 42, 94, 103, 121, 145, 149, 161, 187, 195, 202

Epomeo Green Tuff 31
eruption *passim*
eruptive
　column 14, 16, 17, 18, 175, 183
　styles 3, 105, 153
Eurasian plate 2, 3, 4, 7, 33, 55, 81,
　132, 133, 166
explosion pit 59, 65, 66, 86, 90, 94,
　105, 109, 121, 126, 139, 211
explosive eruption 4, 9, 23, 31, 32,
　37, 40, 47, 66, 82, 88, 105, 136, 137,
　151, 170, 173
extension (crustal) 32, 37, 56, 171
extinct volcano 6, 7, 9, 24, 33, 34,
　44, 71, 137, 162, 171, 200, 203,
　216, 202

fajãs 151
fatalities 18, 20, *20*, 55, 58, 87,
　Plate 1b
fault 2, 3, 4, 9, 12, 26, 30, 31, 33,
　36, 37, 39, 51, 55, 56, 58, 83, 101,
　112, 133, 135, 149, 151, 155, 157,
　158, 170, 174, 175, 196, 197, 201,
　210, 214, 215, 217
final eruption 14, 15, 16, 30, 120,
　207, 208
fissure *passim*
　eruption 64, 66, 67, 102, 108,
　　122, 133, 152, 153, 170, 171,
　　182, 186, 187, 194, 196; *see also*
　　Icelandic eruption
flank eruption 4, 53, 65, 67, 69,
　107
fluid lava 14, 62, 63, 66, 67, 120,
　127, 168, 177
fluid magma 65
fluvioglacial deposits 179
fragments *passim*; *see also* lithic
　fragments
fumarole 3, 6, 9, 14, 16, 27, 33, 47,
　48, 51, 71, 72, 81, 82, 86, 92, 94,
　106, 126, 133, 136, 144, 145, 149,
　173, 175, 182, 183, 197, 216; *see
　also* hydrothermal eruption,
　solfatara

gas 12, 14, 17, 18, 29, 48, 53, 58,
　59, 62, 65, 69, 87, 116, 121, 127,
　133, 147, 148, 180, 187, 196
geyser 12, 27, 173, 184, 185
geyserite 185
gour 210
gravity 26, 120, 126

hawaiite 9, 56, 57, 58, 59, 62, 65,
　70, 101, 103, 106, 122, 134, 135,
　144, 145, 146, 149, 150, 151, 155,
　157, 158, 159, 196, 201, 203, 204,

212; *see also* labradorite, trachy-
　basalt
hazard 17, 21, 22, 65
helluhraun 167, 175, 177; *see also*
　pahoehoe lava
historical time and eruptions
　passim
hornito 62, 63, 65, 109, *113*, 117,
　117, 120; *see also* spatter
hot spring 8, 22, 28, 30, 45, 48, 49,
　71, 82, 93, 94, 114, 135, *135*, *136*,
　138, 184, 185
hotspot 2, 4, 101, 124, 132, 133,
　167, *169*, 194
　plumes 133
hydrothermal eruption 94, **184–5**,
　185; *see also* fumarole, solfatara
hydrovolcanism 14, 15, 18, 24, 27,
　29, 32, 41, 43, 47, 55, 58, 59, *60*, 61,
　65, 66, 67, 84, 92, 94, 105, 106, 120,
　127, 136, 137, 140, 141, 151, 174,
　175, *176*, 204, 205, 208, 210, 214,
　215, 217, *Plate 5b*; *see also* Etna,
　Surtseyan eruption

ice 172, 173, 174, 178, 179, 180,
　182, 183, 184, 194; *see also*
　Icelandic eruptions, subglacial
　eruptions
　age 179
　cap 4, 84, 86, *166*, 167, 169,
　　169, *170*, 171, 178, 179, 182,
　　183, *183*, 184, 194
　core 84, 86
　-filled crater 179
　sheet 167, 168, 169, 172, 174,
　　182
Icelandic eruption 168, 173; *see
　also* fissure eruption
ignimbrite 23, 31, 70, 84, 103, 123,
　145
indicator bed 138, 179, 204, 205,
　211, 214, 216
inflation (crustal) 175, 177
intermediate eruption 14; *see also*
　andesite

jameos 114
jökulhlaup 180, 183, 184

labradorite 201; *see also* hawaiite,
　trachybasalt
Lajes Ignimbrite 145
lake (lac, lago, lagoa) 7, 8, 22, 23,
　24, 27, 28, 69, 120, 136, 137, 138,
　139, 140, 141, 145, 146, 149, 155,
　162, 174, 182, 183, 187, 200, 208,
　210, 212, 214, 216, 217
landslide 11, 46, 104, 124, 127,

135, 210
lapilli 6, 8, 15, 21, 24, 34, 37, 41,
　46, 48, 54, 65, 112, 115, 136, 149,
　153, 159, 174, 189, 207
Late Minoan IA 84, 87
Late Minoan IB 87
latite 6, 7, 23, 30, 38, 40
lava
　accretionary lava ball 106, 109
　delta *110*, *156*
　flow *passim*
　fountain 11, 14, 15, 36, 59, 65,
　　67, 68, 69, 109, 111, 120, 126,
　　161, 170, 171, 172, 173, 175,
　　180, 181, 187, 189, 197
　temperature 27, 48, 49, 56, 61,
　　70, 90, 94, 121, 170, 184, 187,
　　216
　tunnel 114, 115
leucite 7, 8, 214
leucitites 8, 9, 17
lithic fragments 8, 18, 21, 23, 84,
　92, 138, 217

maar 3, 7, 8, 32, 61, 92, 123, 136,
　137, 140, 155, 168, 174, 200, 201,
　203, 208, 210, 211, 214, 215, 216,
　217
magma *passim*
　reservoir 8, 14, 17, 18, 21, 23,
　　24, 25, 33, 48, 172, 173, 178,
　　180, 181, 187, 214, 216; *see also*
　　stratified reservoir
magmatic pressure 14, 18, 30
malpaís 92, 100, *112*, 114, 115,
　116, 117, 120, 122; *see also* aa lava,
　apalhraun, biscoitos
mantle 4, 6, 30, 38, 56, 84, 155,
　171, 174, 204
　plume 174
marine erosion 43, 45, 48, 49, 94,
　95, 103, 114, 124, 125, 133, 135,
　143, 156, 187, 196
microplates 2, 3, 7, 12, 55
mid-ocean ridge 3, 4, 133, 157,
　166, 167, 168, 194
mistério 133, *144*, *145*, 146, 152,
　153, *153*, 154, 155, *157*, 160, 161
móberg 168, 182; *see also*
　palagonite
molten lava 61, 62, 65, 68, 89, 120,
　152, 161, 210
mudflow 15, 17, 21, 84, 94, 103,
　127, 135
mugearite 9, 56, 57, 101, 103, 123,
　134, 135, 139, 141, 144, 145, 149,
　150, 151, 155, 157, 162, 196, 201,
　203, 204; *see also* doreite,
　ordanchite, trachyandesite

narse 211
Neapolitan Yellow Tuff **23–4**
nested cone 207
nuée ardente 3, 6, 10, 15, 16, 17,
17, 18, 19, 21, 27, 30, 36, 40, 57, 58,
70, 71, 84, 92, 103, 104, 123, 135,
136, 137, 139, 140, 152, 183, 204,
215, 216, *Plate 1b*; *see also*
pyroclastic flow

obsidian 33, 39, 40, *40*, 41, 42, 45,
46, 48, 103, 106; *see also* rhyolite
ocean floor 3, 103, 123, 124, 125,
137, 139, 149, 153, 162, 194
oceanic crust 2, 3, 101, 122
olivine 32, 56, 109, 111, 112, 114,
117, 120, 122, 138, 144, 146, 167,
177, 187, 189
ordanchite 201; ; *see also* doreite,
mugearite, trachyandesite

pahoehoe lava 65, 155; *see also*
helluhraun
palagonite 168, 172, 179; *see also*
móberg
palagonitic breccia 179
pantellerite 70, 71, 134, 144, 145
peperini tuff 8
peralkaline rhyolite 123
phonolite 6, 7, 8, 9, 13, 16, 31, 101,
103, 106, 108, 123, 126, 201, 214,
215, 216, *Plate 7a*
picón 112, *112*
pillow lava 182
plagioclase feldspar 56, 57
plateau basalt 100, 168
Plinian
column 3, 18, 21, 57, 84, 87,
138, 181, 216
eruption 10, 14, 16, 17, 18, 21,
23, 64, 83, 84, 95, 138, 157, 173,
178, 216
plinth 50, 51, 57, 92, 132, **153–5**,
159, 195, 200, 201
poisonous gas 116, 154, 171, 180,
189; *see also* carbon dioxide
polygenetic cone 208
potassic series 23
pumice 7, 9, 10, 13, 14, 16, 17, 18,
19, 21, 23, 24, 27, 28, 29, 32, 33, 36,
39, 40, 41, 47, 70, 71, 72, 82, 83, 84,
86, 87, 92, *92*, 94, 95, 103, 104, 105,
106, 108, 134, 135, 136, 137, 138,
139, 140, 141, 142, 144, 145, 157,
173, 179, 180, 184, 195, 201, 202,
204, 214, 215, 216
cone 27, 41, 71, *107*
grey pumice 18, 19, 21
phonolitic 13, 216

white 18, 21, 27, 43, 82, 84, 173,
179, 216
pyroclastic
flow 16; *see also* nueé ardente
surge 16, 32, 47, 65
pyroxene 56

quartz 40, 177

radiocarbon dating 22
rhyolite 4, 6, 33, 34, 40, 41, 44, 45,
47, 70, 92, 94, 101, 168, 178, 180,
201, 202; *see also* obsidian
Richter scale 133, 157, 196
rift 3, 56, 58, 61, 62, 63, 67, 70, 72,
101, 103, 108, 109, 110, 111, 124,
125, 133, 144, 146, 156, 157, *167*,
169, 170, 177, 214
ring fracture 24, 105
rootless cone 174, *176*, 204, 210,
211, *212*

sancyite 201; *see also* benmoreite,
trachyandesite
satellite cone 9, 10, 35, 138, 155
seamount 7, 33, *33*, 37, 44
shallow magma reservoir 30, 101,
184
sheeted dyke 166
shield 51, 70, 82, 83, 103, 122,
125, 128, 135, 145, 167, 168, 177,
178, 182, 187
basal 100, 101, 102, *102*, **103**,
123, 144, 153, 154, 155, 195,
196
volcanoes **177**
shoshonitic lava 9, 45
silica 70, 180, 185, 204
silicic magma 83
solfatara 6, 9, 90, 94, 145, 173, 175,
184; *see also* fumarole,
hydrothermal eruption
spatter 32, 37, 38, 101, 110, 117,
120, 127, *142*, 143, 144, 146, 152,
155, 161, 171, 196
cone 14, 36, 38, 59, 62, 83, *113*,
115, 117, *117*, 120, 121, 122,
144, 146, *146*, 152, 155, 161,
172, 197, *211*, *Plate 2*; *see also*
hornito
steam explosion 211
stratified reservoir 70; *see also*
crystal fractionation,
differentiation
stratosphere 14, 16, 172
stratovolcano *passim*
Strombolian
cone 150, *162*
eruption 36, 69, 206, 208, 210,

215, 216
explosion 14, 30
subduction zone 7, 9, 33
subglacial
eruption 168, 177
table mountain 169, 177
submarine eruption 72, 90, 132,
141, **146–8**, 151, 152, 157, 186,
187; *see also* Surtseyan eruption
suc 200, 201
sulphuric acid 27, 120, 171
Surtseyan eruption 9, 30, 32, 49,
50, 51, 72, 86, 90, 108, 113, 114,
115, 121, 133, 134, 138, 140, 141,
142, 143, 144, 148, 150, 155, 158,
159, 161, 168, 182, 183, 187, 195,
208, 210; *see also*
hydrovolcanism, submarine
eruption

tectonics 3, 18, 26, 35, 36, 38, 39,
55, 58, 65, 87, 104, 151, 167, 194,
197
plate 2, 3, 4, 7, 9, 33, 55, 56, 81,
100, 132, 133, 162, 166, 169,
170, 172
boundaries *4*
tephrite 6, 7, 9, 13, 14, 16, 123,
214, 215
tephritic magma 14
timpa 58
trachyandesite 9, 56, 201; *see also*
benmoreite, doreite, mugearite,
ordanchite, sancyite
trachybasalt 6, 9, 23, 56, 103, 201;
see also hawaiite, labradorite
trachyte 6, 7, 9, 23, 24, 27, 30, 32,
38, 45, 47, 50, 56, 57, 70, 103, 112,
123, 124, 134, 135, 139, 141, 144,
145, 146, 149, 157, 158, 162, 195,
201, 203, 204, 205, 214; *see also*
domite
tree ring 86
tremor see *earthquake*
tsunami 86, 87
tuff
cone 33, 45, 46, 47, 48, 72, 86,
88, 133, 140, 143, 148, 159, 161,
183
ring 23, 24, 32, 41, 90, 92, 120,
123, 126, 127, 136, 137, 168,
174, 183, 204, 208, 210, 214,
215, 217
tumuli 63, 211
twin cone 65, 207

vent *passim*
viscous lava 32, 40, 93, 105, 150,
204

TOPICS AND THEMES

volatile   178
volcanic
  arc   91
  cone   *11*
  field   39, 83, 92, 108
  islands   *4*
  rock   3, 7, 8, 17, 23, 83, 92, 101,
    105, 123, 144, 167
  system   169, *169*, 170, 171, 172,
    174, 177, 178, 179, 182, 183
  zones   *7*
Vulcanian eruption style   6, 33

xenolith   103, 120